Springer Complexity

Springer Complexity is an interdisciplinary program publishing the best research and academic-level teaching on both fundamental and applied aspects of complex systems—cutting across all traditional disciplines of the natural and life sciences, engineering, economics, medicine, neuroscience, social and computer science.

Complex Systems are systems that comprise many interacting parts with the ability to generate a new quality of macroscopic collective behavior the manifestations of which are the spontaneous formation of distinctive temporal, spatial or functional structures. Models of such systems can be successfully mapped onto quite diverse "real-life" situations like the climate, the coherent emission of light from lasers, chemical reaction-diffusion systems, biological cellular networks, the dynamics of stock markets and of the internet, earthquake statistics and prediction, freeway traffic, the human brain, or the formation of opinions in social systems, to name just some of the popular applications.

Although their scope and methodologies overlap somewhat, one can distinguish the following main concepts and tools: self-organization, nonlinear dynamics, synergetics, turbulence, dynamical systems, catastrophes, instabilities, stochastic processes, chaos, graphs and networks, cellular automata, adaptive systems, genetic algorithms and computational intelligence.

The three major book publication platforms of the Springer Complexity program are the monograph series "Understanding Complex Systems" focusing on the various applications of complexity, the "Springer Series in Synergetics", which is devoted to the quantitative theoretical and methodological foundations, and the "Springer Briefs in Complexity" which are concise and topical working reports, case studies, surveys, essays and lecture notes of relevance to the field. In addition to the books in these two core series, the pr[illegible]m also incorporates individual titles ranging from textbooks to major reference work[illegible]

Indexed by SCOPUS, INSPEC, zbMATH, SCImago.

Understanding Complex Systems

Founding Editor: Scott Kelso

Michał Chorowski · Tomasz Gubiec ·
Ryszard Kutner

Anomalous Stochastics

A Comprehensive Guide to Multifractals, Random Walks, and Real-World Applications

Springer

Michał Chorowski
Biomedical Physics Division, Institute of Experimental Physics, Faculty of Physics
University of Warsaw
Warsaw, Poland

Google
Mountain View, USA

Tomasz Gubiec
Biomedical Physics Division, Institute of Experimental Physics, Faculty of Physics
University of Warsaw
Warsaw, Poland

Ryszard Kutner
Biomedical Physics Division, Institute of Experimental Physics, Faculty of Physics
University of Warsaw
Warsaw, Poland

ISSN 1860-0832 ISSN 1860-0840 (electronic)
Understanding Complex Systems
ISBN 978-3-031-80394-9 ISBN 978-3-031-80392-5 (eBook)
https://doi.org/10.1007/978-3-031-80392-5

This Springer imprint is published by the registered company Springer Nature Switzerland AG
The registered company address is: Gewerbestrasse 11, 6330 Cham, Switzerland

To our Families

I (Ryszard Kutner) would like to thank my mentors, Izabela Sosnowska, Kurt Binder, Wolfgang Dieterich, and Klaus Kehr (of late memory), to whom I owe so much
I (Michał Chorowski) want to thank my wife, Allison, for supporting me through every late night of writing this book. I also want to thank my mentor, Ryszard Kutner, for everything he taught me while working together.

Preface

Stochastic dynamics, which encompass a variety of complex systems, can be characterized by different stochastic processes and their corresponding random walks. Among these processes, singular processes are particularly noteworthy due to their universal characteristics. Within this context, we focus on a specific subclass known as anomalous diffusion. This monographic textbook examines processes of this type in detail.

Singular stochastic processes are essential for understanding universality in complex systems. They encompass a variety of phenomena and the dynamics of stochastic processes, particularly in the context of asymptotic behavior. This examination is primarily framed from the perspective of physicists who study critical phenomena, especially dynamics and phase transitions.

An additional rationale for presenting and analyzing singular stochastic processes to a broader audience is their pervasive occurrence not only within the domains of physics and the natural and mathematical sciences but also across diverse fields such as neuroscience, cardiac science, linguistics, climate science and ecology, socio-economic sciences, telecommunications, and infrastructure networks, among others.

We are focusing on processes classified as non-Brownian and/or non-Gaussian, which encompass non-Fickian and/or non-Debye phenomena. These processes extend our understanding beyond the Central Limit Theorem and Bernoulli's Law of Large Numbers. This significant topic is often only superficially addressed in textbooks, particularly in comparison to the more comprehensive analyses found in specialized monographs.

This monographic textbook serves as a condensed and comprehensive rendition of three interdisciplinary physics lectures delivered by one of the authors (RK) to undergraduate and graduate students at the Faculty of Physics, University of Warsaw, covering over two decades. The lectures address the following topics:

- Non-Brownian and/or non-Gaussian chains (referring to discrete time) and processes (referring to continuous time) within the natural sciences, incorporating elements of econophysics and sociophysics,

- An introduction to the physics of complexity, including the statistical physics of complex networks,
- The methods of physics in economics (econophysics): An introduction.

The textbook includes characteristic examples, control questions, exercises, and problems designed to help students learn and consolidate the material. We recommend it to all readers, particularly undergraduate and graduate students and academic lecturers conducting classes (lectures and exercises) supplemented by this textbook.

The textbook includes topic extensions that may benefit early-career scientists, particularly those who recently completed their doctorate. The material is designed to provoke open research questions, especially given the intensive development of singular stochastics and its growing applications in the real world, including physics and other disciplines. This textbook presents characteristic examples of these applications.

In this monographic textbook, we pose several essential and thought-provoking questions, systematically answer them, and discuss their implications. We believe that this systematic approach may also be beneficial not only for the audiences above but also for scientists conducting studies and research in the areas covered by our textbook.

The prerequisites for engaging with the material presented in this textbook are surprisingly straightforward. A fundamental understanding of probability and statistics is essential, along with basic knowledge of differential and integral calculus. Additionally, familiarity with the fundamental properties of Laplace and Fourier transforms is required. However, a deeper comprehension of the textbook's content necessitates more advanced knowledge, such as skills in solving ordinary and partial differential equations, understanding the properties of the Mellin transformation, and the ability to perform integration by residues. However, there is a notable contrast between the simplicity of the mathematical methods employed and the depth of the subject matter in singular stochastics. In terms of required physical knowledge, a background in statistical thermodynamics will be beneficial. Moreover, for a more thorough understanding, familiarity with statistical physics at a foundational level is recommended.

The structure of the textbook's content is outlined as follows: Chap. 1 introduces the subject matter, situating it within a broader context and defining our objectives. In Chap. 2, we highlight a fundamental concept that is pivotal in transitioning stochastic processes from Brownian and Gaussian chains to the domain of singular (including anomalous) chains and processes. This concept serves as a signature theme throughout almost all sections of the textbook. Even at an early stage, specifically in Chap. 3, we critique the Central Limit Theorem and Bernoulli's Law of Large Numbers by illustrating the conditions these classical results hold. We explore the transition to singular stochastic processes and provide examples from canonical physics. In Chap. 4, we present the foundational concepts necessary for understanding multifractality, specifically in the context of non-deterministic fractals. An introduction to multifractal random walks, encompassing topics from bi-fractals to binomial cascades, can be found in Chap. 5. In Chap. 6, we present a category of physical experiments about the anomalous (or non-Debye) relaxation of photocurrents in amorphous materials. The description and understanding of these phenomena are rooted in the Continuous-Time Random Walk model developed by Montroll, Weiss, and Scher. The influence of this model on the singular random walks discussed in Chap. 7 (via Weierstrass flights)—in other words, on non-Fickian diffusion—is invaluable. Another area explored in Chap. 8, inspired by this model, is multifractality. Chapter 8 demonstrates the potential of the Continuous-Time Random Walk (CTRW) formalism in this domain by analyzing two characteristic examples: the substrate generated by the canonical Gaussian distribution and that generated by the generalized Gaussian distribution, specifically the distribution of the stretched exponential type. A defining feature of singular stochastic processes is the prominence of extreme events, a topic discussed in Chap. 9. As such, the CTRW formalism is central to this textbook and has significant connections to various areas within complexity science. Chapter 10 illustrates the presence of Lévy flights in financial markets, resulting from the herd effect and moral hazard, where rare but significant losses or gains can occur. The presence of multifractality in the market of competing companies is discussed in Chap. 11, where we also present a classification of full multifractality. Chapter 12 concludes the textbook by summarizing its most important achievements.

Warsaw, Poland
October 2024

Michał Chorowski
Tomasz Gubiec
Ryszard Kutner

Contents

Chapter 1
Introduction

Abstract In this introductory chapter, we aim to outline the fundamental motivations behind the development of this monographic textbook. Our objective is to convey our fascination with complex systems, particularly through the lens of singular stochastic processes. We strive to present this information in a well-documented and accessible manner, targeting a younger audience.

1.1 Motivation and Goals

We anticipate our textbook will captivate readers interested in random processes and their ubiquity [1–11]. It is aimed at undergraduate and graduate students with a fundamental grasp of differential and integral calculus, probability theory and stochastic processes, and algebra in the context of random walks.[1] In addition, experienced researchers, both theoretical and practitioners, can expect to uncover new, valuable insights and tools. These resources will aid in understanding the pivotal role and implications of extreme events or "black swans" in real-world systems.

[1] In our textbook, we mainly consider the marked/tracer/tagged particle wandering randomly. For example, we often consider tracer or self-diffusion, referring to collective (chemical or bulk) diffusion only when necessary. Simply put, when discussing anomalous diffusion, we mean the former. The distinction between self-diffusion and collective diffusion is important because they are very different [12–14]. In the former type, we track each particle *individually*. In contrast, in the latter type, we only examine the state of a point in space, e.g., the occupation of a network node by *any* particle. Therefore, if we talk about collective diffusion, we will indicate it explicitly, as opposed to self-diffusion.

M. Chorowski et al., *Anomalous Stochastics*, Understanding Complex Systems,
https://doi.org/10.1007/978-3-031-80392-5_1

We delineate the scope of knowledge central to this textbook, focusing on anomalous random walks or, more broadly, *singular stochastic processes*. By singular stochastic processes, we refer to a class of stochastic processes encompassing anomalous and ultra-slow processes, Lévy flights, and particularly the multifractal ones. These processes demonstrate a non-linear/non-Brownian function of time in their asymptotic variance. Our analysis adopts a physicist's perspective at the intersection of physics, complexity science, and applied mathematics. The latter is presented in a manner that is universally applicable across various specific domains.

The rough structure of our textbook is presented in the diagram in Fig. 1.1. We hope this diagram will make it easier to navigate through our textbook.

Until two decades ago, the singular stochastic processes, as a significant branch of stochastic physics, were termed exotic. We can say that today, it is the other way around: statistical physics based on the Central Limit Theorem [15]. As a reminder, CLT applies to a cumulative multi-step random variable $X(n) = \sum_{j=1}^{n} x_j$, where x_j, $j = 1, 2, \ldots, n$, are elementary single-step random variables of the i.i.d. type (i.e., identical, statistically independent distribution). Their variances $\sigma^2 = \langle x_j^2 \rangle - \langle x_j \rangle^2$ (taking into account drift) are identical (i.e., independent of the time step $1 \leq j \leq n$) and finite ($\sigma^2 < \infty$), where $\langle \ldots \rangle$ means the statistical ensemble average on the replicas of the random walk. For sufficiently large n, the PDF of the cumulative random variable, $X(n)$, approximates a Gaussian PDF with mean $\langle X(n) \rangle$ and variance $\sigma_X^2(n) = n\sigma^2$. The Berry-Esséen theorems determine the convergence

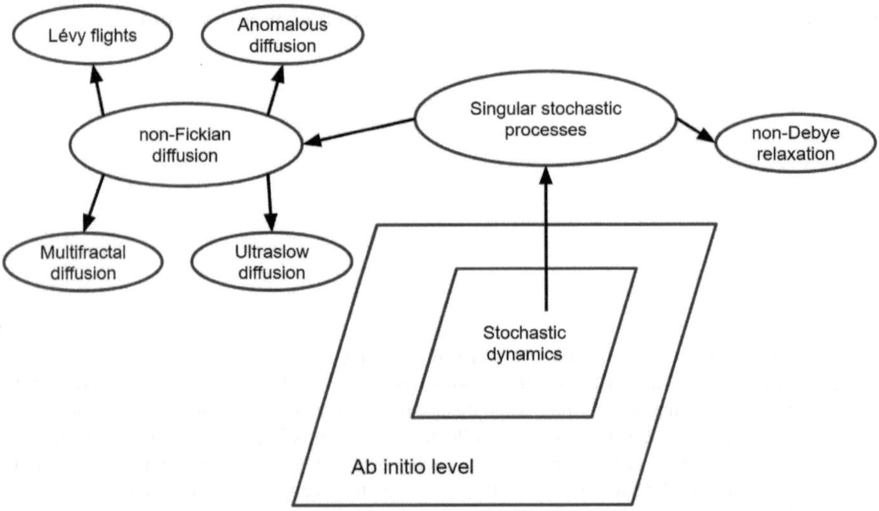

Fig. 1.1 Schematic diagram of the monographic textbook structure

speed [16]. More general variants of CLT exist,[2] such as when elementary random variables x_j are not identically distributed but have different variances $\sigma_j^2 < \infty$.

In mathematical and natural sciences, including biological, socioeconomic, engineering, and complexity science, anomalous stochastic chains and processes, or, more generally, singular stochastics, prevail, challenging and diverging from the CLT and the laws of large numbers.

The singular stochastic processes mentioned above may be rooted in deterministic chaos [18, 19]. This is an intriguing area of ongoing research, currently being analyzed in fragments. However, deterministic chaos is beyond the scope of our textbook and would excessively increase its size.[3]

We regret that due to limited space, we were unable to discuss the important topic of anomalous random walk/diffusion in the presence of percolation phenomena in regular lattices [23, 24] and networks [25–28].

For the same reason, we were unable to discuss the canonical topic of random walks (including anomalous) on complex systems (including networks) [29, 30].

Other important topics omitted include anomalous polymer diffusion/relaxation [31, 32], anomalous or fractional diffusion/relaxation with reaction [33], and diffusion-limited aggregation [34].

From such a broad point of view, we managed to discuss only the very important topics, which were poorly represented at the textbook level. These topics we listed in Sect. 1.3—they may constitute an indispensable introduction to the anomalous diffusion/relaxation domain or, more generally, singular stochastic processes.

[2] Thorough considerations regarding CLT, including the method and speed of convergence to the Gaussian distribution, can be found in the monographic book [17].

[3] The textbook relies on computer simulations utilizing a congruential (pseudo)random number generator/method [20, 21]. There are many types of (pseudo)random number generators—the optimal choice depends on the specific problem, the software used, and the hardware used. The congruential generator is based on a one-dimensional dynamic, iterative map that produces a sequence of (pseudo)random numbers (of independent identically distributed, i.i.d. type), exhibiting characteristics akin to deterministic chaos (this sequence is located in a chaotic window that does not contain any structures). This sequence of random numbers can be treated as a stochastic chain obeying the Central Limit Theorem, leading to Brownian and Gaussian diffusion. Given a generator of (pseudo)random numbers drawn from a uniform distribution defined on the interval [0,1), one can construct a random number generator with an arbitrary distribution (theorem on the inversion of CDFs). Hence, it is possible to produce anomalous diffusion within deterministic chaos [18, 19, 22].

Of course, the topics mentioned above are widely discussed in monographs. Unfortunately, they are not discussed in modern monographic textbooks and even more so in textbooks—we are dealing with a painful gap waiting to be filled.

1.2 Inspiring Questions

In this monographic textbook, we pose several essential and inspiring questions, systematically answer them, and show their consequences.

The first central question is: How does theoretically infinite variance manifest itself in the real world, where everything is, after all, finite? We approach this issue from an analytical perspective [24, 35], which is a very helpful and supporting numerical. For instance, they showed analytically that the empirical variance would be finite for a finite system with a mathematical infinite variance but would grow without bounds (with large fluctuations) as the sample size increases. We visualized/verified this through numerical experiments. In the papers [36, 37], we have demonstrated (indirectly through the behavior of the variance estimate of the cumulative random variable over time) that there is a violation not only of the Central Limit Theorem but also of Bernoulli's Law of Large Numbers (BLLN).

We simulate a random walk with (a mathematically) infinite variance to see how such a variance looks in practice, depending on time. For this purpose, we introduce the Weierstrass wandering,[4] which has (depending on the parameter values) finite or infinite single-step variance. This allows us to compare both complementary worlds. We believe that this comparison will be very inspiring for the reader. The essential message of this book is to answer this central question and to point out its several significant consequences.

The second essential question that complements the one formulated above is about anomalous diffusion having the multifractal property, i.e., governed by the generalized Hurst exponent. This exponent is a function of the order of the moment of the stochastic process.[5] These types of processes are increasingly being detected in the real world. Here, we analyze only two inspiring examples, i.e., amorphous materials and stock markets. Let us emphasize that the textbook considers the stochastic processes studied in the context of monofractals, bifractals, and multifractals (including the parabolic multifractal).

It should be noted that in this textbook, we mainly use non-Brownian and/or non-Gaussian stochastic processes, e.g., Lévy stochastic process and especially Weierstrass wandering,[6] which have heavy/fat/bold tails of probability distributions. In this

[4] By the term wandering, we mean both a walking-type stochastic process and a flight-type process. The former stochastic process is characterized by continuous trajectories and the latter by discontinuous trajectories or instantaneous jumps.

[5] Recall that for monofractal processes, the Hurst exponent does not depend on the order mentioned.

[6] The Weierstrass wandering is a versatile process in the sense that (depending on the process parameters) it may or may not have a fat (algebraic) tail.

context, we also use continuous-time random walk (CTRW) formalism as a working process with an empirical grounding in the area of amorphous/glassy conductive films. It allows us to look at singular stochastic processes from the perspective of stochastic dynamics and fractal diffusion equations. The point is to finally answer the third principal question of how deeply in scale the subject of anomalous diffusion goes and how extensive it is.

We deal here primarily with models that play (or may play) a dual role: they are used in physics and (after reinterpretation and often generalization) to describe phenomena and processes in financial markets on micro- and macroeconomic scales and other complex systems. This leads to a more precise and in-depth interpretation of the aforementioned economic phenomena and processes and sociological ones, especially the type of bubbles and stock market crashes [38]. This approach to economics and sociology falls within the emerging interdisciplinary fields of knowledge known as econophysics and sociophysics, respectively.

1.3 Highlights

In this textbook, we tie together several (partially overlapping) key threads/pillars, the cores of which are multifractality [39, 40]. The pillars are as follows:

(i) Breaking of the Central Limit Theorem and the laws of large numbers[7] [24].
(ii) From *ab intio* fractal and multifractal stochastic dynamics to fractional relaxation and evolution equations [41].
(iii) Generalized master equation and memory [42].
(iv) Anomalous transport & diffusion and ergodicity breaking [43].
(v) Long-range dependence and correlations (containing auto-correlations) [35].
(vi) Heavy/fat tails of probability distributions: broadened distributions [24].
(vii) Fractal wanderings (including the theory of stochastic fractals) [44].
(viii) Multifractal wanderings (including the theory of stochastic multifractals) versus thermodynamic formalism. From bi-fractals to multifractals, including the parabolic multifractals implemented through thermally activated hopping on a Gaussian amorphous substrate/domain/medium [18, 45, 46].
(ix) Elements of Extreme Value Theory and its applications [17, 47–49].
(x) Lévy flights, "black swans" and "dragon kings" [50].

[7] Breaking a theorem means breaking the assumptions of that theorem, i.e., these assumptions (or at least one) are not met. For example, this is the case when the reality that inspired the adoption of certain assumptions has changed, forcing the modification of these earlier assumptions. The strong term "breaking" has been used to emphasize the importance of this situation.

We intricately interlace the principal themes of our textbook, specifically extended anomalous diffusion, the formalism of multifractality & statistical thermodynamics, generalized extreme value analysis, and the augmented formalism of continuous-time random walk (CTRW). We incorporate illustrative examples from the realms of physics and various markets (both financial and corporate). The table of contents demonstrates the integration of these fundamental concepts within statistical physics.

In our textbook, we frequently discuss examples from econophysics, examining pivotal issues in financial markets and free-market economies. We approach these topics through the lens of models utilized to explain physical phenomena and processes. This unconventional approach, which mainly focuses on models accommodating the occurrence of extreme or even super extreme events (i.e., rare and super-rare events), proves promising and intriguing. Such events play an increasingly significant role in our macro and micro-worlds.

1.4 Warning

This textbook illuminates crucial topics that, if incorporated into economic analysis, could potentially avert market disasters like the one in 1998. The idea is to protect against the inappropriate (even despite initial successes) use of random walks to obtain profits on financial markets.

Founded in 1993 by the prominent bond trader Salomon Brothers, John Meriwether, and Nobel Prize winner Myron Scholes (co-author of the Black-Scholes model), the Long-Term Capital Management (LTCM) later welcomed Nobel Prize winner Robert C. Merton into its management. Both Nobel laureates also served on the LCTM board of directors.[8] This famous duo and Fischer Black formulated a canonical option pricing model—promptly dubbed the Black-Scholes-Merton (BSM) formula.

The BSM formula is primarily derived under the assumption that the prices of underlying financial instruments on which options were issued follow geometric Brownian motion, with the considered options being risk-neutral. LTCM embraced highly leveraged asset strategies grounded in this optimistic theoretical framework. Furthermore, the LTCM portfolio held substantial amounts of Russian government short-term commitments (GCO) and treasury bonds. This approach enabled LTCM to achieve a net annualized return of over 20% for three consecutive years.

[8] In brief, the LTCM was a significant hedge fund co-led by Nobel Prize-winning economists and renowned Wall Street traders that collapsed in 1998. The fallout prompted the U.S. government to intervene to prevent a complete financial market collapse.

Nevertheless, between August and September 1998—shortly after the Asian (1997) and Russian (1998) financial crises—opportunities for arbitrage profits, particularly in fixed-income arbitrage with treasury bonds, markedly diminished despite LTCM's active engagement in interest rate swaps. The fund incurred losses of approximately 4.5 billion US dollars, causing severe disruptions in global markets for several months. The LTCM ultimately collapsed in the aftermath.

The company's downfall stemmed from a breach of the theory's fundamental assumptions in evolving market conditions and consistent neglect to validate these assumptions and the corresponding operational strategies. A comprehensive and systematic econophysics analysis of this subject, particularly addressing market risks, was presented in 2001 by Jean-Philippe Bouchaud and Marc Potters in their book [17].

The evolving nature of risks in our society is fascinating and multifaceted. In many aspects, our world has become safer and more secure than in the past. Advances in medicine, technology, and social systems have significantly improved areas like infant mortality rates.

However, the point about the potential rise in the frequency and severity of shocks, crashes, and systemic failures in financial markets is also valid. Risk control and management challenges intensify as systems become more interconnected and complex. The comparison to the Long-Term Capital Management incident is a stark reminder of the consequences of underestimating or mismanaging risks in the financial domain.

Striking a balance between progress and risk mitigation is crucial. The pursuit of a "zero-risk society", as suggested by Sornette and Cauwels [51], may be an ideal example. Still, it's likely an unattainable one. Risks are inherent in dynamic systems, and the adaptability and resilience of these systems play a crucial role in managing and mitigating those risks.

Updated models and risk management strategies are imperative in the financial world. Rapid change, technological advancements, and global interconnectedness require constant vigilance and innovation in risk assessment and control. The reference to "black swans" and "dragon-kings" of Sornette [50] underscores certain events' unpredictable and potentially catastrophic nature (e.g., wars), emphasizing the importance of being prepared for the unexpected.

The key lies in fostering a culture of proactive risk management, continuous learning, and adaptability. It's a complex challenge, but acknowledging the dynamic nature of risks and staying ahead of the curve in understanding and managing them is essential for the stability and sustainability of our increasingly interconnected world.

1.5 Authors' Hope

This textbook, referring, among others, to econophysics and sociophysics, imparts practical knowledge to readers on understanding and mathematically describing extreme and super-extreme events in the peculiar random processes prevalent in physics, markets, industries, and everyday life. Drawing from two decades of lectures and applied research (RK), as well as practical insights into business requirements, risks, and modeling within financial institutions, and data-driven quantitative analysis in the leading technological companies (MCh), this work aims to equip readers with theoretical knowledge and valuable tools and models. The intention/dream is to support practitioners dealing with real-life data.

In the contemporary landscape, complex systems have veered away from the placid waters of normality. Knowledge of extreme events and anomalous processes modeling has become indispensable for any professional or scientist working with real-world data. It's worth noting, however, that in this textbook, our emphasis leans more towards unraveling the mechanisms behind observed facts rather than mnemonic procedures/technical analysis—addressing the latter requires distinct and specific treatment.

As the last section, each chapter of our textbook (except the first and last) contains the control and review questions and exercises. They are intended to highlight the most important elements of each chapter.

In our textbook, we have included only the literature directly related to the subject matter, specifically (i) textbooks, books, monographic books, and journal special/topical issues, (ii) review articles, (iii) milestone and key research articles, and (iv) popular science articles. Including additional references would result in an excessively long bibliography, especially since the cited works already contain essential references mentioned above. Of course, we also quote (v) our articles (including chapters in books), which are the principal elements of this textbook. This is well illustrated by the literature cited in this chapter. The bibliography listed in this chapter constitutes the general basis of our textbook. The bibliography on specific issues is shown in the following chapters.

References

1. Shlesinger, M.F., Zaslavsky, G.M., Frisch, U. (eds.).: Lévy Flights and Related Topics in Physics. Lecture Notes in Physics, vol. 450, Springer, Berlin (1995). https://doi.org/10.1007/3-540-59222-9
2. Pękalski, A., Sznajd-Weron, K. (eds.).: Anomalous Diffusion. From Basics to Applications. Lecture Notes in Physics, vol. 519, Springer, Berlin (1999). https://doi.org/10.1007/3-540-59222-9
3. Metzler, R., Klafter, J.: The random walk's guide to anomalous diffusion: a fractional dynamics approach. Phys. Rep. **339**(1), 1–77 (2000). https://doi.org/10.1016/S0370-1573(00)00070-3

4. Evangelista, L.A., Lenzi, E.K.: An Introduction to Anomalous Diffusion and Relaxation. PoliTO Springer Series. Springer Nature, Switzerland AG (2023). https://doi.org/10.1007/978-3-031-18150-4
5. Seuront, L., Stanley, H.E.: Anomalous diffusion and multifractality enhance mating encounters in the ocean. PANS **111**(6), 2206–2211 (2014)
6. Krapf, D., Lukat, N., Marinari, E., Metzler, R., Oshanin, G., Sulhuber-Unkel, Ch., Squarcini, A., Stadler, L., Weiss, M., Xu, X.: Spectral content of a single non-brownian trajectory. Phys. Rev. **X9**, 011019(13) (2019). https://doi.org/10.1103/PhysRevX.9.011019
7. Ben-Avraham, D., Havlin, S.: Diffusion and Reactions in Fractals and Disordered Systems. Cambridge University Press, Cambridge, UK (2000). https://doi.org/10.1017/CBO9780511605826
8. Paul, W., Baschnagel, J.: Stochastic Processes. From Physics to Finance, 2nd edn. Springer, Berlin (2013). https://doi.org/10.2307/3085855
9. Klyatskin, V.I.: Stochastic Equations Through the Eye of the Physicist. Elsevier, Amsterdam (2005). https://doi.org/10.1016/B978-0-444-51797-5.X5000-X
10. Van Kampen, N.G.: Stochastic Processes in Physics and Chemistry, 3rd edn. Elsevier, Amsterdam (2007)
11. Masoliver, J.: Random Processes. First-Passage and Escape. World Scientific, Singapore (2018). https://doi.org/10.1142/10578
12. Kutner, R.: Chemical diffusion in the lattice gas of non-interacting particles. Phys. Lett. **81**(4), 239–240 (1981). https://doi.org/10.1016/0375-9601(81)90251-6
13. Spohn, H., Yau, H.-T.: Bulk diffusivity of lattice gases close to criticality. J. Stat. Phys. **79**(1–2), 231–241 (1995). https://doi.org/10.1007/BF02179388
14. Van Beijeren, H., Kutner, R., Spohn, H.: Excess noise for driven diffusive systems. Phys. Rev. Lett. Phys. Rev. Lett. **54**(18), 2026–2029 (1985). https://doi.org/10.1103/PhysRevLett.54.2026
15. Elizar, I., Klafter, J.: Anomalous is ubiquitous. Ann. Phys. **326**(9), 2517–2531 (2011). https://doi.org/10.1016/j.aop.2011.07.006
16. Mantegna, R.N., Stanley, H.E.: An Introduction to Econophysics. Correlations and Complexity in Finance. Cambridge University Press, Cambridge, UK (2000). https://doi.org/10.1063/1.1341926
17. Bouchaud, J.-P., Potters, M.: Theory of Financial Risks. From Statistical Physics to Risk Management, 2nd reprint. Cambridge University Press, Cambridge (2001)
18. Schuster, H.G.: Deterministic Chaos—An Introduction, 2nd Revised edn. VCH Verlagsgesellschaft, Weinheim (1988). https://doi.org/10.1002/BBPC.198800268
19. Klages, R.: From Deterministic Chaos to Anomalous Diffusion. In: Schuster, H.G. (ed.) Reviews of Nonlinear Dynamics and Complexity, vol. 3, Chap. 6, pp. 169–227. Wiley-VCH, Wiley Online Library (2010). https://doi.org/10.1002/9783527630967
20. Landau, D.P., Binder, K.: A Guide to Monte Carlo Simulations in Statistical Physics. Cambridge University Press, Cambridge UK, (2000). https://doi.org/10.2277/0521842387
21. Landau, R.L., Páez M.J., Bordeianu, C.C.: Computational Physics. Problem Solving with Computers, 2nd Revised and Enlarged Edition. Wiley-VCCH, Weinheim (2007)
22. Geisel, T.: Lévy walks in chaotic systems: useful formulas and recent applications. In: Shlesinger, M.F., Zaslavsky, G.M., Frisch, U. (eds.) Lévy Flights and Related Topics in Physics. Lecture Notes in Physics, vol. 450, Springer, Berlin (1995). https://doi.org/10.1007/3-540-59222-9
23. Stauffer, D.: Introduction to Percolation Theory. Taylor & Francis, London & Philadelphia (1985)
24. Bouchaud, J.-P., Georges, A.: Anomalous diffusion in disordered media: statistical mechanics, models, and physical applications. Phys. Rep. **195**(4&5), 127–293 (1990). https://doi.org/10.1016/0370-1573(90)90099-N
25. Masuda, N., Mason, A.P., Lambiotte, R.: Random walks and diffusion on networks. Phys. Rep. **716–717**, 1–58 (2017). https://doi.org/10.1016/j.physrep.201
26. Albert, R., Barabási, A.-L.: Statistical Mechanics of complex networks. Rev. Mod. Phys. **74**, 47–97 (2002). https://doi.org/10.1103/RevModPhys.74.47

27. Dorogovtsev, S.N.: Lectures on Complex Networks. Oxford Master Series in Physics, Computational, and Theoretical Physics. Clarendon Press, Oxford (2010)
28. Dorogovtsev, S.N., Goltsev, A.V., Mendes, J.F.F.: Critical Phenomena in complex networks. Rev. Mod. Phys. **80**(4), 1275–1335 (2008). https://doi.org/10.1103/RevModPhys.80.1275
29. Nicolis, G., Nicolis, C.: Foundations of Complex Systems. Emergence, Information and Prediction, 2nd edn. World Scientific, New Jersey (2012). https://doi.org/10.1142/8260
30. Metzler, R., Klafter, J.: The restaurant at the end of the random walk: recent developments in describing anomalous transport by fractional dynamics. J. Phys. A: Math. Gen. **37**, R161–R208 (2004). https://doi.org/10.1016/s0370-15730000070-3
31. Richert, R., Blumer, A. (eds.): Disorder Effects on Relaxational Processes: Glasses. Polymers. Ptoteins. Springer, Berlin (1994)
32. Tokuyama, M., Oppenheim, M. (eds.): Slow Dynamics in Complex Systems. AIP, Woodbury (1999)
33. Baeumer, B., Kovács, M., Meerschaert, M.M.: Numerical solutions for fractional reaction-diffusion equations. Comp. Math. Appl. **55**(10), 2212–2226 (2008). https://doi.org/10.1016/j.camwa.2007.11.012
34. Aharony, A.: Fractal Growth. In: Bunde, A., Havlin, S. (eds.) Fractals and Disordered Systems. Second Revised and Enlarged Edition. Springer, Berlin (1996)
35. Sornette, D.: Critical Phenomena in Natural Sciences. Chaos, Fractals, Self-Organization, and Disorder: Concepts and Tools. 2nd edn. Springer Series in Synergetics. Springer Complexity. Springer, Berlin (2006). https://doi.org/10.1007/3-540-33182-4
36. Kutner, R., Regulski, M.: Hierarchical spatiotemporal coupling in fractional wanderings. (II). Diffusion phase diagram for Weierstrass walks. Physica A **264**, 107–133 (1999). https://doi.org/10.1016/S0378-4371(98)00455-5
37. Gubiec, T., Klamut, J., Kutner, R.: Multi-phase long-term autocorrelated diffusion: stationary continuous-time weierstrass walk versus flight. In: Grech, D., Miśkiewicz, J. (eds.) Simplicity of Complexity in Economic and Social Systems, Chap. 4, Springer Proceedings in Complexity, pp. 55–88. Springer, Berlin (2021). https://doi.org/10.1007/978-3-030-56160-4-4
38. Sornette, D.: Why Stock Markets Crash. Critical Events in Complex Financial Systems. Princeton University Press, Princeton & Oxford (2002)
39. Jiang, Z.-Q., Xie, W.-J., Zhou, W.-X., Sornette, D.: Multifractal Analysis of Financial Markets Review. https://doi.org/10.1088/1361-6633/ab42fb
40. Kutner, R., Masoliver, J.: The continuous time random walk still trendy: fifty-year history, current state, and outlook. Eur. Phys. J. B, Top. Issue (2018)
41. Hilfer, R. (ed.): Applications of Fractional Calculus in Physics. World Scientific, Singapore (2000). https://doi.org/10.1142/3779
42. Haus, J.W., Kehr, K.W.: Diffusion in Regular and Disordered Lattices. Phys. Rep. **150**(5& 6), 263–406 (1987). https://doi.org/10.1016/0370-1573(87)90005-6
43. Kozłowska, M., Kutner, R.: Anomalous transport and diffusion versus extreme value theory. Physica A **357**, 282–304 (2005). https://doi.org/10.1016/j.physa.2005.06.03
44. Kutner, R., Masoliver, J.: The continuous-time random walk, still trendy: fifty-year history, state of the art and outlook. Eur. Phys. J. B **90**, 50 (2017). https://doi.org/10.1140/epjb/e2016-70578-3
45. Bunde, A., Havlin, S. (eds.): Fractals and Disordered Systems. Second Revised and Enlarged Edition, Springer, Berlin (1996)
46. Drożdż, S., Kowalski, R., Oświęcimka, P., Rak, R., Gęębarowski, R.: Dynamical variety of shapes in financial multifractality. Complexity **218**, ID 7015721 (2018). https://doi.org/10.1155/2018/7015721
47. Albeverio, S., Jentsch, V., H. Kantz, H. (eds.): Extreme Events in Nature and Society. (The Frontiers Collections). Springer, Berlin (2006). https://doi.org/10.1007/s10955-007-9389-7
48. Embrechts, P., Klüppelberg, C., Mikosh, T.: Modelling Extremal Events for Insurence and Finance. Springer, Berlin (1997)
49. Malevergne, Y., Sornette, D.: Extreme Financial Risks. From Dependence to Risk Management. Springer, Berlin (2006)

50. Sornette, D.: Dragon-kings, black swans and the prediction of crises. Int. J. Terraspace Sci. Eng. **2**(1), 1–18 (2009). https://doi.org/10.2139/ssrn.1596032
51. Sornette, D., Cauwels, P.: Trapped in "zero-risk" society and how to break free. Swiss Finance Institute Research Paper Series No. 20–78 (2020). https://doi.org/10.2139/ssrn.3684550

Chapter 2
Fundamental Concepts

Abstract We highlight a fundamental concept pivotal in transitioning stochastic processes from Brownian and Gaussian chains and processes to the domain of singular (containing anomalous) chains and processes. This concept serves as a signature across almost all sections of the textbook. We specifically refer to the heavy/fat tail notion in the probability distribution function (PDF) and the principles of long-range dependence and long memory observed in stationary chains/processes. Our discussion encompasses long-range dependence and long-term memory, examining scenarios (mainly) in the presence of a fat-tailed PDF, particularly heavy-tailed probability distribution. The notion of stationarity of a chain/process goes beyond merely indicating an equilibrium state (as understood in the context of the system's statistical equilibrium). It enables the consideration of stochastic chains/processes in the presence of drift- and relaxation-type processes, which hold significance in the real world. Stationary chains/processes, distinguished by their simplicity compared to more complex non-stationary counterparts, are extensively studied. This is particularly advantageous as they can effectively describe a broad spectrum of reality, spanning physical, biological, financial, and social domains.

2.1 Definitions of the Power-Law and Sub-Exponential Distribution Tails

The following sections delve into the properties of distributions featuring a power-law distribution tail—a concept widely employed across various scientific domains.[1,2] However, in this particular section, we present the definition of the algebraic distribution tail along with characteristic examples by introduction. Additionally, this section provides a valuable elucidation of the sub-exponential distribution tail.

---

[1] We remind that the term chain refers to discrete-time and process refers to continuous time.

[2] Both canonical volumes [1, 2] by the mathematician William Feller are very helpful here.

M. Chorowski et al., *Anomalous Stochastics*, Understanding Complex Systems,
https://doi.org/10.1007/978-3-031-80392-5_2

2.1.1 Power-Law Distribution Tail

Let $f(x)$ be the probability density function (PDF). By survival function $\hat{F} = 1 - F$, we denote the complementary cumulative distribution function (CCDF), where F is the cumulative distribution function (CDF) given by the formula,

$$F(x) = \int_{x_0}^{x} f(x')dx', \tag{2.1}$$

where we keep normalization $F(x = \infty) = 1$. We say that the distribution f has a power-law tail at infinity if and only if for $x \to \infty$ we have,

$$\hat{F}(x) \approx L(x)\, x^{-\alpha}, \tag{2.2}$$

where index $\alpha > 0$ and $L(\cdot)$ is a slowly-varying function of x at infinity defined for every $c > 0$ as follows,

$$\lim_{x \to \infty} \frac{L(cx)}{L(x)} = 1, \tag{2.3}$$

and discussed in Appendix A.10 as a member of Tauberin theorems (see also definition (2.5) in a Sect. 2.1.2 below). The PDFs, for which CCDFs satisfy the (2.2) condition are also called *broadened* or *fat tail* PDFs.[3] Note that the term "fat tail" refers loosely to the PDFs that have tails decaying slower than a Gaussian PDF (for financial applications, for instance) or slower than an exponential PDF (for use in recurrence problems, for instance; see also Sect. 2.1.2 below).

It is exciting and essential *the concept of the heavy tail PDF* [3–6], the most used in this textbook. For instance, the Central Limit Theorem (CLT) and Bernoulli Law of Large Numbers (BLLN) breaking are accompanied by heavy-tailed PDFs. We are dealing with heavy-tail PDF if and only if the shape exponent (or stability index) α in Eq. (2.2) is contained in the range $0 < \alpha \leq 2$, which is a particular case of the PDFs with the *power-law/algebraic tail* (in infinity) or, more generally, a particular case of the fat-tail PDFs. In other words, synonymous with the heavy-tailed PDF is a distribution with infinite variance.[4]

There are many PDFs with a fat tail (in infinity)—some are stable, but only one is α-stable. This is the Lévy heavy-tail PDF for $0 < \alpha < 2$, which we consider later. Let's add that the Gauss PDF is stable but not α-stable.[5] The distribution is α-stable

[3] The simplifying assumption that L is a constant function does not affect the understanding of the essence of the definition (2.2).

[4] In the following, we show that all absolute moments of the order $r \geq \alpha$ diverge, with the domain of distribution being unlimited.

[5] The Gauss PDF is obtained by putting $\alpha = 2$ in the characteristic function of the Lévy PDF.

if and only if the sum of i.i.d. random variables, following a heavy-tail PDF with stability index α, follow the heavy-tail PDF with the same index α [7]. The term "stability" means stability due to summing random variables. That is, the shape of the distribution does not change under the influence of the summation transformation.

To complete essential and valuable concepts and theorems, we also discuss further in the textbook:

(i) the relationship of variance (fluctuation) with the autocorrelation of process and noise versus stationarity,
(ii) the relationship of spectral density with autocorrelation function, including the so-called $1/f$ noise, and
(iii) the relationship of large deviations and rare and super-rare events with long-range dependence.

We emphasize that we study anomalous diffusion both in the presence and absence of heavy tails.

2.1.2 Sub-exponential Distribution Tail

Sub-exponential distributions are a class of fat-tailed distributions whose importance is hard to overestimate. Their name comes simply from the definition given below by Eq. (2.4). Therefore, we are dealing with the sub-exponential distribution tail of the probability distribution function $f(x)$ if and only if its complementary cumulative distribution function, CCDF,[6] (denoted here by $\hat{F}(x)$) has a following diverging limit,

$$\lim_{|x|\to\infty} \frac{\hat{F}(x)}{\exp(-a \mid x \mid)} = \infty, \tag{2.4}$$

for every $a > 0$. This definition implies that although large values (fluctuations) of a given random variable appear rarely, they are frequent enough not to be neglected. Moreover, these rare/extreme events ("black swans") are key in the generated statistical/stochastic long-term/range structure. Such a situation makes PDF, which has a sub-exponential distribution tail, a flexible candidate for modeling actual data.

[6] The notation without hat, $F(x)$, refers to the cumulative distribution function, CDF.

An important subclass of the sub-exponential class are *PDFs with regularly varying tails* defined as follows,

$$\lim_{x\to\infty} \frac{R(cx)}{R(x)} = c^{\alpha}, \tag{2.5}$$

for every $c > 0$. As one can see (compare Appendix A.10), for $\alpha = 0$, the R function becomes a slowly varying function L.

2.1.2.1 Characteristic Examples

(i) Truncated (from the bottom) log-gamma PDF,

$$R(x) = \frac{\alpha^{\beta}}{\Gamma(\beta)} \left(\ln\left(\frac{x}{a}\right)\right)^{\beta-1} \left(\frac{x}{a}\right)^{-\alpha-1}, \; 0 \le a \le x, \; \alpha > 1, \; \beta > 0, \tag{2.6}$$

or more generally,

$$R(x) = L\left(\frac{x}{a}\right)\left(\frac{x}{a}\right)^{-\alpha-1}, \; 0 \le a \le x. \tag{2.7}$$

(ii) Pareto PDF,

$$R(x) = \frac{\alpha}{a} \frac{1}{\left(1 + \frac{x}{a}\right)^{\alpha+1}}, \; x \ge 0, \; a, \alpha > 0. \tag{2.8}$$

(iii) Burr PDF,

$$R(x) = \tau\alpha a^{\alpha} \frac{x^{\tau-1}}{(a + x^{\tau})^{\alpha+1}}, \; x \ge 0, \; a, \tau, \alpha > 0. \tag{2.9}$$

In this textbook, we discuss both sub-exponential and super-exponential classes of PDFs; super-exponential PDFs decay to infinity faster than exponential. The second class includes such essential processes as normal fractional Brownian motion (nfBm).[7]

[7] The nfBm is a Gaussian process where the process variance is a non-linear function of time (i.e., sub- or super-linear).

The classifications considered in the textbook help us to discuss singular stochastic processes from different points of view, i.e., both from the point of view of the nfBm and fBm. The fractional Brownian motion (fBm) is a more general process than the nfBm as a non-Gaussian PDF can also characterize it. As a part of the fBm, we pay special attention to such stochastic processes whose PDFs are based on scaling functions and stochastic processes based on complex statistics. It is a look inspired by the processes taking place in the real world. We add that we treat the stochastic processes based on distributions with heavy tails as a reference case.

2.1.3 Additional Remarks

We emphasize that Gaussian distributions and processes (including those meeting, e.g., Central Limit Theorem (CLT)) are treated here only as a necessary reference point,[8] because Gaussian processes are insufficient to describe the reality surrounding us. This type of approach we can justify by the fact that the non-Gaussian stochastic processes can be non-ergodic, i.e., they can be governed by rare events, which are most likely the basis of both the bubbles mentioned above and stock market crashes. They are also the basis of the slow, non-Debye type relaxation of photocurrents in amorphous materials, that underlies nonergodic laser cooling, which is a necessary experimental step leading to the Bose-Einstein condensate [8, 9]. We must replace the well-known CLT with the so-called Lévy-Khinchine Generalized Central Limit Theorem (GCLT) for stable distributions with infinite variances.

The interdisciplinary physics, which we deal with in this textbook, is based mainly on the observation that the stochastic processes by which we try to describe the physical or socio-economic reality around us (e.g., stochastic population development in the presence of an external source, epidemic spread, migrations, also the evolution of a stock investor's portfolio) are often multiplicative-additive. It means they contain both multiplicative noise as well as additive noise [10].

[8] We are talking here about Gaussian PDF and processes in the narrow sense, i.e., those that satisfy the CLT. In other words, the variance is a linear function of time (discrete or continuous). Thus, we do not mean, for example, normal fractional Brownian motion (nfBm)—this we will discuss later.

With the help of the multiplicative-additive type of process, by introducing a fair competition (i.e., an appropriate balance) of both types of noise, it is possible to reproduce asymptotic random variables from the Gaussian distribution, log-normal distribution, and especially power-law distribution, which play a fundamental role in the natural and economic and social sciences. Of course, in our considerations, we will not omit the stretched exponential distribution of both the random variable and its logarithm, which we use to describe relaxation in vitreous systems [4, 11–14].

We have added to the book sections devoted to viscoelastic materials such as biopolymers, in which the relaxation of deformation under the influence of applied stress is described by fractal dynamics, i.e., various types of fractal relaxation equations [15]. As it turns out, solutions to such equations are based on H-Fox functions, which have interesting properties from the financial market's point of view. For example, they decay asymptotically according to the power law, while they behave as a stretched exponent for short periods. It allows recreating trends (rising and falling), forming local maxima of time series of daily stock indices [16].

2.2 Relative Character of the Fat Tail and Subexponential

We discuss two inspiring examples: a fat-tailed PDF and a log-normal one. We show how to obtain them formally, i.e., through appropriate nonlinear transformations of random variables. The point is that the random variables before and after such a transformation have physical interpretations in these examples.

2.2.1 *From the Weibull Distribution to the Power Law*

Let's ask an instructive question: What transformation do we have to use (if any) to turn the Weibull distribution into a power distribution? So let's start with the Weibull distribution of the variable ε,

$$W(\varepsilon) = \frac{\eta}{\bar{\varepsilon}} \left(\frac{\varepsilon}{\bar{\varepsilon}}\right)^{\eta-1} \exp\left(-\left(\frac{\varepsilon}{\bar{\varepsilon}}\right)^{\eta}\right), \quad 0 \le \varepsilon < \infty, \ \bar{\varepsilon}, \ \eta > 0, \tag{2.10}$$

where we are looking for the distribution of the transformed variable, $\tau(\varepsilon)$, given in the form of the stretch exponential function,

$$\tau(\varepsilon) = \tau_0 \exp\left((\beta\varepsilon)^{\eta}\right), \tag{2.11}$$

where β (like η) is a parameter of the non-linear transformation $\varepsilon \Rightarrow \tau$. The ε variable can be treated, for example, as the depth of the potential minimum in the energy landscape of a one-dimensional amorphous substrate (due to the random nature of these minima having a Weibull distribution). Between these minima, a single charge jumps cross-barrier (a thermally activated jump), where $\tau(\varepsilon)$ can be the average charge lifetime at a given potential minima.

Thus, we are looking for the probability distribution $P(\tau)$. As a scalar, this distribution satisfies the invariance condition (here for right-handed orientation) of the form,

$$P(\tau)d\tau = W(\varepsilon)d\varepsilon \equiv P(\tau) = W(\varepsilon(\tau))\left(\frac{d\tau(\varepsilon)}{d\varepsilon}\right)^{-1}. \tag{2.12}$$

Using the definition (2.11) we get,

$$\frac{d\tau(\varepsilon)}{d\varepsilon} = \eta\,\beta(\beta\varepsilon)^{\eta-1}\tau(\varepsilon) \tag{2.13}$$

and

$$W(\tau) = \frac{1}{(\beta\bar{\varepsilon})^{\eta}}\frac{1}{\tau}\exp\left(-\alpha(\beta\varepsilon)^{\eta}\right) = \frac{\alpha}{\tau}\exp\left(-\alpha\ln\frac{\tau}{\tau_0}\right) = \frac{1}{\tau_0}\frac{\alpha}{(\tau/\tau_0)^{1+\alpha}}, \tag{2.14}$$

where the exponent $\alpha = (\beta\bar{\varepsilon})^{-\eta}$ and along the way we used the definition (2.11).

As one can see, thanks to the non-linear transformation given by Eq. (2.11), we transformed a distribution that does not have a fat tail for $\eta > 1$ into one that does. So the fat tail property is relative—it depends on the variables in which we represent the distribution, and we are talking about variables that have physical interpretations, not just any variables.

2.2.2 *From Gaussian to Log-Normal PDF*

It is easy to check that the Gaussian distribution has no fat tail. However, it is possible to find a nonlinear transformation of a random variable that transforms the Gaussian distribution into a log-normal distribution with a fat tail.

So, let a one-dimensional Gaussian half-distribution be given,

$$G(\varepsilon) = \frac{2}{\sqrt{2\pi\sigma^2}}\exp\left(-\frac{\varepsilon^2}{2\sigma^2}\right), \tag{2.15}$$

where the random variable ε can be interpreted in the same way as in the previous example.

Then, we introduce a non-linear transformation,

$$\varepsilon \Rightarrow \tau = \tau(\varepsilon) = \tau_0 \exp(\beta\varepsilon). \tag{2.16}$$

Our goal is to find (as in the previous example) the probability distribution of $P(\tau)$ of the variable τ. Using the equality (2.12) (where W is replaced with G) we get,

$$\begin{aligned} P(\tau) &= G(\varepsilon(\tau)) \left(\frac{d\tau(\varepsilon)}{d\varepsilon}\right)^{-1} \\ &= \frac{1}{\tau_0}\frac{2}{\sqrt{2\pi}}\frac{1}{\beta\sigma}\frac{1}{\tau/\tau_0} \exp\left(-\frac{1}{2(\beta\sigma)^2}\ln^2\left(\frac{\tau}{\tau_0}\right)\right), \end{aligned} \tag{2.17}$$

i.e., the distribution we are looking for. Using the definition (2.4), it is easy to check that it has a fat tail.

We emphasize that how we obtained the distributions in both examples is only one way—these distributions can also be obtained in other ways. However, in this subsection, we wanted to show the role of the non-linear transformation of random variables and their distributions. One can't always guess this transformation so easily—the hint came from statistical physics here.

The target distributions (power and log-normal) presented in this subsection are also derived below in Sect. 2.3 for stochastic chains and processes with direct physical interpretations.

2.3 Multiplicative Chains: Log-Normal PDF

The log-normal distribution appears when solving probabilistic problems, such as the type of crushing (grinding) of coal, stone, or ore. Following Kolmogorov [4], such chains and processes are multiplicative.[9] So far, we have discussed additive chains and processes, i.e., they are the sum of sequential, single, independent random variables through successive filters. It leads to the log-normal PDF, thus giving its physical interpretation.

Suppose that as a result of multistage sequential crushing (in sieves/filters with smaller and smaller mesh diameters), the linear grain size X creates a sequence of random variables $(X_0, X_1, X_2, \ldots, X_{n-1}, X_n)$, where X_n is the linear size of a single grain at n-th stage (degree, level) of crushing,

[9] We talk about stochastic chains when time is discrete. When time is continuous, we speak about stochastic processes.

$$0 < X_n < X_{n-1},\ n = 1, 2, \ldots. \tag{2.18}$$

Of course, X_n is a random variable with a certain (finite, usually small) statistical scatter around average grain size $\langle X_n \rangle$[10] at a given stage (i.e., after some time) crushing. Subsequent difference $X_n - X_{n-1} (< 0)$ is some random fraction of the output grain size X_{n-1},

$$X_n - X_{n-1} = R_n \cdot X_{n-1},\ n = 1, 2, \ldots, \tag{2.19}$$

where R_n is a random variable in the range $-1 < R_n < 0$[11] with a distribution (for simplicity that does not change the essence of things) independent of n. From the above, it follows (by recursively using recursion (2.19) for successive n) that the random variable X_n has the following multiplicative representation,

$$x_n \overset{def.}{=} \frac{X_n}{X_0} = \prod_{j=1}^{n} p_j,\ n = 1, 2, \ldots, \tag{2.20}$$

where $p_j = 1 + R_j = X_j / X_{j-1}$ is a random variable from the range $0 < p_j < 1$ (of course, also distributed independent of j). Thus, taking logarithms of this expression, we get the additive representation of the Eq. (2.20)

$$\ln x_n = \sum_{j=1}^{n} y_j,\ n = 1, 2, \ldots, \tag{2.21}$$

where $y_j = \ln p_j < 0,\ j = 0, 1, 2, \ldots,$ is a random variable with a finite average value $\langle y_j \rangle = \langle \ln p_j \rangle < 0$[12] and finite variance, $\sigma^2(y_j) = \langle y_j^2 \rangle - \langle y_j \rangle^2 = \langle (\ln p_j)^2 \rangle - \langle \ln p_j \rangle^2$. For simplicity, we assume that both quantities are now independent of j, i.e., $\langle y_j \rangle = \langle y \rangle = \langle \ln p_j \rangle = \langle \ln p \rangle$ and $\sigma^2(y_j) = \langle y^2 \rangle - \langle y \rangle^2 = \langle (\ln p)^2 \rangle - \langle \ln p \rangle^2 = \sigma^2$.

Note that the Eq. (2.21) can now be rewritten as,

$$X_n = \exp(-n \overline{|\ln p\,|}) X_0, \tag{2.22}$$

---

[10] This average should be understood as the average overall grains at a given nth stage of crushing. Note that at each crushing step, n, a certain steady state is created, in which the grain size is no longer reduced. This is because the grains have flown onto a sieve with a smaller hole diameter, thus moving on to the next crushing stage, $n + 1$. They are subject to further crushing until a new steady state is formed, etc.

[11] This double inequality follows immediately from the double inequality (2.18) expressed as $0 < X_n < (1 + R_n) X_{n-1} < X_{n-1}$.

[12] The angles' brackets, $\langle \ldots \rangle$, mean, as usual, the average over the statistical ensemble of replicas of the considered stochastic chain (here, for given j).

which shows that arithmetic average, $\overline{| \ln p |} \stackrel{def.}{=} \frac{1}{n}\sum_{j=1}^{n} | \ln p_j |$, acts as the rate of asymptotic decay of the variable X_n, i.e., how fast the process $\ln x_n$ moves **infinitely** from its initial state $\ln(x_0 = 1) = 0$ to its left side.

As one can see, *random variable*, $\ln x_n$, *has no lower bound*, and is bounded by 0 from the top. It follows that only when the effect of this constraint is negligible can it be assumed that the total random variable $\sum_{j=1}^{n} y_j$ satisfies CLT, i.e., that the random variable $\ln x_n$ has an asymptotically normal distribution,

$$\mathcal{P}(\ln x_n))d\ln x_n \approx \frac{1}{\sqrt{2\pi\sigma^2(n)}} \exp\left(-\frac{(\ln x_n - \mu_n)^2}{2\sigma^2(n)}\right) d\ln x_n \tag{2.23}$$

where the average overall grains at a given stage of crushing,

$$\mu_n \stackrel{def.}{=} \langle \ln x_n \rangle = \sum_{j=1}^{n} \langle y_j \rangle = n\langle y \rangle = n\langle \ln p \rangle \tag{2.24}$$

and the sum variance,

$$\sigma^2(n) = \sigma^2(\ln x_n) = \sum_{j=1}^{n} \sigma^2(y_j) = n\sigma^2, \tag{2.25}$$

both independent of j. In the above notation, the condition of fulfilling the CLT can be written as follows,

$$| \mu_n | \gg \sigma(n) \equiv \langle y \rangle \gg \frac{\sigma}{\sqrt{n}}, \tag{2.26}$$

which is also a condition for moving the PDF center far from its upper limit.

From PDF (2.23) we get directly log-normal PDF for variable x_n,

$$P(x_n) \approx \frac{1}{\sqrt{2\pi\sigma^2(n)}} \frac{1}{x_n} \exp\left(-\frac{(\ln x_n - \mu_n)^2}{2\sigma^2(n)}\right), \tag{2.27}$$

or for variable $x_n/\hat{x}_n$, where $\ln \hat{x}_n \stackrel{def.}{=} \mu_n (= \langle \ln x_n \rangle)$,

$$P\left(\frac{x_n}{\hat{x}_n}\right) \approx \frac{1}{\sqrt{2\pi\sigma^2(n)}} \frac{1}{x_n/\hat{x}_n} \exp\left(-\frac{(\ln(x_n/\hat{x}_n))^2}{2\sigma^2(n)}\right), \tag{2.28}$$

which form the basis of our further discussion in this chapter. Of course, neither the (2.23) nor the (2.27) and (2.28) distributions are asymptotically stationary, but their usefulness is hard to overestimate. Let us emphasize once again that *underlies the full (unbounded) distribution log-normal was that there were no spatial constraints on the random variable multiplicative stochastic chain.*

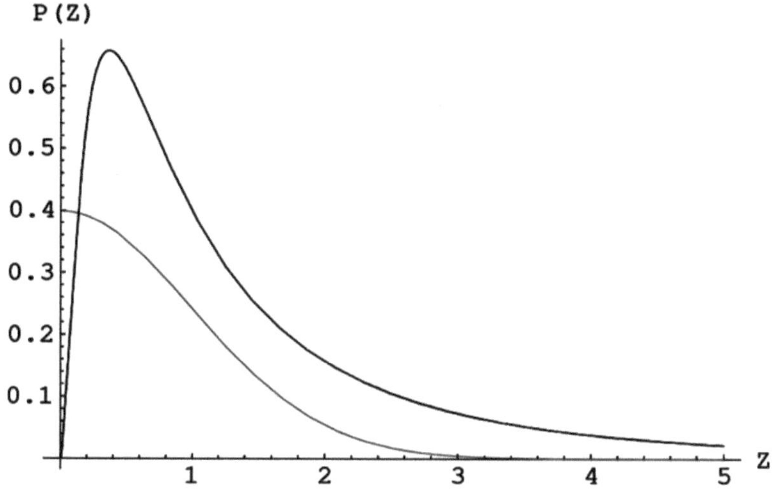

Fig. 2.1 Comparison of the log-normal PDF (the one with a local maximum) with the corresponding Gaussian PDF, both for $Z \geq 0$

It is easy to check (as a simple exercise) that the distribution $P\left(\frac{x_n}{\hat{x}_n}\right)$ has a maximum for

$$\frac{x_n^{\max}}{\hat{x}_n} = \exp(-\sigma^2(n)) \tag{2.29}$$

and it is asymmetric.

Figure 2.1 shows this distribution for the relative variable $Z \overset{def.}{=} x_n/\hat{x}_n$ and unit dispersion $\sigma(n) = 1$—one can see how it differs from the corresponding normal (Gaussian) PDF with zero expected value and unit variance.

Note that the maximum of the $\ln(x_n^{\max})$ distribution (2.27) shifts to increasingly negative numbers as n grows or is "pushed" away from zero in the negative direction because the average $\langle \ln p \rangle$ is less than zero.

The log-normal PDF has one more characteristic property: the knowledge that even all moments of this distribution do not define it unambiguously. It can be easily checked (in the form of an exercise) that the PDF of the form,

$$P(Z) = \frac{1}{\sqrt{2\pi\sigma^2(n)}}\frac{1}{Z}\exp\left(-\frac{\ln^2(Z)}{2\sigma^2(n)}\right)[1 + a\sin(2\pi\ln(Z))],\ -1 < a < 1, \tag{2.30}$$

has the same moments as the corresponding log-normal PDF (2.28). This property can be called a *mathematical camouflage*.

2.3.1 From Log-Normal to Power-Law Distribution

We now show the important property of the log-normal PDF that mimics the power-law PDF over a wide range of independent/explanatory, x-variable. For this, we note that the following identity holds,

$$\exp(a \cdot \ln^2 x) \equiv x^{a \cdot \ln x}; \tag{2.31}$$

using it, we can rewrite the log-normal distribution (2.28) as:

$$P(x_n) = \frac{1}{\sqrt{2\pi\sigma^2(n)}} \left(\frac{x_n}{\hat{x}_n}\right)^{-1-\alpha(x_n)}, \tag{2.32}$$

where $\alpha(x_n)$ is a slowly varying function as $\alpha(x_n) \overset{def.}{=} \frac{1}{2\sigma^2(n)} \ln\left(\frac{x_n}{\hat{x}_n}\right)$. As one can see when the condition is met

$$\mid \alpha(x_n) \mid \ll 1 \iff \mid \ln(x_n/\hat{x}_n) \mid \ll 2\sigma^2(n), \tag{2.33}$$

then we are dealing with a power-law PDF with an exponent approximately equal to 1, a source of noise of the $1/f$ type, which is discussed below.

In the Fig. 2.2 we compared the log-normal PDF (2.28) that satisfies condition (2.33) (because we assumed that $\sigma(n) = 10$ and $1 \leq Z(= x_n/\hat{x}_n) \leq 100$) with the corresponding distribution (i.e., having the exact value of the parameter $\sigma(n)$) exponential by simply putting the exponent $\alpha = 0$ in the formula (2.32).

As can be seen, distinguishing between the two distributions in these conditions is even more complicated when considering the inevitable scattering of experimental points (not shown in the figure).

2.3.2 Stochastic Multiplicative Chain Versus Repulsive Barrier

Often, in multiplicative stochastic processes with which we try to describe reality, random variables have a natural bottom-up limit—they are positive. For example, the share price on the stock exchange or the size of a particular animal population in a given reserve territory, in fact, never drops to zero because when the lower limit is reached (either the value of the stock or the size of the population in the reserve), there is an external intervention that prevents the threshold from being exceeded. Thus, reality often imposes a lower limiting boundary condition on stochastic dynamics. That is, a barrier (threshold) appears that blocks the decrease of these characteristics. This has consequences, which we discuss below.

Suppose then that the Eq. (2.19) describes the stochastic price dynamics of the stock, $X_n(> 0)$, at time n [13]. Let us rewrite this equation in the logarithmic form,

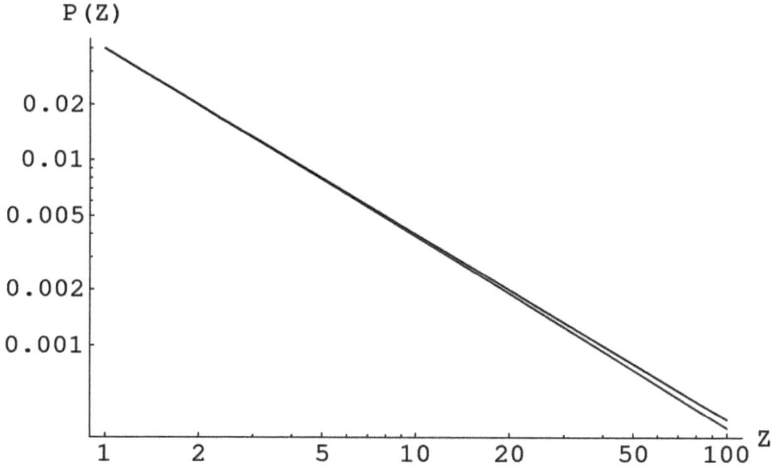

Fig. 2.2 Comparison of the log-normal PDF (the one located slightly lower) with the corresponding power-law PDF with an exponent of 1 on a log-log scale. As can be seen, both curves are difficult to distinguish in the Z range satisfying Eq. (2.33)

$$Y_{n+1} = Y_n + l_n, \ n = 0, 1, 2, \ldots, \tag{2.34}$$

which indicates that we are dealing here with a random walk of the Brownian motion type, in discrete time n, where the current position $Y_n = \ln X_n$ and the current displacement $l_n = \ln p_n$. We continue to assume (as before) that p_n (and therefore l_n) is drawn from a distribution that is independent of n and that $\langle \ln p \rangle < 0$ which doesn't always mean $p_n < 1$—sometimes it can happen here that $p_n \geq 1$, which distinguishes this situation from the previous one.[13] Note that the latter condition allows for occasional (due to fluctuations) $\ln p_n > 0$. However, after a while, there is a drift towards the blocking barrier located to the left of Y_0 at the distant point Y^{bar}, i.e., $Y^{bar} \ll Y_0$. Taking into account the existence of the blocking barrier, Eq. (2.34) can be rewritten in an algorithmic form,

$$Y_{n+1} = \begin{cases} Y_n + l_n, & \text{for } Y_n > Y^{bar} \\ 0, & \text{for } Y_n \leq Y^{bar}. \end{cases} \tag{2.35}$$

It should be emphasized that the existence of the blocking barrier leads to accumulating all the displacements before it. Its absence would lead to the realization of the unlimited process to the left of the barrier. In this way, in the vicinity of the barrier, for asymptotically large n, a significant repulsive thermodynamic force (chemical, proportional to the chemical potential gradient and hence to the probability density gradient) can arise, leading to diffusion in the positive direction (to the right of the barrier). Existence of two opposing currents: drift $j_{drift} = V \mathcal{P}_{stat}(Y_n)$

[13] Of course, in both cases the displacements of l_n are by definition statistically independent quantities.

towards the barrier, where $V = - \mid V \mid$ is the (dimensionless) drift velocity with $\mid V \mid = \mid \langle \ln p \rangle \mid$, $\mathcal{P}_{stat}(Y_n)$ is the desired stationary probability distribution,[14] and diffusion $j_{diff} = -D \frac{d\mathcal{P}_{sat}(Y_n)}{dY_n}$ in the opposite direction caused by the existence of this barrier, where $D = [\langle (\ln p)^2 \rangle - \langle \ln p \rangle^2]/2$ is the (dimensionless, one-dimensional) diffusion coefficient, can lead to a steady state[15] of the process, in contrast to the situation discussed previously (also in this Sect. 2.3).

From the condition of the total flux disappearing in the stationary state,

$$j = j_{drift} + j_{diff} = 0, \tag{2.36}$$

we get,

$$\frac{d\mathcal{P}_{stat}(Y_n)}{dY_n} = -\frac{\mid V \mid}{D} \mathcal{P}_{stat}(Y_n), \tag{2.37}$$

from which it follows directly (with a good approximation) that

$$\mathcal{P}_{stat}(Y_n) = \begin{cases} \alpha \exp(-\alpha(Y_n - Y^{bar}), & \text{for } Y^{bar} < Y_n \leq Y_0 \\ 0, & \text{for } Y_n \leq Y^{bar}, \end{cases} \tag{2.38}$$

where we introduced the notation $\alpha = \frac{|V|}{D}$ and used the normalization condition,

$$1 \approx \int_{Y^{bar}}^{\infty} \mathcal{P}_{stat}(Y_n) dY_n. \tag{2.39}$$

Using the Eq. (2.38), we return in the PDF that occurs there to the output variable X_n, i.e., we use the invariance of distributions, obtaining,

$$P_{stat}(X_n) = \mathcal{P}_{stat}(Y_n(X_n)) \left| \frac{dY_n}{dX_n} \right| \Rightarrow P_{stat}(X_n) = \frac{\alpha}{X^{bar}} \frac{1}{(X_n/X^{bar})^{1+\alpha}}, \tag{2.40}$$

where X^{bar} is the coordinate of the barrier, and n is the time step for the system to reach the steady state; we got the Pareto PDF we discussed initially. As one can see, stationary PDF is driven by the exponent defined by the signal-to-noise ratio (SNR)[16]. A stationary power-law distribution can be formed even for $\alpha \ll 1$

---

[14] The concept of stationarity is defined in many different ways in the literature because we deal with many types of it. In our textbook, each time we talk about a specific type of stationarity, we will specify what we mean. The stationarity we are dealing with here belongs to the category of strong stationarity because it concerns PDF—here $\mathcal{P}_{stat}$.

[15] The stationary state and steady state we consider as synonym.

[16] The SNR is a key parameter characterizing the transmission of, for example, an electrical signal in various types of devices.

because it is asymptotic. We continue these considerations below in the context of the significant multiplicative-additive stochastic chain.

Let us emphasize: the emergence of the power-law PDF was possible thanks to introducing *the lower bound to the value of the log-normal distribution.*

2.3.3 Multiplicative-Additive Stochastic Chain: A Useful Property

As we have seen, every (purely) multiplicative chain, driven by a positively defined noise and a positive initial value of the chain, corresponds to an additive chain. Conversely, there is a multiplicative chain to every (purely) additive chain. The relationship between such chains is built using the logarithm function; we obtain the corresponding chain additive by taking the logarithm of a multiplicative chain. Now, we deal with the multiplicative-additive chain and indicate under what circumstances it leads to a stationary (steady) state.

Thus, let be given a stochastic multiplicative-additive chain defined by the discrete-time equation of stochastic dynamics,

$$X_{n+1} = p_n \cdot X_n + b_n, \; n = 0, 1, 2, \ldots, \tag{2.41}$$

where the independent multiplicative p_n and additive b_n are positive and drawn from two independent distributions, respectively.[17] It should be emphasized that no additional conditions have been imposed on the above problem, e.g., boundary conditions (in this sense, it is a free problem). As it is known [17], the above equation has the exact solution in the form,

$$X_n = X_0 \exp(n\langle \ln p \rangle) + \sum_{j=1}^{n} b_j \prod_{l=j+1}^{n} p_l, \tag{2.42}$$

We continue to discuss the case $\langle \ln p \rangle < 0$, which (as we will show) is analogous to the one discussed previously for the (purely) multiplicative process, with the role of the "repulsive force" in the above solution played by a positive ("soft") inhomogeneity built of both types of noise, and not, as before, by a (rigid) repulsive barrier. In other words, the rivalry of both noises can lead to an asymptotically stable solution of the form,

[17] Note that the process (2.41) is also called an affine map and the variable X_n is called in mathematics the Kesten variable.

$$X_{n\to\infty} = \sum_{j=1}^{n\to\infty} b_j \prod_{l=j+1}^{n\to\infty} p_l. \tag{2.43}$$

To find a plateau for this process, we average both sides of the expression (2.42) over this chain's statistical ensemble of replicas. This gives us the plateau we are looking for,

$$\langle X \rangle = \frac{\langle b \rangle}{1 - \langle p \rangle}, \tag{2.44}$$

where (recall) $\langle p \rangle < 1$. Thus, it is the average distance that the test particle will reach after an arbitrarily long time.

The subsection below presents other essential properties of the Eq. (2.41).

2.3.4 The Langevin and Fokker-Planck Equations and Power-Law PDF

2.3.4.1 General Representation

The recurrence Eq. (2.41) can be reduced to the Langevin equation for an asymptotically long time, i.e., for $n \gg 1$. To do this, first, we will average them over the statistical set of replicas of this process, with a fixed value of the multiplicative noise p_n, i.e., over the additive noise b_n, obtaining the equation,

$$\bar{X}_{n+1} = p_n \cdot \bar{X}_n + \langle b \rangle, \; n = 0, 1, 2, \ldots, \tag{2.45}$$

where $\bar{..}$ is the mean over additive noise. Now, we rewrite (2.45) in a more useful form as

$$\frac{\bar{X}_{n+1} - \bar{X}_n}{\bar{X}_n} = \frac{\langle b \rangle}{\bar{X}_n} + p_n - 1, \tag{2.46}$$

from which we get the equation we are looking for

$$\frac{dY(t)}{dt} = F(Y(t)) + V + \eta(t),$$
$$F(Y(t)) \stackrel{def.}{=} \langle b \rangle \exp(-Y(t)), \tag{2.47}$$

where for convenience,

- we replaced the discrete time n by the continuous (dimensionless) t, replacing the unit of time by the infinitesimal short time interval dt, So we went from a chain to a stochastic process;

- we used a new variable $Y(t) \stackrel{\text{def.}}{=} \ln \bar{X}(t)$;
- we introduced a (dimensionless) drift velocity $V = \langle p \rangle - 1 \approx \langle \ln p \rangle$, which in this case can be both positive and negative; below we will introduce a more precise velocity, V, constrain;
- here we defined δ-self-correlated white noise $\eta(t) = p(t) - \langle p \rangle$, having a vanishing average and a (dimensionless) variance $2D = \sigma^2 = \langle \eta^2 \rangle = \langle p^2 \rangle - \langle p \rangle^2 \approx \langle \ln^2 p \rangle - (\langle \ln p \rangle)^2$.

As one can see, $F(Y(t))$ acts as a force that is stronger the lower $Y(t)$ is, preventing the process from reaching negative $Y(t)$ values. We are dealing here not only with the additive noise $\eta(t)$ (which can take both positive and negative values) but also with the positive stochastic force $F(Y(t))$ playing a vital role in this problem.

The Fokker-Planck equation corresponding to the Langevin equation (2.47) takes the form (see general derivation provided in Appendix A.14),

$$\begin{aligned} \frac{\partial \mathcal{P}(Y,t)}{\partial t} &= -\frac{\partial j(Y,t)}{\partial Y}, \\ j(Y,t) &= [V + F(Y)]\mathcal{P}(Y,t) - D\frac{\partial \mathcal{P}(Y,t)}{\partial Y}. \end{aligned} \tag{2.48}$$

For this reason, we are looking for an equilibrium solution, i.e., an answer to the question of under what conditions the current density $j(Y,t)$ disappears.

2.3.4.2 Stationary Solution

The stationary solution is obtained assuming the vanishing current density (2.48), analogously to how we did it in Sect. 2.3.2. It leads to the following equation independent of time,

$$\frac{d \ln \mathcal{P}_{stat}(Y)}{dY} = \frac{V + F(Y)}{D}, \tag{2.49}$$

Solving the above equation, we get, after returning to the $\bar{X}$ variable, PDF $P_{stat}(\bar{X})$ in the form of time-independent inverse gamma PDF,

$$P_{stat}(\bar{X}) = \frac{const}{\bar{X}^{1-V/D}} \exp\left(-\frac{\langle b \rangle}{D}\frac{1}{\bar{X}}\right), \tag{2.50}$$

where we assume that there is a constraint on the drift velocity in the form $V < D$.

Note that the exponent here is identical to the multiplicative process with a repulsive barrier. This is because, under the conditions mentioned above, the role of additive noise is vanishing. After all, we neglected the fluctuations of this noise. This means we have given the conditions under which the present problem reduces to the one considered in the Sect. 2.3.2.

As one can see, the PDF given by the Eq. (2.50), for $\bar{X} \gg \langle b \rangle / D > 0$, goes to the power distribution of the variable $\bar{X}$. We obtained an in-depth explanation of the circumstances in which multiplicative-additive stochastic dynamics generate a power-law PDF.

2.4 Breaking of the Central Limit Theorem by the Long-Range Autocorrelations

2.4.1 Definition of the Long-Range Autocorrelations

We consider a stationary stochastic chain x_n, $n = 0, 1, 2, \ldots$, and a bilinear autocorrelation function[18] for this chain defined as follows,

$$C_j = \langle x_n \, x_{n+j} \rangle, \tag{2.51}$$

where $\langle \ldots \rangle$ is the average over all process executions and $\langle x_n \rangle = 0$ (i.e., the case of no drift). We say that we are dealing with long-range autocorrelations of index η if and only if for $j \to \infty$,

$$C_j \approx L_j \, j^{-\eta} \tag{2.52}$$

where L_j is a slowly-varying function of j and the exponent $\eta = 2 - 2H(> 0)$, while H is the Hurst exponent.

2.4.2 General Relation Between Variance and Autocorrelation

We now show that one of the reasons for CLT failure is long-term autocorrelations of the stochastic process occurring between single-process (test particle) erratic movement. The existence of autocorrelation means that individual displacements of the test particle are statistically dependent on each other. A typical example of such a situation may be the stray ("heads") of the stiffened polymer depending on the orientation of the starting monomer, called "directed random walk," "self-avoiding random walk," or correlated channel walk (one-dimensional), erratic ions

[18] A given function $L(x, y)$ is bilinear if it is linear for each of the arguments x and y separately. (Here, $L = C_j$, $x = x_n$, $y = x_{n+j}$).

traveling in the crystal lattice (accompanied by "back-jump correlations" or "feed-back" correlations).

Consider the sum autocorrelation function that appears in the variance $\sigma_X^2(n)$ (e.g., one-dimensional random walk without drift) of the cumulative (n-step) displacement of the process $X(n) = \sum_{j=1}^{n} x_j$, where $x_j,\ j = 1, 2, \ldots, n$, are single-step (elementary) displacements. Namely,

$$\sigma_X^2(n) = \langle [X(n)]^2 \rangle = n\sigma_x^2 + 2\mathcal{K}(n), \tag{2.53}$$

where $\langle \ldots \rangle$ means the average over the statistical ensemble of n-step replicas of the process, the elementary variance $\sigma_x^2 = \langle x_j^2 \rangle$ is independent of j due to time uniformity (or stationarity of the process), and sum autocorrelation function,

$$\mathcal{K}(n) = \frac{1}{2} \sum_{i \neq j}^{n} K(i, j) = \sum_{i<j} K(i, j), \tag{2.54}$$

where the autocorrelation function,

$$K(i, j) \overset{\text{def.}}{=} \langle x_i \cdot x_j \rangle, \tag{2.55}$$

applies to two arbitrarily selected different single-step displacements of the test particle.

The formula (2.54) can be rewritten as

$$\mathcal{K}(n) = \sum_{i=1}^{n-1} \sum_{j=i+1}^{n} K(i, j) = \sum_{i=1}^{n-1} \sum_{j-i=1}^{n-i} K(i, j), \tag{2.56}$$

which will continue to transform. Assuming that all autocorrelation functions are homogeneous/uniform in a discrete time[19] and does not exhibit asymptotic oscillations, it can be written,

$$K(i, j) = K(j - i) \tag{2.57}$$

which together with (2.56) yields,

$$\mathcal{K}(n) = \sum_{i=1}^{n-1} \sum_{m=1}^{n-i} K(m). \tag{2.58}$$

[19] This type of time dependence defines the category of weak stationarity because it concerns only secondary characteristics and their dependence solely on differences between given moments. That is, it is homogeneous/uniform from the point of view of time.

Table 2.1 A summary of all elements of the sum $\sum_{i=1}^{n-1}\sum_{m=1}^{n-i} K(m) + \sum_{j=1}^{n} jZ(j)$

$i = 1$	$K(1)$	$K(2)$	...	$K(n-2)$	$K(n-1)$	$Z(n)$
$i = 2$	$K(1)$	$K(2)$	...	$K(n-2)$	$Z(n-1)$	$Z(n)$
...	...	...	...	...	...	...
$i = n-1$	$K(1)$	$Z(2)$	...	$Z(n-2)$	$Z(n-1)$	$Z(n)$
$i = n$	$Z(1)$	$Z(2)$	...	$Z(n-2)$	$Z(n-1)$	$Z(n)$

It should be emphasized that the homogeneity/uniformity property is a common feature- it is observed not only in equilibrium or, more generally, stationary states of the system but even in the case of relaxation. Using the expression (2.58), we show below that the sum autocorrelation function $\mathcal{K}(n)$ can be transformed into a convenient form,

$$\mathcal{K}(n) = \sum_{j=1}^{n}(n-j)K(j) = n\sum_{j=1}^{n} K(j) - \sum_{j=1}^{n} jK(j). \tag{2.59}$$

The derivation of the formula (2.59) is simply a compilation of all components of the double sum (2.58) supplemented with the auxiliary sum $\sum_{j=1}^{n} jZ(j)$. The components $K(j)$ and $Z(j)$, $j = 1, 2, \dots, n$, have been set in a properly ordered manner in the Table 2.1 (for now, the elements $Z(j)$, $j = 1, \dots n$, are arbitrary).

Now, summing the elements of the Table 2.1 and assuming that $Z(j) = K(j)$, $j = 1, \dots, n$, we get,

$$n\sum_{j=1}^{n} K(j) = \sum_{i=1}^{n-1}\sum_{m=1}^{n-i} K(m) + \sum_{j=1}^{n} jK(j), \tag{2.60}$$

and hence, (basing on (2.58)) the searched expression (2.59).

2.4.3 *Variance and a Specific Autocorrelation Function*

Finding the sum variance, $\sigma_X^2(n)$, versus (discrete) time steps n for large enough n requires knowing the sum autocorrelation function, $\mathcal{K}$, for large n and hence the dependence of the autocorrelation function K on j and m for large n.

There are three different cases of asymptotic (when $j,\ n \to \infty$) and monotonic decay of the autocorrelation function,

(1) fading faster than $1/j$, i.e., $K(j) \approx Cj^{-(1+\alpha)}$, $\alpha > 0$,
(2) decaying slower than $1/j$, i.e., $K(j) \approx Cj^{-\gamma}$, $0 < \gamma < 1$,
(3) $K(j) \approx C/j$,

where $C(> 0)$ is a constant. Case (1) defines *short-term autocorrelations*, and case (2) *long-term autocorrelations*, while case (3) defines a *marginal* (transitional) case.

Let us now represent the sum autocorrelation function as the sum of the following two components

$$\mathcal{K}(n) = \mathcal{K}(n_0) + \mathcal{K}_>(n), \tag{2.61}$$

where

$$\mathcal{K}_>(n) = n \sum_{j=n_0+1}^{n} K(j) - \sum_{j=n_0+1}^{n} jK(j), \tag{2.62}$$

here n_0 is the smallest natural number above which both sums in (2.62) can be replaced integration. Therefore,

$$\mathcal{K}_>(n) \approx n \int_{n_0}^{n} K(j)dj - \int_{n_0}^{n} jK(j)dj; \tag{2.63}$$

as one can see, $\mathcal{K}(n_0)$ is a constant (negligible for further considerations).

We can now determine the asymptotic behavior of the sum autocorrelation function $\mathcal{K}(n)$ for the three cases mentioned above.

2.4.3.1 Case 1

Let us first consider the situation where $0 < \alpha \neq 1$. Then, from (2.61) and (2.63) after simple integration, it follows that for $n \to \infty$,

$$\mathcal{K}_>(n) \approx Cn \int_{n_0}^{n} \frac{1}{j^{1+\alpha}} dj - C \int_{n_0}^{n} \frac{1}{j^{\alpha}} dj = C_0 + C_1 n^{1-\alpha}, \tag{2.64}$$

where $C_0,\ C_1$ are constants (easily related to the constant C and the exponent α).

Let's consider the same case but for the special situation where $\alpha = 1$. Then, similarly as before, from (2.61) and (2.63) we obtain,

$$\mathcal{K}(n) \approx C_0 - C_1 \ln(n). \tag{2.65}$$

As one can see, each of the expressions (2.64) and (2.65) divided by n is, in the case of short-range autocorrelations, a decreasing function of n. Thus, from the equation (2.53) for the variance of the cumulative displacement of the process, we get for asymptotically large n,

$$(\sigma_X(n))^2 \approx n(\sigma_x)^2. \tag{2.66}$$

We can state that *short-term autocorrelations do not asymptotically change the Brownian nature of random walk.*

2.4.3.2 Case 2

Performing the calculations in the same way as in the previous case, we get

$$\mathcal{K}(n) \approx C_0 + C_1 n^{2-\gamma}, \tag{2.67}$$

where, as one can see, the exponent $2 - \gamma > 1$. Combining the above expression with (2.53) we get for asymptotically large n,

$$(\sigma_X(n))^2 \approx n(\sigma_x)^2 + (C_1 + C_2)n^{2-\gamma} \sim n^{2-\gamma}. \tag{2.68}$$

Thus, the sum displacement dispersion grows superlinearly with the total number of displacements. This result means that the *long-range autocorrelations increases the universality class of random walk, leading to its persistence or superdiffusion.*

Let us note that the extreme/borderline case of superdiffusion is the ballistic diffusion corresponding to the exponent $\gamma \to 0$, which physically means that all single displacements are identical (i.e., of equal length and facing the same direction)—this can describe the extreme case of "diffusion" of a 100% polymer stiffness. In other words, it can be, for example, uniform motion (with a constant speed) of a rigid rod.

2.4.3.3 Case 3

As in the previous cases, with (2.61) and (2.63) we get (for asymptotically large n) after simple calculations,

$$(\sigma_X(n))^2 \approx n((\sigma_x)^2 - C_1) + C_2 n \ln(n) \sim n \ln(n). \tag{2.69}$$

As one can see, this results from a different nature than the previous ones—**slowly (according to Zipf law) decaying autocorrelations enrich the broadly understood diffusion.**

It should be emphasized that long-range autocorrelations generally lead to boundary distributions having asymptotically non-Brownian variance [24].

2.5 CLT and Power-Law Decay: The Collision of Two Random Walk Worlds

We present a prominent example to get an idea of how CLT works when a given PDF decays asymptotically according to power-law, i.e., it has an algebraically vanishing tail at a finite variance [12, 13].

It is a different situation from the one we deal with, for example, in the case of the truncated Lévy PDF [11], where the random variable follows the Lévy distribution in a finite interval of the independent variable, beyond which the distribution quickly vanishes (e.g., exponentially like the Gamma PDF).

2.5.1 Gaussian Distribution and Power-Law Distribution in One

In this subsection, we consider such ranges of the domain of a given PDF in which it takes the form of either a Gaussian PDF or a power-law PDF, both in the context of CLT and large deviations in the sum of variables, while introducing the vital concept of "infinite divisibility."

As an initial step, let us consider an explicit single-step PDF in the form of flexible one-dimensional Student's (i.e., W. S. Gosset's) t-distribution with $\nu = 1, 2, 3, \ldots,$ degrees of freedom,

$$p_\nu(x) = \frac{1}{\sqrt{\pi}} \frac{\Gamma\left(\frac{1+\nu}{2}\right)}{\Gamma\left(\frac{\nu}{2}\right)} \frac{\sigma^\nu}{(\sigma^2 + x^2)^{(1+\nu)/2}}, \tag{2.70}$$

where σ measures its width.

One can see that for $x^2 \gg \sigma^2$ the above PDF takes the power-law form,

$$p_\nu(x) \approx \frac{1}{\sqrt{\pi}} \frac{\Gamma\left(\frac{1+\nu}{2}\right)}{\Gamma\left(\frac{\nu}{2}\right)} \frac{\sigma^\nu}{\mid x \mid^{1+\nu}}. \tag{2.71}$$

Let's add that the absolute moment the order of m of the PDF (2.70),

$$\langle \mid x \mid^m \rangle = \infty, \tag{2.72}$$

for exponent $m \geq \nu$.

The beneficial Fourier transform of the Student's PDF takes the form [18],

$$\tilde{p}_\nu(k) = 2\left(\frac{\sigma \mid k \mid}{2}\right)^{\nu/2}\left[\Gamma\left(\frac{\nu}{2}\right)\right]^{-1} K_{\nu/2}(\sigma \mid k \mid), \tag{2.73}$$

where $K_{\nu/2}(\sigma \mid k \mid)$ is the modified Bessel function.

Our goal is to show that satisfying CLT means that as the number of components in the sum of n of (single-step) variables increases, the size of the domain of the Gaussian PDF of this sum increases. Thus, the size of the remaining domain (that is, for another PDF—in our case, power-law PDF) is decreasing. This can be clearly shown already for three degrees of freedom (i.e., $\nu = 3$). Then, Eq. (2.70) reduces to the much simpler form,

$$p_{\nu=3}(x) = \frac{1}{\pi}\frac{2\sigma^3}{(\sigma^2 + x^2)^2}, \tag{2.74}$$

which is characterized by a finite variance of simply σ^2.

The Fourier transform of the above PDF takes (thanks to Eq. (2.73)) a form convenient for further manipulation. Hence, we get, after taking into account Eq. (2.74), that

$$\tilde{p}_{\nu=3}(k) = (1 + \sigma \mid k \mid)\exp(-\sigma \mid k \mid). \tag{2.75}$$

Consider (for completeness) the case when $(x/\sigma)^2 \ll 1$,

$$p_{\nu=3}(x) \approx \frac{1}{\pi}\frac{2}{\sigma}\left[1 - 2\left(\frac{x}{\sigma}\right)^2\right] \approx \frac{1}{\sqrt{\pi/2}} \cdot p_G(x; \sigma/2), \tag{2.76}$$

where $p_G(x; \sigma/2)$ is the (zero-centered) Gaussian PDF of the standard deviation random variable x equals $\sigma/2$.

Figure 2.3 presents a very instructive graphic summary of the above considerations for $\nu = 3$—it compares both the exact PDF (2.74) and the other approximate one (2.71), and two in the expression (2.76).

2.5.1.1 How CLT Is Built: The Competition Between Gauss and Power-Law PDFs

We now turn to **the primary step** of our considerations, i.e., multi-step PDF for $\nu = 3$. Note that the original PDF's finite variance, σ^2, (2.74) allows the Central Limit Theorem to be applied. For this, we will express the density probabilities for

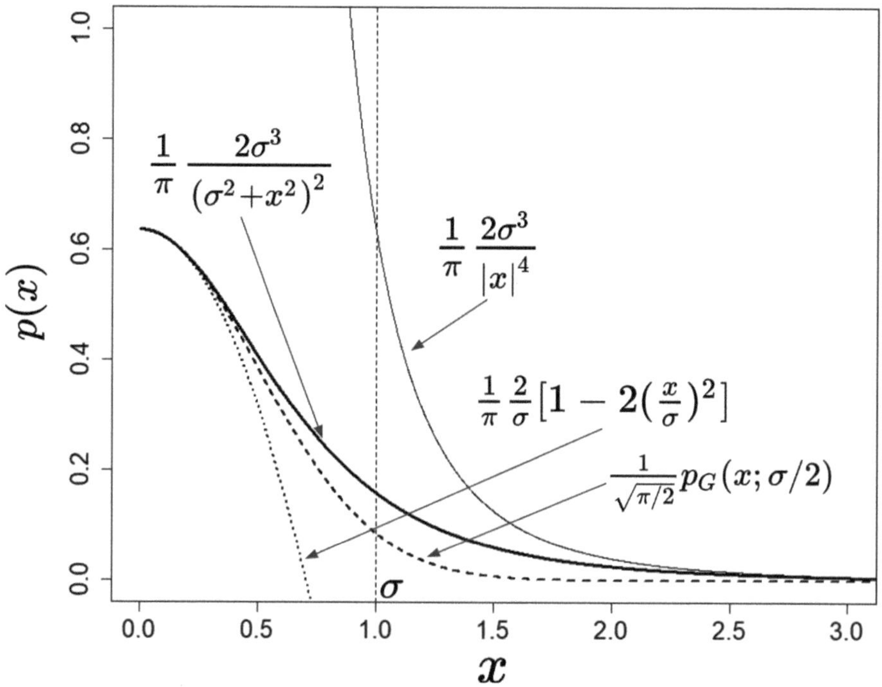

Fig. 2.3 The comparison (for $\nu = 3$ and $\sigma = 1$) of the exact PDF (2.74) with the other approximate distributions—one given by Eq. (2.71) and two others from (2.76). We obtain a similar result for negative values of x because all functions here are even. The dominance of the power-law and Student's PDFs over the Gauss PDF in the entire range of values of the random variable x is visible. This is an instructive starting point for understanding the functioning of CLT, as presented below. The plot is inspired by Fig. 3.12 in the book of Sornette [13]

the sum random variable $X_n = x_1 + x_2 + \cdots + x_n,\ n = 1, 2, 3, \ldots,$ in the form of the following convolution,[20]

$$P_n(X_n) = \overbrace{(p \otimes p \otimes \ldots \otimes p)}^{n}(X_n) =$$
$$= \int_{-\infty}^{\infty} dX_{n-1} \int_{-\infty}^{\infty} dX_{n-2} \ldots \int_{-\infty}^{\infty} dX_2 \int_{-\infty}^{\infty} dX_1$$
$$p(X_n \mid X_{n-1})p(X_{n-1} \mid X_{n-2}) \ldots p(X_2 \mid X_1)p(X_1), \tag{2.77}$$

where $p(X_j \mid X_{j-1})$ is the conditional probability density of the specified sum variable X_j given the occurrence of the sum variable $X_{j-1},\ j = 1, 2, 3, \ldots, n$. Still, using the condition of homogeneity (stationarity) of the process,

$$p(X_j \mid X_{j-1}) = p(X_j - X_{j-1}),\ j = 2, 3, \ldots, n, \tag{2.78}$$

[20] It is a general definition of the convolution and not for the particular stationary case and $\nu = 3$.

where $X_j - X_{j-1} = x_j$, and we get from (2.77) that its Fourier transform (for $\nu = 3$),

$$\tilde{P}_n(k) = [\tilde{p}_{\nu=3}(k)]^n = (1 + \sigma \mid k \mid)^n \exp(-n\sigma \mid k \mid) \approx (1 + n\sigma \mid k \mid) \exp(-n\sigma \mid k \mid), \tag{2.79}$$

the (latter) approximate equality holds for $n\sigma \mid k \mid \ll 1$. It is identical to that given equation (2.75), except here, σ has been replaced by $n\sigma$. The second equality in (2.79) indicates that the $\tilde{P}_n(k)$ is the infinitely divisible characteristic function of the PDF $P_n(X)$[21] as it is the nth power of the characteristic function, $\tilde{p}_{\nu=3}(k)$, of the elementary single-step probability density $p_{\nu=3}(x)$ (where the former, $\tilde{p}_{\nu=3}(k)$, is given by Eq. (2.75), while the latter, $p_{\nu=3}(x)$, by Eq. (2.74)).

Generally, the PDF that can be expressed as the PDF for the sum of any number of independent and identically distributed (i.i.d.) random variables is called *infinitely divisible*. Infinitely divisible distributions need not be stable. The distributions can change shapes when arguments go from a single random variable to a sum of random variables. Notably, the closed form of the PDF with the characteristic function in the form of Eq. (2.79) is not known. Therefore, we are forced to use only appropriate approximations. By the way, let's add that stable distributions must be infinitely divisible.

Expanding the *action* function $\ln \tilde{P}_n(k)$ in series around the point $k = 0$ and limiting it to terms of order $\mid k \mid^3$, the characteristic function (2.79) can be approximated as follows,

$$\tilde{P}_n(k) \approx 1 - \frac{1}{2} n\sigma^2 k^2 + \frac{1}{3} n\sigma^3 \mid k \mid^3 \tag{2.80}$$

that is, also up to terms of $\mid k \mid^3$. Let us emphasize that terms of this type depend on the modulus k, which means that the third derivative of the characteristic function $\tilde{P}_n(k)$ as a function of k does not exist at $k = 0$. There are only left- and right-hand derivatives, which differ. Therefore, for $k = 0$, the third derivative jump occurs, here $4n\sigma^3$.

Thus, there is no third moment of the distribution, as is the case for the power decay distribution with the exponent 4. Note that regardless of the value of $n = 1, 2, \ldots,$ the term $\propto \mid k \mid^3$, which we can call singular, is always present, i.e., the characteristic function is stable due to this term or, in other words, the distribution's algebraically vanishing tail is stable under n. It follows (by analogy to Eq. (2.71)) that for any n, there is always X large enough that,

$$P_n(X) \approx \frac{2n\sigma^3}{\pi \mid X \mid^4}. \tag{2.81}$$

[21] We continue to omit the n index of the sum variable X variable for the sake simplicity.

One can easily check using Eq. (2.79), the variance $\sigma^2(n)$ of the distribution $P_n(X)$ is

$$\sigma^2(n)\left(=-\frac{d^2\tilde{P}_n(k)}{dk^2}\mid_{k=0}\right)=n\sigma^2, \tag{2.82}$$

which is finite for finite n. Based on the CLT, we get for a sufficiently large $n(\gg 1)$ that,

$$P_n(X)\approx\frac{1}{(2\pi n)^{1/2}\sigma}\exp\left(-\frac{X^2}{2n\sigma^2}\right), \tag{2.83}$$

a one-dimensional Gaussian PDF asymptotically approximates $P_n(X)$ with a variance given by Eq. (2.82). This approximation is the better, i.e., it holds for the greater the range of the random variable X, the larger n is. Thus, we can expect the existence of such a characteristic value of $X_{tr}(n)$ monotonically increasing with increasing n, that only for $X\le X_{tr}(n)$ the Gaussian approximation (2.83) is satisfied. In contrast, for $X\ge X_{tr}(n)$, the proper approximation of $P_n(X)$ is PDF given by Eq. (2.81). A simple task is to find the dependency of threshold $X_{tr}(n)$ on n, i.e., the size of the Gaussian domain from n.

We are looking for the Gaussian domain's left and right border/transition points, $X_{tr}(n)$. These are the intersection points of the Gaussian curve with the power-law curve. Therefore, comparing Eq. (2.81) with Eq. (2.83)

$$\begin{aligned}\frac{2n\sigma^3}{\pi X_{tr}(n)^4}&=\frac{1}{(2\pi n)^{1/2}\sigma}\exp\left(-\frac{X_{tr}(n)^2}{2n\sigma^2}\right)\Leftrightarrow\\ \frac{2}{\pi\sqrt{n}\,Y_{tr}(n)^4}&=\frac{1}{(2\pi)^{1/2}}\exp\left(-\frac{Y_{tr}(n)^2}{2}\right),\end{aligned} \tag{2.84}$$

where $Y\overset{\text{def.}}{=}\frac{X}{\sigma\sqrt{n}}$ is a Gaussian scaling variable. We emphasize that the left sides of both equations in (2.84) were obtained assuming $\mid X_{tr}(n)\mid\gg\sigma$ or $Y_{tr}(n)\gg\frac{1}{\sqrt{n}}$, respectively.

Taking the logarithm of both sides of the second equation in (2.84) and rearranging them we get a transcendental equation in the form,

$$Y_{tr}(n)^2-8\ln Y_{tr}(n)=\ln n-3\ln 2+\ln\pi, \tag{2.85}$$

which has, as needed, two solutions (let's denote them by $Y_{tr}^1(n)$ & $Y_{tr}^2(n)$ and $X_{tr}^1(n)$ & $X_{tr}^2(n)$, respectively).

In a very special situation, when we limit ourselves to $Y^2-\ln n\gg 8\ln Y-\frac{3}{2}\ln 2+\ln\pi$, as a result, get an approximate solution,

$$\mid Y_{tr}^{approx}(n)\mid=\sqrt{\ln n}\Leftrightarrow\mid X_{tr}^{approx}(n)\mid=\sigma\sqrt{n\ln n}, \tag{2.86}$$

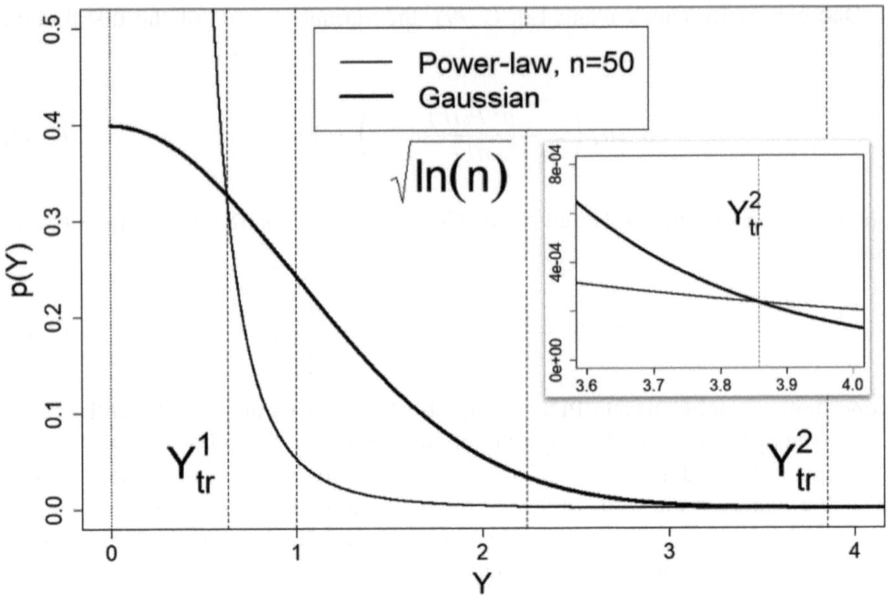

Fig. 2.4 Illustration of the transition from the Gaussian PDF to power-law one. The Gauss PDF dominates inside the $Y^1_{tr}(n) \le Y(n) \le Y^2_{tr}(n)$ domain, while the power-law PDF dominates outside it. In Fig. 2.5 below we show that when n increases, the range of the domain of the Gaussian distribution increases, and the range of the domain of the power distribution decreases. Thus, only the Gaussian PDF remains in the limit $n \to \infty$, as it should be according to CLT. Let us say for completeness of explanation that the second vertical dashed line was drawn at the inflection point $\sigma = 1$ of the standardized Gaussian curve $(2\pi)^{-1/2}\exp(-Y^2/2)$ (the thick solid curve). However, the third vertical dashed line was drawn at the characteristic point $Y^{approx}_{tr}(n) = \sqrt{\ln n}$ (see explanations above). A thin solid curve is given by a power law distribution $\frac{2}{\pi\sqrt{n}\,Y(n)^4}$. The plot is inspired by Fig. 3.12 in the book of Sornette [13]

where we neglected the logarithmic correction $8\ln Y_{tr}(n)$ and a constant $-3\ln 2 + \ln\pi$ in Eq. (2.85). Threshold $X^{approx}_{tr}(n)$ increases (slowly) superlinearly with n, while $Y^{approx}_{tr}(n)$ sublinearly. However, this has only indicative/preliminary significance for our further consideration.

An exact approach requires a numerical solution of the transcendental equation (2.85) for variable $Y_{tr}(n)$—Figs. 2.4 and 2.5 well illustrate the situation. It leads to the conclusion that an increase in n leads to an expansion of the area of dominance of Gauss PDF over power-law PDF located in between $Y^1_{tr}(n)$ & $Y^2_{tr}(n)$.

A similar situation occurs for the negative Y values.

We can write it as follows:

$$P(\mid Y \mid \ge Y_{tr}) = P(X \ge Y_{tr}) + P(Y \le -X_{tr}) = \frac{4}{\pi\sqrt{n}} \int_{\sqrt{\ln n}}^{\infty} \frac{1}{Y^4} dX = \frac{4}{3\pi} \frac{1}{\sqrt{n \ln^3 n}}, \quad (2.87)$$

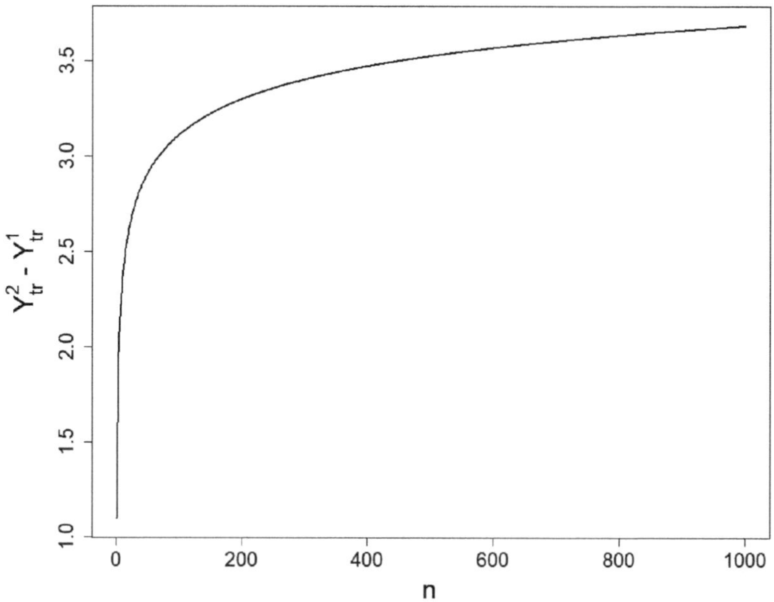

Fig. 2.5 Dependence of the difference $Y_{tr}^2(n) - Y_{tr}^1(n)$ on n. As can be seen, this difference is an increasing function of n, as it should be. It is also worth noting that its growth slows as n increases. Of course, the same applies to difference $X_{tr}^2(n) - X_{tr}^1(n)$. To understand the functioning of CLT, increasing this difference by increasing n is crucial

or equivalently

$$P(|\ X\ |\geq X_{tr}) = \frac{4n\sigma^3}{\pi}\int_{\sigma\sqrt{n\ln n}}^{\infty}\frac{1}{X^4}dX = \frac{4}{3\pi}\frac{1}{\sqrt{n\ln^3 n}}, \tag{2.88}$$

tending to zero when $n \to \infty$. Thus, the CCDF $P(|\ Y\ |\geq Y_{tr})$ or $P(|\ X\ |\geq X_{tr})$ approach zero as slow as $1/\sqrt{n\ln^3 n}$. Just speaking at the beginning of this subsection about the flexibility of Student's PDF, we meant the situation presented in Fig. 2.4.

In this subsection, a typical example shows the scope of CLT's applicability.

2.5.2 Stationary Power-Law Distribution on the Stock Exchange

As it is well-known from the physical bibliography [13, 19, 20], Langevin's variant of the canonical model of stochastic dynamics of stock price,[22] $S(t)$, in the stock market, is a model of Brownian motion with drift. That is, variable $\ln(S(t)/S(0))$ subject to the arithmetic (ordinary/standard) Brownian motion with drift, described by the stochastic differential equation (SDE),[23]

$$dX(t) = \mu dt + \sigma dW(t), \tag{2.89}$$

where $X(t) = \ln(S(t)/S(0))$, i.e., $S(t)$ follows a geometric Brownian motion with drift or growth rate μ and volatility σ. Increment $dW(t)$ is the stochastic differential of the i.i.d. Wiener process for the arithmetic Brownian motion. The Eq. (2.89) is a simplified version of the Langevin equation because we assumed that the coefficients μ and σ are constant. This makes it easier to show the power-law nature of the stationary PDF given by the equation (2.100) below without losing this nature. Moreover, in this case, the representation of Itô and the representation of Stratonovich of the Langevin equation become identical [19].

The solution of equation (2.89) is of the form,

$$S(t) = S(0)\exp\left(\mu t + \sigma W(t)\right), \tag{2.90}$$

where $W(0) = \langle W(t)\rangle = 0$ and $\langle W(t)^2\rangle = t$, while $\langle\ldots\rangle$ means the average over statistical ensemble of all realizations of the stochastic process $W(t)$. The Eq. (2.90) is the well-known time dependence of the underlying financial instrument $S(t)$, which is the basis of the canonical Black-Scholes model for pricing a derivative financial instrument exposed to that underlying instrument.

It is worth realizing that the equation of stochastic dynamics (2.89) can be written in an equivalent form as a simplified noninertial/massless Langevin equation[24] of stochastic dynamics with a drift,

---

22 The price, $S(t)$, of the underlying financial instrument plays an analogous role in the market as the position, $X(t)$, of a stray Brownian particle in suspension. We continue to consider the noninertial (or massless/overdamped) variant of the Langevin model, in which the component $m\frac{d^2X(t)}{dt^2} \to 0$, where m is the mass of the Brownian particle, has been neglected in comparison with the other components of the Langevin equation, i.e., the resistance force of the medium, the external force causing the drift and the stochastic force.

23 In the SDE there are stochastic differentials and not the simple ordinary differentials found in ordinary differential equations. However, we will still not explicitly use the properties of the former. The appearance of stochastic differentials results from erratic random walk trajectories being continuous but non-differentiable, i.e., they contain points of sharp turns of the trajectories.

24 More generally, the Langevin equation with no acceleration (that is, no second derivative term $\frac{d^2X(t)}{dt^2}$) is also called overdamped.

$$\frac{dX(t)}{dt} = \mu + \sigma\eta(t) \tag{2.91}$$

where between the Wiener process, $W(t)$, and the white noise $\eta(t)$ there is a relation in the form,

$$W(t) = \int_0^t \eta(t')dt' \Leftrightarrow dW(t) = \eta(t)dt. \tag{2.92}$$

The right-hand sides of the equations (2.89) and (2.91) have two different components. The first describes the systematic/deterministic drift of the random variable $X(t)$, and the second represents the contribution to the change of this random variable coming from the fluctuations of the medium/environment in which the $X(t)$ process occurs.

It can be shown[25] from the Itô lemma [21] applied to the PDF, $P(X, t)$, that the Eq. (2.91) is equivalent to the Fokker-Planck (FP) equation of the form,

$$\frac{\partial P(X,t)}{\partial t} = -\mu \frac{\partial P(X,t)}{\partial X} + \frac{\sigma^2}{2}\frac{\partial^2 P(X,t)}{\partial X^2} = -\frac{j(X,t)}{\partial X}, \tag{2.93}$$

where $\sigma^2/2$ plays the role of the diffusion coefficient and $j(X, t)$ is the probability density current defined by the following constitutive equation,

$$j(X,t) = j_{drift}(X,t) + j_{diff}(X,t), \tag{2.94}$$

while

$$\begin{aligned} j_{drift}(X,t) &= \mu P(X,t) \\ j_{diff}(X,t) &= -\frac{\sigma^2}{2}\frac{\partial P(X,t)}{\partial X} \end{aligned} \tag{2.95}$$

The Eq. (2.93) is the continuity equation for PDF, $P(X, t)$.[26] The system is in a steady macrostate[27] if and only if the total probability density current

$$j(X,t) = 0, \tag{2.96}$$

that is, vanishes.[28] It means that the drift probability density current $j_{drift}(X, t)$ is oriented opposite to the diffusion density current $j_{diff}(X, t)$, so this is the situation

[25] See derivation in Appendix A.14.

[26] Equation (2.93) is also called the Smoluchowski equation, the Kolmogorov equation, or the general diffusion equation.

[27] Terms "steady macrostate" and "stationary macrostate" are synonyms.

[28] The system would be in statistical equilibrium if both partial probability density currents $j_{drift}(X, t)$ and $j_{diff}(X, t)$ disappeared separately.

that we already discussed in Subsect. 2.3.2 earlier. In such a situation, there is an equality obeyed,

$$\frac{\partial P(X,t)}{\partial X} = \frac{2\mu}{\sigma^2} P(X,t). \tag{2.97}$$

The solution of this linear homogeneous first-order partial differential equation is the function,

$$\Pi(X) = \Pi(0) \exp\left(\frac{2\mu}{\sigma^2} X\right) \tag{2.98}$$

where the constant $\Pi(0)$ is determined below. As can be seen, the spatial dependence of the solution is governed by the ratio of the growth rate to the square of the volatility.

The function given by Eq. (2.98) is a stationary solution if and only if it is a function of the variable X decreasing to zero at infinity, i.e., if $\mu < 0$. Now we reduce $\Pi(X)$ to the form explicitly dependent on S/S_0,

$$\Pi(X = \ln(S/S_0)) = \frac{2|\mu|}{\sigma^2} \exp\left(-\frac{2|\mu|}{\sigma^2} \ln(S/S_0)\right) = \frac{\frac{2|\mu|}{\sigma^2}}{(S/S_0)^{\frac{2|\mu|}{\sigma^2}}}, \tag{2.99}$$

where the constant $\Pi(0) = \frac{2|\mu|}{\sigma^2}$ was determined from the normalization condition, $\int_0^\infty \Pi(X)dX = 1$, for a stationary situation, where σ is a measure of price volatility. Using probability invariance as a scalar, we get the stationary PDF we are looking for, $P_{stat}(S)$,

$$P_{stat}(S) = \Pi(X)\frac{dX}{dS} = \frac{(1-\alpha)}{S_0} \frac{1}{(S/S_0)^{1+\frac{2|\mu|}{\sigma^2}}}. \tag{2.100}$$

It is the normalized stationary solution of the equation (2.97) in the variable S/S_0, which is the Pareto distribution with the Pareto index $\frac{2|\mu|}{\sigma^2}$ (see Eq. (2.8) in Sect. 2.1.2 for comparison). It is important to realize that this stationary solution appeared in a very specific situation of current balancing. In general, this does not happen.

The stationarity mentioned above is because there is a lower, impassable (barrier) price of securities, $S_0 > 0$, which leads to balancing the diffusion current directed from this barrier and the drift towards this barrier. The obtained result, given by the Eq. (2.100), is (as expected) analogous to Eq. (2.40) in Sect. 2.3.2; the role of the SNR is performed by the exponent $\frac{2|\mu|}{\sigma^2}$.

The power-law solution (2.100) is a good reference case for the asymptotic/long-range dynamics of the stock market. It is helpful because shape exponent $\frac{|\mu|}{\sigma^2}$ is explicitly expressed by the ratio of the two most important parameters characterizing this stock market model and is not an *ad hoc* parameter. Let us add that for $\frac{|\mu|}{\sigma^2}$, the PDF given by Eq. (2.100) has the heavy tail, while in the opposite case, $\frac{|\mu|}{\sigma^2}$, only the fat tail with finite variance.

More generally, the shape exponent can be expressed by parameters characterizing the system and the impact of this system's environment on itself. In the next part of the textbook, we present a model in which thermodynamic parameters characterize the shape of the exponent.

2.6 Random Hops on Random Path

The random hops/flights of a particle on a linear chain is a principal reference case of statistics, and the result for the mean square displacement after n hops, $\langle X^2(n)\rangle \propto n$, is commonly known from the Central Limit Theorem. In this section, we consider modified random hops of a particle on a linear chain that does not extend uniformly in one direction but has been constructed by a random flight description [22]. The particle randomly wandering on this folded chain will have a reduced mean square displacement after n hops. This section will show that the mean square displacement is proportional to $n^{1/2}$, i.e., we are dealing here with anomalous diffusion with the Hurst exponent $H = 1/4$, which is antipersistent random wandering.

Physical situations, which can be described by the random wandering of a particle along random paths, are easily visualized. For instance, a particle in an amorphous substance might find an irregular path along which it can hop relatively easily. A particle tagged in a linear chain occupied by a finite concentration c of particles can perform random hops as far as other particles allow. Since the other particles are allowed to move, the tagged particle experiences a varying range in which it can move. Thus, one expects a similarity to random flight on random paths. Richards [23] has numerically studied the mean-square displacement of a tagged particle in a concentrated linear chain and found $\langle X^2(t)\rangle \propto t^{1/2}$, where time t is related to the number of steps n through a characteristic time scale. De Gennes [24] has considered the diffusive motion of a polymer in the presence of fixed obstacles, which provide a tube of random configuration to it. This "reptation" of a polymer chain is similar to the random flight on random paths, although it is even more complicated. In this context, de Gennes has given some results for the random flight on random paths in the Gaussian limit.

2.6.1 Mean Square Displacement and Excess Kurtosis

The usual random flight procedure will generate the linear chain in which the particle performs random hops. Random wandering starts with $X = 0$ and goes from X to the left or right with the same probability $1/2$.

The bottom plot in Fig. 2.6 shows a projection of a particular realization/trajectory of such a random wandering as a function of the number of hops ν on the $(X, \nu, n = 0)$ plane. At the same time, the top plot in the same figure puts this trajectory in a broader, three-dimensional context. Right from the start, we note the dual role of the auxiliary

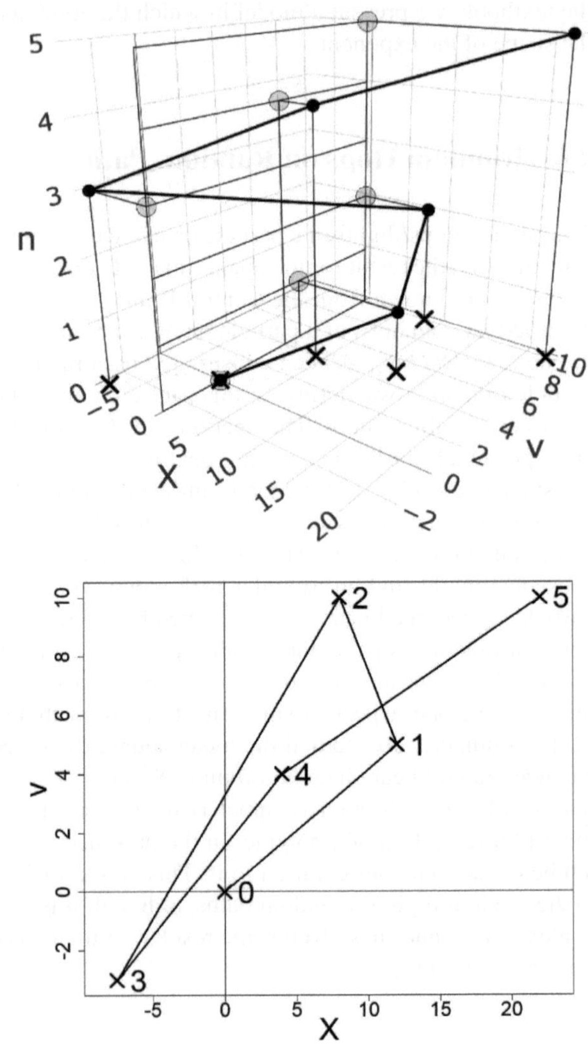

Fig. 2.6 A schematic fragment of an example random walk trajectory defined by Eq. (2.101). In the top plot, there is a three-dimensional random walk space with marked turning points (black circles) and their projection onto the $(X, \nu, n = 0)$ plane (crosses) and the $(X = 0, \nu, n)$ plane (gray circles). In the bottom plot, there is a projection of the random walk trajectory implemented in the three-dimensional space (shown in the top plot) onto the $(X, \nu, n = 0)$ plane. The numbers next to the crosses correspond to the subsequent steps of the random walk. The crosses on both plots mark the same points. There are $2n + 1$ turning points (X, ν, n) of the random walk with fixed coordinates X and n, because the variable $-n \leq \nu \leq n$ may be changed

variable ν. It is imposed on the one hand by the position of particle X and on the other by the total number of hops it performs (which is also a step measure of discrete time).

Trajectories X versus ν are described by PDF $P_\nu(X)$ and the type ν versus n by PDF $P_n(\nu)$. The first PDF is the probability of finding a value X after ν hops, while the second PDF defines the probability of finding ν after n hops. The probability of finding the particle at X after n hops is given by the combination (complex statistics),

$$W_n(X) = \sum_{\nu=-n}^{n} P_\nu(X) P_n(\nu). \tag{2.101}$$

These are the statistics that we can also call superposed. We emphasize that we are interested in asymptotic situations when $n \to \infty$. Moreover, summing over the variable ν is still treated as equal to integrating over ν (i.e., extending this variable).

To better visualize the wandering generated by the above statistics, we present a single realization of the trajectory in three-dimensional space (X, ν, n) on the top plot in Fig. 2.6. On the bottom plot in Fig. 2.6, we depict the projection of this curve onto the plane $(X, \nu, n = 0)$. The evolution in discrete time n is illustrated frame by frame vertically, creating a movie where each frame represents a separate plane for a selected moment $n = 1, 2, 3, \ldots$. Each empty circle, drawn from the distribution $P_n(\nu)$ for a given moment $n \geq 0$, represents a point in this frame and consequently in the vertical plane (ν, n). These empty circles are arranged chronologically relative to the variable n with randomly selected coordinate values ν.

Contrastingly, the black points lie on subsequent planes (x, ν) and are associated with the corresponding empty circles based on joint coordinates ν. Originating from the $P_\nu(X)$ distribution, these black points define the turning points of the $X(n)$ process trajectory in the three-dimensional space (X, ν, n), as presented on the top plot in Fig. 2.6. It is noteworthy that the trajectory constructed in this manner is not inherently chronological in the random variable ν. Achieving this chronology independently can be pursued through three approaches: (i) combining the black points chronologically concerning ν, (ii) projecting the black points onto a single, arbitrarily selected plane, and then organizing them chronologically based on the ν variable, or (iii) selecting any plane and subsequently drawing the value of the process ν. However, achieving simultaneous chronology over the n and ν variables is not typically possible in this approach on average.

However, as the ν variable is stochastic, a spontaneous chronology for this variable is conceivable, synchronized with the chronology of the n variable within its finite interval (occasionally smaller and occasionally larger; nonetheless, in the instance depicted in Fig. 2.6, such an interval was absent). In particular, all trajectory types are included when summing/integrating the current variable ν in Eq. (2.101). It is crucial to emphasize that our approach focuses on positioning turning points (black and empty circles) within the (X, ν, n) space, which remains independent of the connecting curves.

We continue to assume for simplicity that both distributions appearing under the sum in Eq. (2.101) the data are the corresponding Gaussian distributions of the form:

$$P_\nu(X) = P_{|\nu|}(X) = \frac{1}{\sqrt{2\pi \mid \nu \mid}} \exp\left(-X^2/2 \mid \nu \mid\right) \tag{2.102}$$

and

$$P_n(\nu) = \frac{1}{\sqrt{2\pi n}} \exp\left(-\nu^2/2n\right). \tag{2.103}$$

Please take note that the trajectories in Fig. 2.6 have been extrapolated to negative values of ν due to the assumed mirror symmetry of the distribution $P_\nu(X)$ concerning ν.[29]

Our goal is to calculate the mean square displacement for $n \to \infty$,

$$\langle X^2(n)\rangle = \lim \int_{-\infty}^{\infty} X^2 W_n(X) dX = \int_{-\infty}^{\infty} \mid \nu \mid P_n(\nu) d\nu = n^{1/2}, \tag{2.104}$$

where along the way, we used equality,

$$\langle X^2(\nu)\rangle = \frac{1}{\sqrt{2\pi \mid \nu \mid}} \int_{-\infty}^{\infty} X^2 \exp\left(-X^2/2 \mid \nu \mid\right) dX = \mid \nu \mid. \tag{2.105}$$

As evident, we have derived a subdiffusive dependence of mean square displacement, $\langle X(n)^2\rangle$, on discrete time n. This outcome aligns with expectations, considering a spontaneous self-limiting influence on the propagation of the absolute value of the process, $\mid X \mid$, due to the tendency of forward jumps to be counteracted by backward jumps (and vice versa). According to the Central Limit Theorem (CLT), this results (in the absence of long-term autocorrelations) in a linear (indirect) correlation between $\langle X(\nu)^2\rangle$ and the internal time ν. Furthermore, ν has additional significance; specifically, it serves as a metric for indirect propagation, giving rise to random/folded channels or fundamental trajectories for the complex process $X(n)$ versus n (not restricted to $X(\nu)$ vs. ν).

Based on the above results, one can easily calculate the fourth-order moment $\langle X(n)^4\rangle$ and hence the excess kurtosis, obtaining,

$$\kappa = \frac{\langle X^4(n)\rangle - 3\langle X^2(n)\rangle^2}{\langle X^2(n)\rangle^2} = \frac{2\sum_{\nu=0}^{\infty}(3\nu^2 - 2\nu)P_n(\nu) - 3n}{n} = -\frac{2}{n^{1/2}}. \tag{2.106}$$

As observed, a non-zero excess kurtosis κ indicates a departure from a Gaussian distribution. Consequently, it is pertinent to inquire about the specific probability density function (PDF) characterizing the distribution in question.

---

[29] While a similar symmetry assumption for time n is conceivable, we opt not to delve into this further as we are specifically examining the forward time evolution of n relatives.

2.6.2 Complex Statistics

The composite/complex statistics, $W_n(X)$, for our process was defined using Eq. (2.101), where we defined its Gaussian components $P_\nu(X)$ and $P_n(\nu)$ by Eqs. (2.102) and (2.103), respectively. Hence,

$$W_n(X) = \frac{1}{\pi n^{1/2}} \int_0^\infty \frac{1}{\nu^{1/2}} \exp\left(-\frac{\nu^2}{2n} - \frac{X^2}{2\nu}\right) d\nu, \tag{2.107}$$

which in the Fourier domain gives its characteristic function in the form,

$$\tilde{W}_n(k) = \frac{1}{(2\pi n)^{1/2}} \int_0^\infty \exp\left(-\frac{\nu^2}{2n} - \frac{k^2\nu}{2}\right) d\nu = \exp(k^4 n/8)\, \mathrm{erfc}[k^2(n/8)^{1/2}], \tag{2.108}$$

and the above result is exact. Its asymptotic behavior is given by,

$$\tilde{W}_n(k) \overset{n\to\infty}{\longrightarrow} \frac{1}{k^2}\left(\frac{8}{\pi n}\right)^{1/2}. \tag{2.109}$$

As evident, the process (decomposition) under consideration is not infinitely divisible [16], thus lacks stability in shape characteristic of stable distributions such as Gaussian or Lévy distributions.

It should be noted that, despite being governed by complex statistics, the process does not exhibit multifractality. Even when extending this process to continuous time by replacing n with t, the fundamental nature remains unchanged, resulting in an asymptotically square root dependence of the mean square displacement on time t. However, such an extension necessitates transitioning to the more intricate Continuous-Time Random Walk (CTRW) formalism.

Control and Review Questions and Exercises

2.1 Exercise
Define PDF in the sub-exponential class. Give an example.

2.2 Exercise
List the differences between nfBm and fBm.

2.3 Exercise
Propose a nonlinear transformation of random variables that converts the Weibull PDF to the power-law PDF. Interpret the resulting exponent.

2.4 Exercise
Propose a nonlinear transformation of random variables that transforms the Gaussian PDF to a log-normal PDF. Interpret the variance of this distribution.

2.5 Exercise
Give at least one stochastic process with its physical interpretation leading to the log-normal PDF.

2.6 Problem
Show that knowing all moments of the log-normal PDF does not uniquely define this distribution. Hint: Give an example of a PDF different from the log-normal one with all moments the same as the log-normal PDF.

2.7 Problem
Show to what extent the independent variable log-normal PDF is similar (to a good approximation) to the power-law PDF.

2.8 Problem
Show that the existence of an equilibrium state for a multiplicative process (e.g. due to the introduction of a repulsive barrier) is described by power-law PDF. What physical meaning does the power-law exponent of this PDF have, then?

2.9 Problem
What is a multiplicative-additive stochastic process? Give its relationship to the Langevin equation and the Fokker-Planck equation. Give the conditions under which a stationary state exists for such a process. Derive the PDF for this state. Does an expectation value and variance exist for this PDF? If so, derive them. When does this stationary distribution become a power-law PDF?

2.10 Questions
What effect does the existence of autocorrelation have on the variance of a stochastic process? What is superdiffusion and how does it relate to slowly decaying autocorrelations? Can they lead to subdiffusion?

2.11 Exercise
Does the equality of all moments of any two PDFs mean these distributions are identical? Under what conditions does this approximately occur?

2.12 Question
Under what conditions does the Student distribution become a Gaussian distribution, and under what conditions does it become a power distribution?

2.13 Exercise
Explain why the unbounded growth of the difference $Y_{tr}^2(n) - Y_{tr}^1(n)$ is crucial for understanding the functioning of CLT, i.e., the growing role of the multi-step Gaussian PDF.

2.14 Exercise
Give the path by which stationary geometric Brownian motion leads to a power law distribution. Interpret the power-law exponent defined here.

2.15 Exercise
Give the interpretation of variables x, ν, n for the process random hops on the random path (also called random walk on a random walk, RWRW). Note the dual role of variable ν.

2.16 Questions
Explain why the RWRW process is subdiffusional/antipersistent. Is the process auto-correlated? If so, are they long-term?

References

1. Feller, W.: An Introduction to Probability Theory and Its Applications, vol. I, 3rd edn. Wiley, New York (1967)
2. Feller, W.: An Introduction to Probability Theory and Its Applications, vol. 2, 2nd edn. Wiley, New York (1971)
3. Bouchaud, J.-P., Georges, A.: Anomalous diffusion in disordered media: statistical mechanics, models, and physical applications. Phys Rep **195**(4&5), 127–293 (1990)
4. Montroll, E.W., Shlesinger, M.F.: On the wonderful world of random walks. In: Lebowitz, J.L., Montroll, E.W. (eds.): Nonequilibrium Phenomena II. From Stochastics to Hydrodynamics. Studies in Statistical Mechanics, vol. XI, pp. 1–121. North-Holland, Amsterdam (1984)
5. Montroll, E.W., Lebowitz, J.L.: Studies in Statistical Mechanics, vol. VII. Fluctuation Phenomena. North-Holland, Amsterdam (1979)
6. Prigogine, I., Rice, S.A.: Advances in Chemical Physics. Wiley, New York (1983)
7. Janicki, A., Weron, A.: Simulation and Chaotic Behavior of α-Stable Stochastic Processes. Marcel Dekker, New York (1994)
8. Phillips. W.D.: Nobel lecture: laser cooling and trapping of neutral atoms. Rev. Mod. Phys. **70**, 721–741 (1998). https://doi.org/10.1103/RevModPhys.70.721
9. Bardou, F., Bouchaud, J.-P., Aspect, A., Cohen-Tannoudji, C.: Lévy Statistics and Laser Cooling. How Rare Events Bring Atoms to Rest. Cambridge University Press, Cambridge (2002)
10. Sornette, D.: Linear stochastic dynamics with nonlinear fractal properties. Physica A **250**, 295–314 (1998). https://doi.org/10.1063/1.1341926
11. Mantegna, R.N., Stanley, H.E.: An Introduction to Econophysics. Correlations and Complexity in Finance. Cambridge University Press, Cambridge UK (2000). https://doi.org/10.1016/S0378-4371(97)00543-8
12. Bouchaud, J.-P., Potters, M.: Theory of Financial Risks. From Statistical Physics to Risk Management. 2nd reprint. Cambridge University Press, Cambridge (2001)
13. Sornette, D.: Critical Phenomena in Natural Sciences. Chaos, Fractals, Self-Organization, and Disorder: Concepts and Tools. 2nd edn. Springer Series in Synergetics. Springer Complexity. Springer, Berlin (2006). https://doi.org/10.1007/3-540-33182-4
14. Sornette, D.: Multiplicative processes and power law. Phys. Rev. E **57**, 4811–4813 (1998). https://doi.org/10.1103/PhysRevE.57.4811
15. Nonnenmacher, T.F., Metzler, R.: Applications of fractional calculus techniques to problems in biophysics. In: Hilfer, R. (ed.) Applications of Fractional Calculus in Physics. World Scientific, Singapore (2000). https://doi.org/10.1142/9789812817747-0002
16. Kozłowska, M., Kutner, R.: Dynamics of the warsaw stock exchange index as analyzed by the nonhomogeneous fractional relaxation equation. Acta Phys. Pol. B **37**(11), 3027–3038 (2006)

17. Brand, L.: Differential and difference equations. Wiley, New York (1966)
18. Bateman, H., Erdélyi, A.: Tables of Integral Transforms, vol. 1. McGraw-Hill, New York (1954)
19. Van Kampen, N.G.: Stochastic Processes in Physics and Chemistry, 3rd edn. Elsevier, Amsterdam (2007)
20. Kubo, R., Toda, M., Hashitsume, N.: Statistical Physics II. Nonequilibrium Statistical Mechanics. Springer, Berlin (1985)
21. Paul, W., Baschnagel, J.: Stochastic Processes. From Physics to Finance. 2nd edn. Springer, Berlin (2013). https://doi.org/10.2307/3085855
22. Kutner, R., Kehr, K.W.: Random walk on a random walk. Physica A **110**, 535–549 (1982). https://doi.org/10.1016/0378-4371(82)90067-X
23. Richards, P.M.: Theory of one-dimensional hopping conductivity and diffusion. Phys. Rev. B **16**, 1393–1408 (1977). https://doi.org/10.1103/PhysRevB.16.1393
24. de Gennes, P.G.: Reptation of a polymer chain in the presence of fixed obstacles. J. Chem Phys. **55**, 572–579 (1971). https://doi.org/10.1063/1.1675789

Chapter 3
Singular Stochastic Processes

Abstract We present both the consequences of scaling symmetry and those of the Generalized Central Limit Theorem, including applications that describe selected elements of the dynamics of financial markets, such as log-periodicity. Additionally, in the context of viscoelastic materials (e.g., non-Newtonian fluids), we explore the phenomenon of slowed (or non-Debye) relaxation. In this context, we introduce the concept of fractal relaxation and the corresponding evolution equation with memory effects. However, the key element of this chapter is anomalous diffusion: we address its definition, the most commonly employed probability density functions (PDFs) leading to anomalous diffusion, and the introduction of singular processes as a generalization of this phenomenon. This framework lays the foundation for defining the conditions necessary to emerge multifractal stochastic processes—a broad category of singular stochastic processes. We discuss the stochastic process known as normal fractional Brownian motion and its generalization, fractional Brownian motion. We emphasize one of the most important properties of singular processes: long-range (and/or long-term) autocorrelations. Furthermore, we relate this type of process to the Langevin equation of stochastic dynamics, which we treat here at an *ab initio* level. At the end of this chapter, we discuss the monofractal subdiffusion process as a reference case.

3.1 "Prehistorical" Outline: Power-Law Distributions

The history of power-law distribution, including Pareto-Lévy, is exciting and worth mentioning. Namely, it may be surprising that it is commonly believed that the power-law distribution, and more precisely this type of relationship, was discovered in the real world only in 1897 by the Italian economist and sociologist Vinifredo Pareto, while power-law relaxation (i.e., the analogous dependence only on time) was already observed in 1729 by the physicist and engineer B. G. Buelfinger, as discussed in Sect. 3.4.

Moreover, at the beginning of the second half of the 19th century, the French mathematician and physicist Baron Augustyn L. Cauchy introduced and analyzed the PDF in the form,

M. Chorowski et al., *Anomalous Stochastics*, Understanding Complex Systems,
https://doi.org/10.1007/978-3-031-80392-5_3

$$p(x) \sim \frac{1}{1+x^2}, \tag{3.1}$$

known today as the Cauchy distribution or the Lorentzian after the Dutch physicist Hendrik A. Lorentz, who was the first to apply this distribution in spectroscopy to describe the shapes of spectral lines and this despite its infinite variance. The distribution discovered by Pareto is essential, if only because of its increasing number of applications in various branches of science—from mathematical and natural to economic and social.

Vilfredo Pareto empirically studied the increase in the wealth of individuals in various societies. He noticed that the number of $y(x)$ units whose income is not less than x could be described for relatively large income values through a power-law complementary cumulative distribution function (CCDF) of the form,

$$y(x) \sim \frac{1}{x^\alpha},\ x > 0, \tag{3.2}$$

where exponent α is the only (besides the pre-exponential normalizing factor) parameter defining the distribution. Pareto empirically determined this exponent; it was roughly 1.5 for societies as diverse as England, Ireland, Germany, Italy, and even Peru. The formula (3.2) can be expressed in terms of PDF, $f(x) \sim 1/x^{1+\alpha}$,

$$y(x) \sim \int_x^{x_{max}} dx' f(x') \sim \int_x^{x_{max}} dx' \frac{1}{x'^{1+\alpha}}, \tag{3.3}$$

where $1 \ll x \ll x_{max}$ (here, x_{max} is the unit's maximum achievable income); the $f(x)$ density is a measure of the relative number of individuals with an income of exactly x. Thus, V. Pareto observed an asymptotic power-law PDF for positive x,

$$f(x) \sim \frac{1}{x^{1+\alpha}}. \tag{3.4}$$

Based on this distribution, it was possible to formulate in the 1950s (within socio-economic sciences) the so-called Pareto 80/20 rule, stating that in a market economy, 20% of the population owns 80% of the capital/wealth. We will return to the power laws later, where we will discuss their scaleless and thus immeasurable nature and other related properties. The point is that the power law describes the self-similar nature of the system, i.e., at each scale, the system looks similar, which makes it impossible to choose a characteristic/physical scale because such a scale does not exist.

The extension of the distribution (3.4) to negative x values and its systematic analysis was published only in 1926 by the French mathematician Paul Lévy and independently by the English physicist, mathematician, meteorologist, and psychologist Lewis F. Richardson, who applied this distribution to describe the motion of objects (the so-called "passive scalars") in the atmosphere in the presence of turbulence [1]. It should be noted that the Pareto-Lévy distribution is still verified. For example, a study presented in 2001 at the Royal (British) Society Scientific Meeting [2] fully confirms the observations of V. Pareto, with α =1.90 (where the error is in the second decimal place) for net income above 100 $k£/year$. Conducting similar research on emerging market societies and capitalist economies would be of great interest.

Research has been conducted on the incomes of many countries and the wealth of individuals. These studies were first performed based on aggregate data obtained from tax offices by the Italian economist and sociologist Corrado Gini in 1922; he found that these incomes are subject (to a good approximation) to power laws with significantly different exponents.

Of particular interest were studies conducted at the end of the last century on the Japanese economy [3], which showed that the income of Japanese enterprises is subject to Zipf's power law[1] (nearly four decades from an income of over 10 million yen to about 10^5 million yen).

Furthermore, the income of enterprises within individual branches of the economy was better described by the power law the more a given branch participated in the free market game, i.e., the less there was state intervention in a given industry (with the exponent α is in the range $0.72 \leq \alpha \leq 1.13$ and its specific value varies from branch to branch). For example, the construction industry, subject to almost entirely free competition, can be described by the power law with the exponent $\alpha = 1.13$ over three decades, while the energy sector, subject to significant state protection, behaves paradoxically in a way that is difficult to describe by any model.

3.2 Scaling and Log-Periodicity Versus Bubbles and Stock Market Crashes

The power-law function, $y(x)$, given by Pareto Eq. (3.2) satisfies the following homogeneous singular scaling equation,[2]

$$y(\lambda x) = g(\lambda)y(x), \tag{3.5}$$

where the factor g scaling the function $y(x)$ depends on λ and is given by,

$$g(\lambda) =\mid \lambda \mid^{\nu}; \tag{3.6}$$

[1] Zipf's power law is asymptotically equivalent to the Cauchy-Lorentz distribution for positive x, i.e., the cumulative probability distribution built on it has a long-range "tail" vanishing with the exponent 1.

[2] More precisely, the scaling relation (3.5) also applies to the Pareto-Lévy i.e., distribution $y(x) \sim \frac{1}{|x|^{\alpha}}$, where the random variable x can take both negative and positive values, and not only positive as in the case of Pareto distribution.

we still assume that the exponent ν can be positive or negative (because it depends on what physical quantity we are dealing with). It should be emphasized that the scaling Eq. (3.5) is general and describes the behavior of substances as diverse as magnetics, double alloys, or gas and liquid in the phase transition region near the critical point (therefore called the scaling area/region or critical area/region) [4].

As experience shows, in this area, most of the physical quantities (describing the phase transition, let's denote them by $\mathcal{F}$) change depending on the temperature T according to the power law[3]

$$\mathcal{F} \sim\mid T - T_c \mid^{\alpha}, \tag{3.7}$$

where α is the name of the critical index or exponent, which assumes universal values (i.e., independent of the type of substance), and T_c is the critical temperature (which, for example, in the case of a ferromagnetic–paramagnetic phase transition, is additionally called the Curie temperature).

Now, let us ask a question about the most general solution of the scaling equation (3.5). It is easy to check the function of the form.

$$y(x) =\mid x \mid^{\alpha} Q\left(\frac{\ln \mid x \mid}{\ln \mid \lambda \mid}\right), \tag{3.8}$$

is the solution of this equation, as it turns out, the most general (as we will show later), where $Q(u)$ is a periodic function of the argument u with a period equal to 1, while the exponent takes the form,

$$\alpha = \frac{\ln g(\lambda)}{\ln \mid \lambda \mid}. \tag{3.9}$$

Substituting the specific form of the function $g(\lambda)$ (here given by the expression (3.6)) we get that $\alpha = \nu$.

We now expand the periodic function $Q(u)$ into a Fourier series of the variable u,

$$\begin{aligned} Q\left(\frac{\ln \mid x \mid}{\ln \mid \lambda \mid}\right) &= \sum_{n=-\infty}^{\infty} c_n \exp\left(2\pi i n \cdot \frac{\ln \mid x \mid}{\ln \mid \lambda \mid}\right) \\ &= c_0\left[1 + 2\sum_{n=1}^{\infty} \frac{c_n}{c_0} \cos\left(\frac{2\pi n}{\ln \mid \lambda \mid} \cdot \ln \mid x \mid\right)\right], \end{aligned} \tag{3.10}$$

where the expansion coefficients c_n are given here in the form of Fourier integrals,

[3] The point is that there are also physical quantities such as, e.g., specific heat that may have a logarithmic singularity at a critical point as in the two-dimensional Ising model; phase transitions in which it occurs do not fully lend themselves to Ehrenfest's classification of phase transitions [5], although in many respects (e.g., lack of latent heat and continuity of the order parameter as a function of temperature) resemble a second-order phase transition.

$$c_n = \frac{1}{2}\int_{-1}^{1} du Q(u) \exp(-2\pi i n u) = \int_{0}^{1} du Q(u) \cos(2\pi n u), \; n = 1, 2, .. \quad (3.11)$$

for simplicity, we assumed that $Q(u)$ is an even function of u, which results in the fact that each Fourier expansion coefficient is the even function of n, i.e., $c_n = c_{-n}$. This property was used in the second equality in (3.10). In addition, we limit ourselves here to the real function Q, which means that the c_n coefficients are also real (the parameter λ which scales a random variable is not arbitrary for the discrete scaling relation (3.5), which we will consider later [6]). Due to its internal structure, the function $Q(\ldots)$ is called the log-periodic function.

Using the renormalization group, we will prove that log-periodicity is present not only in the distributions describing the statistics of random variables, e.g., stock indexes but also in the very stochastic equations describing the dynamics of these variables [7–9].

3.2.1 Market Application

Figure 3.1 shows the quotations of the Dow Jones index on the New York Stock Exchange (NYSE) containing October 1929, i.e., the great crash on Wall Street - the largest stock market crash that affected the world, including the United States of America, in the twentieth century. It was the bursting of the huge stock-market speculative bubble[4,5] on September 3, 1929, closing at 381.17 USD. It declined until July 8, 1932, when it closed at 41.22 USD. The index did not reach its 1929 highest value again until November 23, 1954. The depth of this crash is evidenced by the fact that in the unlucky week: (opening) Wednesday, October 23 - (closing) Tuesday, October 29, the NYSE index lost about 30% of its value.[6]

For comparison, the fit to the empirical data of Eqs. (3.12) and (3.13) is shown in Fig. 3.1 (black dashed and solid curves, respectively). These equations (presented below) are derived from Eq. (3.8) supported by Eq. (3.10). Moreover, we used substitution here $x \stackrel{\text{def}}{=} a \cdot (t_c - t)$, where t is the time (counted in months), and t_c is the time of the crash, i.e., a time of sudden reorientation of the trend in the peak (here

[4] Such a bubble says that it is an increase in share prices occurring in a relatively short period (compared to the entire period under consideration), which significantly deviates from the fundamental valuation of shares, but still falls within the estimates resulting from existing models, as opposed to purely speculative bubbles, which are purely irrational/emotional beyond any modeling.

[5] The general definition of the so-called rational stock bubble comes from [10, 11].

[6] It should be noted that a similar crisis happened, for example, in October 1987 in the week of October 14 (opening) to October 19 (closing); fortunately, its effects were not as dramatic as in 1929.

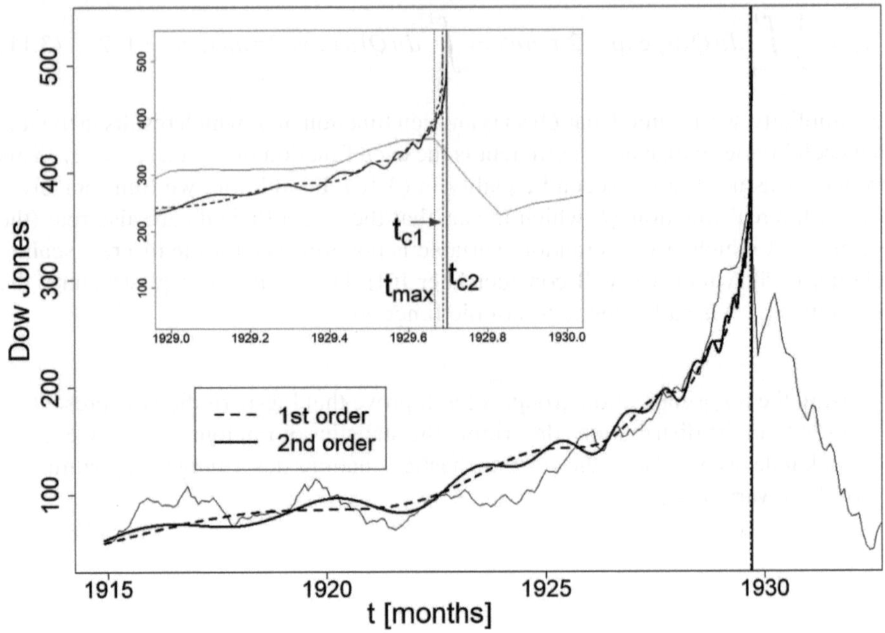

Fig. 3.1 The monthly quotations of the Dow Jones index on the New York Stock Exchange (NYSE) containing October 1929 (the empirical data marked with the gray solid curve), i.e., the great crash on Wall Street. The world's largest stock market crash, including the United States of America, affected the world in the 20th century. We fitted the log-periodic predictions to the data for time $t \leq t_{max} < t_c$, where t_{max} is the date of the highest pre-crisis peak; here $t_{max} = 1929-09-01$. This local maximum of empirical data is marked by the dashed vertical line. The dashed curve derived from the first-order log-periodic formula (3.12) was fitted to the empirical data similarly to the second-order log-periodic formula (3.13) fit represented by the solid curve. The fitting parameters of both curves are presented (accurate to two significant places after the decimal point) in Tables 3.1 and 3.2, respectively. It can be seen that Eq. (3.13) describes the course of quotations a little bit better than the (3.12). (The plot was inspired by Johansen and Sornette [12])

Table 3.1 Fit parameters for Eq. (3.12)

$t_c = t_{c1}\,[10^3]$	$A\,[10^2]$	$B\,[10^2]$	$C\,[10^{-2}]$	$\alpha\,[10^{-1}]$	ω	$\phi\,[10^{-1}]$
1.93	7.16	–4.88	–1.46	1.13	7.45	–3.85

of the cusp type; a is a proportionality constant). As can be seen, adding another harmonic improved the fit a bit.

The mentioned above formulas versus time containing only the first two harmonics (i.e., the zero and first) have been considered,

$$y(t_c - t) \approx A + B \cdot \mid t_c - t \mid^{\alpha} \cdot [1 + C \cos(\omega \ln \mid t_c - t \mid - \phi)], \qquad (3.12)$$

Table 3.2 Fit parameters for Eq. (3.13)

$t_c = t_{c2}\,[10^3]$	$A\,[10^2]$	$B\,[10^2]$	$C\,[10^{-2}]$	$C'\,[10^{-2}]$	$\alpha\,[10^{-1}]$	ω	$\phi\,[10^{-1}]$
1.93	7.16	–4.88	–1.29	1.50	1.13	8.0	7.06

where more convenient notation here than for Eq. (3.8) supported by Eq. (3.10) is used $B = c_0 \cdot a^{\alpha},\ C = 2c_1/c_0,\ \omega = 2\pi/\ln \mid \lambda \mid,\ \phi = -\omega \ln a$. The additive constant A has been added to the function $y(x)$ given by Eq. (3.8) *ad hoc*, representing a reference time-independent shift.

In addition, we have a formula showing the role of the zero, first, and second harmonics of the log-periodic correction present in the expression,

$$y(t_c - t) \approx A + B \cdot \mid t_c - t \mid^{\alpha} \times \left[1 + C \cos\left(\omega \ln \mid t_c - t \mid - \phi\right) + C' \cos\left(2\omega \ln \mid t_c - t \mid - 2\phi\right)\right], \tag{3.13}$$

where $C' = 2c_2/c_0$. Adding the second harmonics significantly changed the phase of ϕ compared to the poorer variant (3.12). We emphasize that both expressions (3.12) and (3.13) do not (practically speaking) change either the local maximum $t_{max} \approx t_{c1} \approx t_{c2}$, or the shape exponent α (as well as the reference level A being the theoretical height of the peak, and the relative amplitude of quotes B). This suggests the modal stability of these parameters, which is particularly important for stock investors.

Furthermore, the above expression (including the second harmonic in addition to the zero and the first harmonics) opens the possibility of examining subtle effects in stock exchange quotations.

Equations (3.12) and (3.13), obtained from the second equality in Eq. (3.10) combined with Eq. (3.8), prepared the comparative Fig. 3.2. Due to the importance of Eqs. (3.12) and (3.13) in the technical analysis of stock-market quotes, we figure the dependence of y on time t and its ingredients: the power law and log-periodic harmonics. The curves' subtle structure (i.e., the second harmonics) is visible in the lower panels.

Considerations like those discussed in this section can also be applied to the declining phases of stock market bubbles. Specifically, one can examine not only the dynamics of bubble formation and their subsequent bursting, i.e., crashes, but also the so-called'after effect' scenarios. Such analyses were initiated in the publication [13], which serves as a foundation for exploring the mechanisms that contribute to the recovery of financial markets and, more broadly, market economies following crashes.

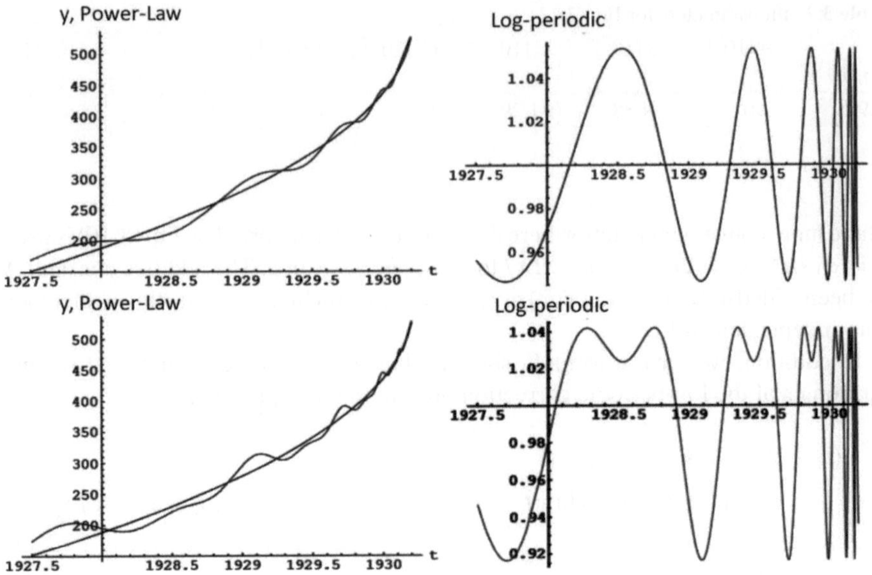

Fig. 3.2 Various components of the log-periodic oscillations versus time t (counted in days) given by Eq. (3.8) supported by Eq. (3.10). We applied them to the region before the crash at t_c, i.e., for $t \leq t_c$. In the top row, we have included results containing only the zero and first harmonics (in square brackets), based on the formula (3.12) fitted to the empirical data presented in Fig. 3.1. In addition, in the bottom row, we have included results containing additionally the second harmonic (based on the formula (3.13)). As can be seen, we are dealing with quasi-periodic cases, where a period is time-dependent, decreasing as we approach t_c. This behavior reflects the mood swings of stock market investors before the crash. (i) The left upper panel: Time-dependence of the function $y(t_c - t)$ given by the oscillating curve described by Eq. (3.12) and its power-law ingredient $A + B \cdot \mid t_c - t \mid^{\alpha}$ (non-oscillating curve). (ii) The right upper panel: the log-periodic component of the $y(t_c - t)$ function, i.e., the expression inside the square brackets in the formula (3.12). (iii) The left lower panel: An oscillating curve described by the Eq. (3.13), where, for example, we put the free parameter $C' = -0.03$. The non-oscillating curves here and in the left upper panels are identical. (iv) The right lower panel: a separated curve describing the first three harmonics i.e., described by the expression in square brackets in the formula (3.13) (built for the same parameter values as the curves presented in the lower left panel)

3.3 Properties of Lévy PDF and Its Financial Market Application

Lévy was primarily interested in the class of stable distributions, i.e., those that do not change their shape after transformation from a single to a sum random variables; they can, for example, undergo appropriate flattening and stretching. Such classes include Gauss and Cauchy (Lorentz) [14]. These two distributions differ in one essential feature: the Gaussian distribution has a finite second moment, while the Cauchy distribution has an infinite one.

Since distributions with a finite second moment need not be stable (for example, consider CLT), Lévy posed a vital question of purely mathematical nature: **what is the most general definition of the entire class of stable distributions**? Lévy formulated the answer to this question and later clarified by Khinchine in the form of the so-called Generalized Central Limit Theorem (GCLT) [15–17].

The most general form of the stable distribution, now called Pareto-Lévy—discussion of GCLT, including this distribution and its various applications in statistical physics and econophysics, is one of the main goals of this textbook.

3.3.1 Significant Properties

Figures 3.3(the upper and lower panels) show indicatively the symmetric Pareto-Lévy distribution, on a linear scale and a log-log scale, respectively, i.e., the most general form of a symmetric stable probability distribution function,[7]

$$P(X) = \frac{1}{\pi}\int_0^{\infty} exp(-\gamma\,\Delta t \mid q \mid^{\alpha})\cos(qX)dq,\ 0 < \alpha < 2,\ \gamma > 0, \quad (3.14)$$

where $\Delta t(> 0)$ is a time interval, fixed for four characteristic values of the Pareto-Lévy index. One can see fundamentally different behavior of $P(X)$ for $\alpha > 1$ and $\alpha \leq 1$ when the expected value of the random variable x exists or does not exist, respectively. Notably, for $\alpha = 2$, Eq. (3.14) becomes a Gauss PDF with variance $\sigma^2 = 2\gamma\Delta t$.

From Eq. (3.14) we obtain the return-to-origin probability density,

$$P(X = 0) = \frac{1}{\pi(\gamma\,\Delta t)^{1/\alpha}}\mathbf{\Gamma}\left(1 + \frac{1}{\alpha}\right), \quad (3.15)$$

as a function of Δt, which decreases according to power law with exponent equals $-1/\alpha$; $\mathbf{\Gamma}(\ldots)$ denotes the Euler-gamma function.

Using the series expansion of the exponent function (standing under the integral in Eq. (3.14)), then substituting the current variable in the expression (3.14) and integrating term by term using the definition of Euler's gamma function (i.e., Euler integrals of the second kind), the Pareto-Lévy PDF can be expressed for $\mid x \mid > 0$ as the following instrumental series,[8]

---

[7] We remind that in the textbook, the capital letter X refers to a multi-step displacement and the lowercase letter x refers to a single-step displacement.

[8] See Appendix A.15 for details.

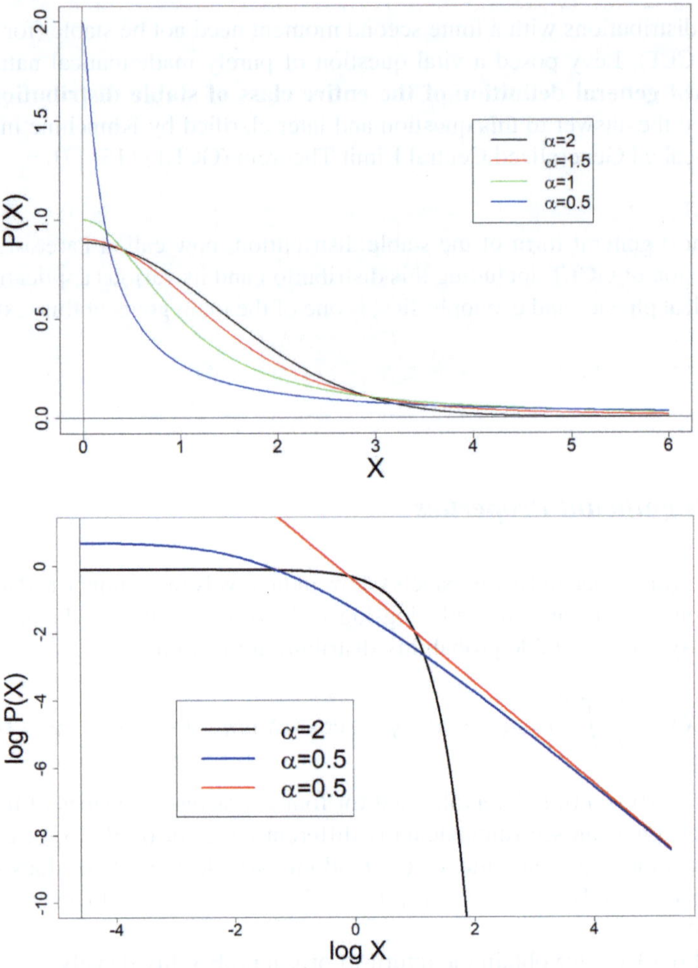

Fig. 3.3 The plot of the Pareto-Lévy probability density function (PDF) given by Eq. (3.14) for four characteristic values of the Pareto-Lévy index in the linear scale (the upper panel) and the log-log scale (the lower panel) for the value of the scaling parameter $\gamma\Delta t = 1$, and $X \geq 0$. It is described as follows: (i) The reference black curve corresponds (on both panels) to the index $\alpha = 2$, representing the Gaussian PDF centered at zero with a dispersion of $\sigma = \sqrt{2\gamma\Delta t} = \sqrt{2}$, and (ii) the blue curve represents (on both panels) the distribution with the index $\alpha = 0.5$. As seen in the lower panel, this curve asymptotically approaches the red straight line given by the power law (3.17). The remaining curves (located in the upper panel) are: (iii) the red curve represents the Pareto-Lévy PDF with index $\alpha = 1.5$, and (iv) the green curve represents the distribution with the index $\alpha = 1$, which is the Cauchy-Lorentz PDF. It is important to note a significant difference between distributions with index $\alpha \geq 1$ and those with index $\alpha < 1$, particularly evident in their behaviors near $X = 0$ and for the asymptotic values of X (the lower panel)

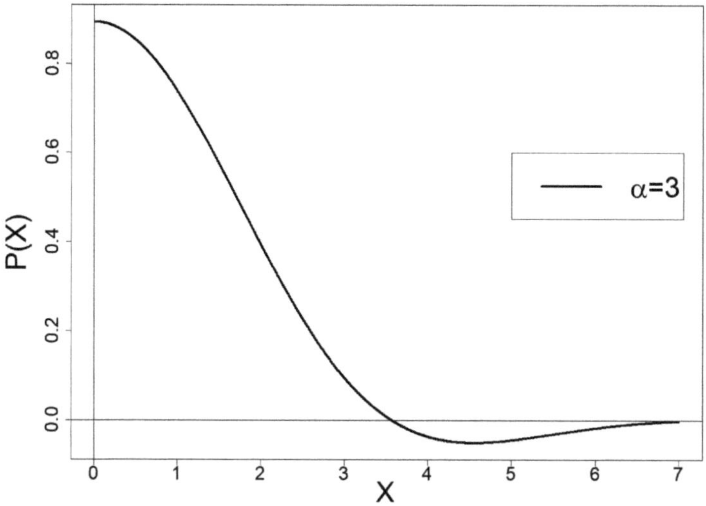

Fig. 3.4 The dependence of $P(X)$ on the variable X given by the formula (3.14) for the parameter $\gamma\Delta t = 1$ and the index $\alpha = 3$. This is a typical relationship when the index α exceeds 2. As can be seen, $P(X)$ takes both positive and negative values, which disqualifies it as a candidate for PDF

$$P(X) = -\frac{1}{\pi}\sum_{j=1}^{\infty}\frac{(-1)^j}{j!}(\gamma\ \Delta t)^j \mathbf{\Gamma}\ (1+\alpha j)\sin\left(\frac{j\pi\alpha}{2}\right)\frac{1}{\mid X\mid^{1+j\alpha}}, \tag{3.16}$$

which shows that for $\mid X\mid\to\infty$ the first term (with $j = 1$) of this series dominate,

$$P(X\to\infty) = \mathbf{\Gamma}\ (1+\alpha)\sin\left(\frac{\pi\alpha}{2}\right)\frac{1}{\mid X\mid^{1+\alpha}}; \tag{3.17}$$

it is visible in the log-log scale in Fig. 3.3 (upper panel), where we compare results of Eq. (3.14) for $\alpha = 2$ (black curve) and $\alpha = 1/2$ (blue curve).

At first glance, the existence of an upper bound on the index of the Pareto-Lévy distribution in the formula (3.14) may seem surprising—Fig. 3.4 explains this question. The point is that when $\alpha > 2$, then $P(X)$ given by formula (3.14) becomes negative for some values of the independent variable X, which was first noticed in 1919 by mathematician Felix Bernstein.

At the end of this section, we give for example, the closed form of the symmetric Pareto-Lévy distribution: in addition to the Gaussian and Cauchy PDFs we have already discussed, there is also the Zolotarev PDF or the Pareto-Lévy PDF with index $\alpha = 2/3$, which is expressed by the Whittaker function $W_{1/2,1/6}$,

$$P(X) = \frac{1}{\sqrt{2\pi}}\frac{1}{\mid X\mid}W_{1/2,1/6}\left(\frac{4}{27}\frac{(\gamma\Delta t)^{2/3}}{X^2}\right)\exp\left(-\frac{2}{27}\frac{(\gamma\Delta t)^3}{X^2}\right); \tag{3.18}$$

as one can see, the Zolotaryov distribution is expressed by a non-elementary function (unlike the other two).

Of course, the three examples above are not the only closed modes of the Lévy PDF.

3.3.2 *Petersburg Paradox and Consequences*

Two brothers, Nicolaus and Daniel Bernoulli, drew attention to infinite average values in probability calculus at the beginning of the 18th century, using the example of the so-called **Petersburg Paradox** (PP). It is a gambling game in which the banker flips a symmetrical coin n times. The player participating in this game wins 2^n units if the following $n-1$ rolls come up on the first side of the coin (avers) before the second side (reverse) comes up on the next $n^{th}-$ roll.[9] The results are presented in Table 3.3.

We now consider three main consequences of the Petersburg Paradox:

(i) Lack of physical scale of phenomena and processes
(ii) Its generalization
(iii) Introduction of the utility function.

They all opened the way to modern theories of scale-free phenomena and processes and modern utility theory.

As one can see, the total win (or expected value) equals the sum of all the numbers in the last column in Table 3.3, i.e., $p(1) \cdot w(1) + p(2) \cdot w(2) + p(3) \cdot w(3) + \cdots = 1 + 1 + 1 \cdots$, diverges as the number of rolls increases to infinity. Thus, for the bet that a player entering the game could propose, his win will sooner or later be more likely than a loss. Even so, the player will not agree to the infinite stake required by fair play because he cannot play indefinitely. On the other hand, the banker will not agree to any player's finite initial payment because (as we said above) he will lose it sooner or later. We are dealing here with an unsolvable conflict; indeed, for this

Table 3.3 Winning table

Roll number n	Winning $w(n)$	Probability $p(n)$	Expected value $p(n) \cdot w(n)$
1	2^1	1/2	$2 \cdot 1/2 = 1$
2	2^2	1/4	$4 \cdot 1/4 = 1$
3	2^3	1/8	$8 \cdot 1/8 = 1$
...	...	...	...
n	2^n	$1/2^n$	$2^n \cdot 1/2^n = 1$
...	...	...	...

[9] Let's agree that the player's undeserved win for $n = 1$, i.e., when the averse has not fallen even once, is supposed to encourage him to join this game.

reason, games with infinite expected value are not suitable for use in casinos. What is the cause of this conflict?

The banker and the player cannot agree on any compromise payment/rate because such a characteristic rate (or, in the language of physics, scale) does not exist.

3.3.3 *Generalization of the Petersburg Paradox*

Weierstrass's generalization of the Petersburg Paradox (PP) is that the player now flips a coin that is (generally) asymmetric. This means that now $p(n) \stackrel{\text{def}}{=} \left(1 - \frac{1}{N}\right) \frac{1}{N^{n-1}}$, where $N > 1$. Of course, when $N = 2$, we return to the expression for probability defined in the PP framework. We still express the reward as a function $w(n) \stackrel{\text{def}}{=} \nu^n$. As one can see, when $\nu = 2$, we return to the PP case. We now introduce the (normalized) PDF in the form,

$$\psi(x) \stackrel{\text{def.}}{=} \sum_{n=1}^{\infty} p(n)\delta(x - w(n)) = \left(1 - \frac{1}{N}\right) \sum_{n=1}^{\infty} \frac{1}{N^{n-1}} \delta(x - \nu^n), \tag{3.19}$$

$x \geq 0$ is a random variable related to the reward taking the values $w(n)$, $n = 0, 1, 2, \ldots$, while $\delta(\ldots)$ is the delta Dirac function. The last equality in (3.19) is the so-called Weierstrass semi-function. We will use the beneficial Weierstrass function later in this textbook.

With the above expression, the expected value can now be expressed in the general form,

$$\begin{aligned} \langle x \rangle &= \int_0^{\infty} x\psi(x)dx \\ &= \left(1 - \frac{1}{N}\right) \sum_{n=1}^{\infty} \frac{\nu^n}{N^{n-1}} = \left(1 - \frac{1}{N}\right) \nu \sum_{n=0}^{\infty} \left(\frac{\nu}{N}\right)^n. \end{aligned} \tag{3.20}$$

The infinite sum of terms of a geometric sequence appearing in the above expression may converge or diverge. It is convergent when $\nu < N$ and divergent when $\nu \geq N$. It can be seen that the PP case corresponds to the marginal situation when $\nu = N = 2$. In other words, we deal with disjoint worlds depending on whether the fraction ν/N is less than, equal to, or greater than 1.

Let us also note that in the case of $\nu \geq N$ also the second moment given by the formula,

$$\langle x^2 \rangle = \int_0^\infty x^2 \psi(x) dx = \left(1 - \frac{1}{N}\right) \sum_{n=1}^{\infty} \frac{\nu^{2n}}{N^{n-1}} = \left(1 - \frac{1}{N}\right) \nu^2 \sum_{n=0}^{\infty} \left(\frac{\nu^2}{N}\right)^n . \tag{3.21}$$

is divergent/infinite because always $\langle x \rangle^2 \le \langle x^2 \rangle$.

The case of $\nu < N$ is more complicated because the second moment, $\langle x^2 \rangle$, can be either finite or infinite. This depends on whether $\nu^2 < N$ or, conversely, $\nu^2 \ge N$. The latter situation means (by Eq. (3.21)) that the variance $\sigma_x^2 = \langle x^2 \rangle - \langle x \rangle^2$ is infinite, i.e., we are dealing here with infinite fluctuations. We continue this matter in the Sect. 3.3.4 subsection below and a few other places in our textbook, especially in Sect. 6.4.8.

3.3.4 Inspiring Digression

To extend the above-defined conflict, let's consider a surprising property of the Pareto-Lévy that has kept physicists from using it for decades—the second-moment divergence, or infinite dispersion of this distribution.

Note that the following formula holds for a moment of order (degree) $m = 0, 1, 2, \ldots$, of a random variable x,

$$\langle x^m \rangle = (-i)^m \frac{d^m}{dk^m} \tilde{P}(k) \mid_{k=0}, \tag{3.22}$$

where $\tilde{P}(k)$ is the Fourier transform, also called the characteristic function of the PDF $P(x)$. As can be seen, the characteristic function of the symmetric Pareto-Lévy distribution (3.14) is given by the expression,

$$\tilde{P}(k) = \exp(-\gamma \mid k \mid^{\alpha}),\ 0 < \alpha < 2,\ \gamma > 0, \tag{3.23}$$

which in combination with the formula (3.22) leads to the conditions,

$$\langle x^m \rangle \begin{cases} < \infty, \text{ for } m \le \alpha \\ = \infty, \text{ otherwise.} \end{cases} \tag{3.24}$$

Thus, the variance of $\sigma_x^2 = \langle x^2 \rangle$ of the distribution (3.23) (that is, the case $m = 2$) is infinite.

Generally speaking, infinite variance means the lack of a characteristic scale of statistical fluctuations; in other words, all scales are equal here. The phenomena described by this distribution occurring in different scales are similar (in the mathematical sense), i.e., connected by the scaling relation, which is discussed in detail later in the textbook.

3.3.4.1 Utility Function

Now, we can answer the significant question. Namely, how to modify the win rate, $w(n)$, in each step, n, of the generalized PP, for the case $v \geq N$, so that the expected value of the modified win is finite, making the game functional. The modification proposed by Daniel Bernoulli is simple,

$$u(n) \stackrel{\text{def}}{=} \ln w(n) = n \ln v. \tag{3.25}$$

The expected value of the modified win, $u(n)$, we define as follows,

$$U \stackrel{\text{def}}{=} \langle u(n) \rangle = \sum_{n=1}^{\infty} p(n)u(n) = (N-1)\ln v \sum_{n=1}^{\infty} \frac{n}{N^n} = \left(1 - \frac{1}{N}\right)^{-1} \ln v. \tag{3.26}$$

As one can see, it is a finite quantity. For this reason, U is called the utility function.

It is the values of the single-step utility functions $u(n)$, $n = 0, 1, 2, \ldots$, which can make sense of the size of the stakes won in this game - stakes that are "flattened" for large values of n, i.e., their increase as n increases are slowed down, as it should be (as the richer may not care as much about winning as the poorer). The approach described above (based on the Petersburg Paradox) paved the way for the emergence of the modern utility theory.

3.4 Slowed Relaxation: The Physical Motivation

Like it or not, the number of experiences that can be described utilizing slowed relaxation, also called long-term (i.e., the anomalous, non-Debye, or non-exponential) is evergrowing. These are primarily phenomena related to the relaxation in an amorphous, disordered environment ("random materials"). Of these, the most famous are those concerning relaxation [18] of,

- photocurrents in amorphous glassy films, including relaxation of photocurrents in xerographic experiments [19];
- magnetization relaxation below the critical temperature;
- spin glasses;
- percolating systems;
- aging glasses and polymers;
- viscoelastic relaxation;
- recombination relaxation in epitaxial semiconductors;
- instabilities in the granular matter (e.g., avalanches in the self-organizing criticality of a sand mound or silo with grain, cement, etc.)

and also concerning

- anomalous hydrogen diffusion in amorphous transition metals [20];
- laser cooling, which is an essential element of the method that allows trapping atoms in the form of a Bose-Einstein condensate, where the lifetime statistics of these atoms in the trap are subject to Lévy distribution with the exponent $\alpha = 1/2$ [21].

The relaxation functions in the abovementioned phenomena form a catalog of functions present (to varying degrees) in other phenomena, such as anomalous transport and diffusion. Below, we present the most characteristic ones used in rheology.

A canonical example within rheology is the viscoelastic or slowed-down relaxation function, $f(t)$, of the Kohlrausch-Williams-Watts (KWW) law,

$$f(t) = \exp\left(-\left(\frac{t}{\tau}\right)^{\alpha}\right), \tag{3.27}$$

where $0 < \alpha < 1$. The expression (3.27) is also called a 'stretched exponent.' Also using Nutting's power law of relaxation

$$f(t) = \frac{1}{(1 + t/\tau)^n}, \tag{3.28}$$

where exponent $0 < n < 1$. As well as, in general, functions that have power-law decay for the asymptotic-long time,

$$f(t) \sim \frac{1}{t^{\alpha}}, \tag{3.29}$$

(where $\alpha > 0$) we will discuss it later.

Figure 3.5 compares three types of relaxation functions on a linear (left panel) and log-log (right panel) scales: (ordinary) exponential, i.e., (standard) Debye's law of relaxation (dotted curve),

$$f(t) = \exp(-t/\tau), \tag{3.30}$$

and the KWW (solid curve, Eq. (3.27)) and Nutting (dashed curve, Eq. (3.28)) laws for the parameters $\tau = 1$ and $\alpha = 0.75$. As one can see, the Debye function relaxes the fastest, and the Nutting function relaxes the slowest.

As we said at the outset, B.G. Buelfinger first observed power-law relaxation in 1729 in studying stress relaxation in materials such as steel and stone. Since 1888, it has been called Bach's law of elasticity.

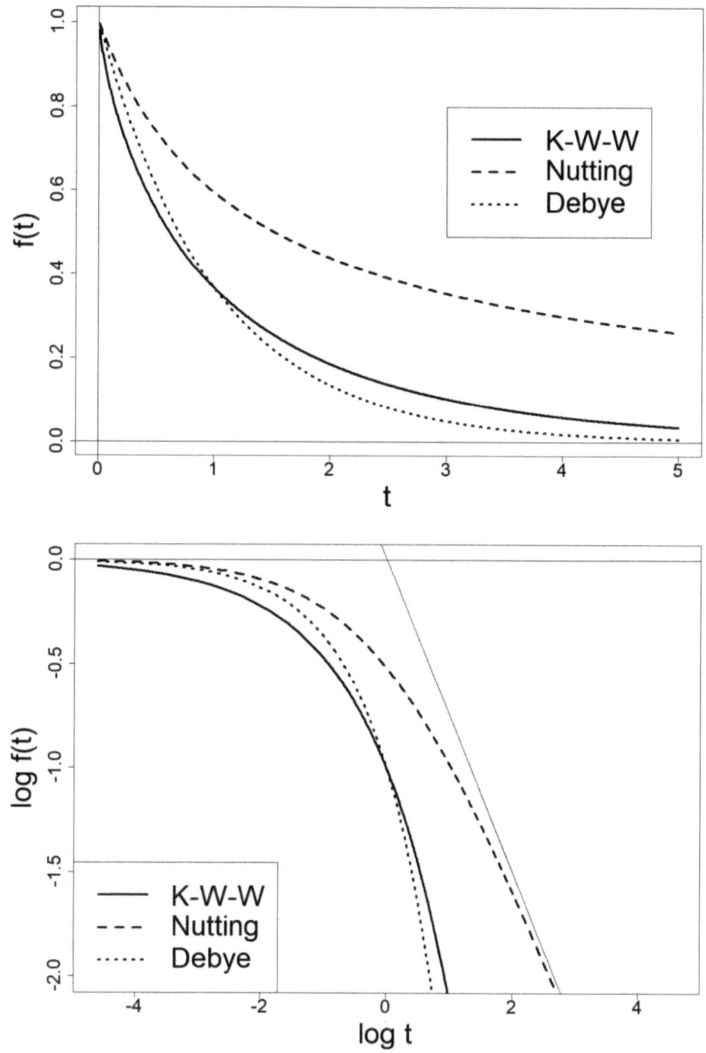

Fig. 3.5 Comparison on the linear (upper panel) and the log-log scales (lower panel) of three types of relaxation functions: Debye's (the dotted curve), Kohlrausch-Williams-Watts (solid curve), and Nutting (dashed curve) relaxation laws for parameters $\tau = 1$ and $\alpha = n = 0.75$. For additional comparison with the Nuttig law, we put the straight thin solid line from Eq. (3.29) for the same $\alpha = 0.75$; therefore, the dashed curve comes very close to this straight line

Roughly speaking, slow relaxation may be the result of the existence of many highly coupled and differently (asynchronously) relaxing subsystems in a complex system, which may lead to retardation or, more generally, to memory and thus to the non-Markovian nature of the relaxation process [22, 23]. This leads to the fractional (or fractal) relaxation equation, whose solution is the Fox function decaying, as is known, to a power-law function. The main goal of this textbook is to discuss these issues.

3.5 Fractal Relaxation

We now introduce a fractal equation for the relaxation function thanks to, which will allow us to reproduce the power law decay introduced earlier. As one knows, the exponential relaxation function (3.30) is a solution to a differential equation,

$$\frac{df(t)}{dt} = -\frac{1}{\tau} f(t),\ t > 0, \tag{3.31}$$

with the given initial condition $f(t = 0) = f_0$. The integral equivalent of the Eq. (3.31) is the following,

$$f(t) - f_0 = -\frac{1}{\tau}\frac{d^{-1} f(t)}{dt^{-1}} \overset{def.}{=} -\frac{1}{\tau}\int_0^t dt' f(t'). \tag{3.32}$$

Since integration is the inverse operation of ordinary differentiation, we labeled it as the negative derivative. Of course, this derivative has a different character than the positive one. Namely, it is of the global type (as opposed to the positive one, which is local).

The formal generalization of the Eq. (3.32) consists in replacing the negative differentiation occurring there with a negative fractional differentiation,

$$f(t) - f_0 = -\frac{1}{\tau^\alpha}\frac{d^{-\alpha} f(t)}{dt^{-\alpha}},\ \alpha > 0, \tag{3.33}$$

where the negative derivative of degree $-\alpha$ is simply the Riemann-Liouville integral operator,

$$\frac{d^{-\alpha} f(t)}{dt^{-\alpha}} \overset{def.}{=} \begin{cases} \frac{1}{\Gamma(\alpha)}\int_0^t dt' \frac{f(t')}{(t-t')^{1-\alpha}} & \text{for } \alpha > 0 \\ f(t) & \text{for } \alpha = 0 \end{cases} \tag{3.34}$$

(see Appendix A.1 for detail). By differentiating the integral equation (3.33) by sides, we get the fractal relaxation equation, which is an integrodifferential equation,

$$\frac{df(t)}{dt} = -\frac{1}{\tau^\alpha}\frac{d^{1-\alpha}f(t)}{dt^{1-\alpha}}, \ \alpha > 0, \tag{3.35}$$

Equation (3.35) is a direct generalization of the ordinary first-order differential equation (3.31) that describes the Debye relaxation. The solution to this equation uses a Fox H-function of type E_α [23, 24],

$$\begin{aligned} f(t) &= \frac{f_0}{\alpha}H^{11}_{12}\left[\frac{t}{\tau}\middle| \begin{matrix}(0, \frac{1}{\alpha}) \\ (0, \frac{1}{\alpha}),\ (0, 1)\end{matrix}\right] = f_0 H^{11}_{12}\left[\left(\frac{t}{\tau}\right)^\alpha \middle| \begin{matrix}(0, 1) \\ (0, 1),\ (0, \alpha)\end{matrix}\right] \\ &= f_0 E_\alpha\left(-\left(\frac{t}{\tau}\right)^\alpha\right) = f_0\sum_{j=0}^{\infty}\frac{\left(-\left(\frac{t}{\tau}\right)^\alpha\right)^j}{\Gamma(1+\alpha j)}. \end{aligned} \tag{3.36}$$

This solution exhibits the following boundary behaviors,

$$f(t) \approx f_0 \begin{cases} \exp\left(-\frac{(t/\tau)^\alpha}{\Gamma(1+\alpha)}\right) & \text{dla } t \ll \tau \\ \frac{1}{\Gamma(1-\alpha)}\frac{1}{(t/\tau)^\alpha} & \text{dla } t \gg \tau \text{and } 0 < \alpha < 1, \end{cases} \tag{3.37}$$

the first of which is nothing else than the law of relaxation mentioned earlier—the Kohlraush-Williams-Watts law and the second is the power decay law with exponent $0 < \alpha < 1$. No asymptote function is obtained within this formalism for other values of α.

This subsection is our first encounter with the fractal integro-differential equation of system evolution. Here, the fractalizing variable is time. Later, we will discuss fractal diffusion equations, where the fractalizing variables are time and/or space. This type of system evolution equation constitutes an important mathematical tool for analyzing singular stochastic processes.

3.6 The Role of Memory in Relaxation

We now look at relaxation from a perspective that allows us to analyze the role of memory. So let's consider the following differential-integral equation with kernel $K(t)$,

$$\frac{df(t)}{dt} = -\int_0^t dt' K(t-t')f(t'). \tag{3.38}$$

It should be emphasized that the above equation was obtained using the Zwanzig projective operator technique [25]. The above equation allows us to derive all the types of relaxation discussed so far by selecting the appropriate integral/memory kernel $K(t)$.

(A) No memory. Suppose there is no memory in the system, i.e., an integral kernel

$$K(t) = \frac{1}{\tau}\delta(t); \tag{3.39}$$

substituting the above expression into the equation (3.38), we get the exponential relaxation function as its solution,

$$f(t) = f_0 \exp\left(-\frac{t}{\tau}\right), \tag{3.40}$$

previously given by the model (3.30).

(B) Constant/uniform memory. Now assume that the memory kernel is time-independent, i.e.,

$$K(t) = \omega^2, \tag{3.41}$$

then, substituting (3.41) into (3.39), we get an oscillating solution for the relaxation function,

$$f(t) = f_0 \cos(\omega t), \tag{3.42}$$

where ω is a constant greater than zero.

(C) Slowly-varying memory. Suppose the integral kernel, $K(t)$, is slowly-varying function of time and initially (for $t \ll \tau$) it increases in a power-law manner,

$$K(t) = K_0 t^\gamma, \ \gamma > 0. \tag{3.43}$$

Hence and from the Eq. (3.38), we get the relaxation function as a solution,

$$f(t) \approx f_0 \exp\left(-\left(\frac{t}{\tau}\right)^{\gamma+2}\right), \ \tau^{-1} = K_0^{1/(\gamma+2)} \tag{3.44}$$

in the form of a stretched exponential function.

(D) Long-term memory. Now, we assume that the memory kernel is for $t > 0$ an algebraic decreasing function of time,

$$K(t) = \frac{K_0}{t^{2-\alpha}}, \ K_0 > 0, \ 1 < \alpha \le 2; \tag{3.45}$$

substituting, as usual, the above form of the kernel into the Eq. (3.38), we get,

$$\frac{df(t)}{dt} = -\frac{1}{\tau^\alpha}\frac{d^{1-\alpha}f(t)}{dt^{1-\alpha}}, \quad \tau^{-\alpha} \overset{def.}{=} K_0\Gamma(\alpha - 1) \tag{3.46}$$

where along the way, we used the definition of a negative fractal derivative (3.34). As one can see, the integral kernel of type (3.45) leads to the fractal relaxation equation (3.35) with the exponent range α limited to $1 < \alpha \leq 2$.

Now the question arises about the relationship between the fractal relaxation equation (3.35) and the equation with memory (3.38) for $0 < \alpha \leq 1$. In other words, it is about the answer to whether for $0 < \alpha \leq 1$ there is an integral kernel $K(t)$, and if so, what is it?

3.7 Anomalous Diffusion Useful Definitions and Expressions

This section shows how easy it is to build the drift-free anomalous stochastic process. Use a scaling function for a probability distribution that depends on the non-Gaussian scaling variable.[10] Then, it can be seen that Gaussian processes are of null measure in the space of all stochastic processes and even in the space of stable PDFs.

3.7.1 Definition of Anomalous Stochastic Process and Initiating Examples

Consider the stochastic process $X(t)$. We say that a process is anomalous if its variance, $\sigma^2(X(t))$, for asymptotically long times, t, is a nonlinear function of time, that is,

$$\sigma^2(X(t)) = \langle X(t)^2\rangle - \langle X(t)\rangle^2 \propto \mid t \mid^\eta + CS(t), \ \eta > 0, \tag{3.47}$$

where angle brackets $\langle\ldots\rangle$ means the average over the statistical ensemble, and $CS(t)$ is a correction to scaling increasing in time slower than t^η. In the special case of the Brownian motion process (i.e., $\eta = 1$), the correction $CS(t)$ is a sublinear function of time, e.g., increasing logarithmically with time [26]. We provide this correction

[10] By Gaussian scaling variable, we mean a compact/scaling random variable of the form $\frac{|X|}{|t|^{1/2}}$. On the other hand, a non-Gaussian random variable can be of the form $\frac{|X|}{|t|^\alpha}$, where $\alpha \neq 1/2$.

because it determines the velocity autocorrelation function for the Brownian motion process (especially the long-term function). Without the correction to scaling, the velocity autocorrelation function vanishes (what we show in the following). We decided to add the anomalous diffusion correction to scaling to the definition to distinguish between these two situations: disappearance or non-disappearance of the process velocity autocorrelation function. This only plays a role in Brownian motions.

We continue to assume that space is isotropic (drift-free process, i.e., $\langle X(t)\rangle = 0$ but $\langle | X(t) | \rangle \neq 0$). Furthermore, we assume that time is homogeneous/uniform at the level of variance (or more generally, second-order statistics)—thus, the process has the property of weak stationarity (We already talked about this in Sects. 2.3.2, 2.3.4, and 2.4.2). That is, we identify this weak stationarity with the property of homogeneity/uniformity in time during the stochastic process.[11]

In addition, we assume that the variance is a symmetric function of time due to the change of sign (orientation) of time from t to $-t$.

Processes satisfying (3.47) can be classified as follows according to the value of the exponent η.

(i) When $\eta < 1$ we are dealing with subdiffusion, i.e. antipersistent stochastic process;
(ii) when $\eta > 1$ we are dealing with superdiffusion, i.e. a persistent stochastic process.

However, if $\eta = 1$, we would deal with a Brownian diffusion process. We do not impose additional constraints on the η exponent, and we will address the problem of η dependence on time in the following.

We want to announce that in Sect. 3.8.3, we will extend the definition of anomalous diffusion (3.47) to include singular processes.

3.7.1.1 Inspiring Examples

As an inspiring example, we consider the probability distribution (PDF) of a process of the form,

$$P(X, t) = \frac{1}{| t |^H} F\left(\frac{| X |}{| t |^H}\right), \tag{3.48}$$

where $F(| Y |)$ is the normalized scaling function of the (compact) scaling variable $Y = X/ | t |^H$. From the equation above,

$$\langle X(t)^2\rangle = \langle Y^2\rangle\, t^{2H}, \tag{3.49}$$

[11] Our definition of weak stationarity, which is convenient for us, is different from the definition of weak stationarity usually found, e.g., in econometric literature where weak stationarity means that for a time series: mean, variance and bilinear autocorrelations are constant over time, but higher-order moments and autocorrelations can vary.

where we assume that the time-independent variance $\langle Y^2 \rangle = \int_{-\infty}^{\infty} dY Y^2 F(|\, Y \,|) < \infty$ and $H = \eta/2$ is the Hurst exponent. As a simple illustration, let's add that the scaling function can be non-Gaussian, e.g., $F(|\, Y \,|) = \exp(-\mid X \mid / \mid t \mid^H)$.

The above example can be extended to the second example regarding the time-dependent Hurst exponent, the Hölder exponent $H(t)$, which is a slowly varying function of time t or a function of constant intervals in time. Then, the expression (3.48) generalizes as follows,

$$P(X,t) = \frac{1}{\mid t \mid^{H(t)}} F\left(\frac{\mid X \mid}{\mid t \mid^{H(t)}}\right). \tag{3.50}$$

Hence, the Eq. (3.49) takes the more general form,

$$\langle X(t)^2 \rangle = \langle Y^2 \rangle \, t^{2H(t)}. \tag{3.51}$$

As one can see, this leads to generalized anomalous diffusion, where the exponent $\eta = \eta(t) = 2H(t)$, i.e., is a time-dependent function, e.g., a slowly-varying function of time.

Let us also derive the dependence of the absolute moment of order q on time t, where q is any real number. Namely, using Eq. (3.50), we get,

$$\langle \mid X(t) \mid^q \rangle = t^{qH(t)} \int_{-\infty}^{\infty} \frac{dX}{t^{H(t)}} \left(\frac{\mid X \mid}{t^{H(t)}}\right)^q F\left(\frac{\mid X \mid}{t^{H(t)}}\right) = t^{qH(t)} \langle \mid Y \mid^q \rangle, \tag{3.52}$$

where $\langle \mid Y \mid^q \rangle < \infty$ for each value of q. The linear dependence of the exponent on q here shows that we are dealing with monofractal wandering, even though it is wandering governed by the Hölder function.

Often, it is more convenient to use the scaling function in the form,

$$F = F\left(-\left(\frac{\mid X \mid}{\mid t \mid^{H(t)}}\right)^{\gamma}\right), \tag{3.53}$$

Then the expression (3.50) becomes

$$P(X,t) = \frac{1}{\mid t \mid^{H(t)}} F\left(-\left(\frac{\mid X \mid}{\mid t \mid^{H(t)}}\right)^{\gamma}\right). \tag{3.54}$$

Thus, formally speaking, the expression (3.51) does not change, but here $\langle Y^2 \rangle = \int_{-\infty}^{\infty} dY Y^2 F(-\mid Y \mid^{\gamma}) < \infty$. The probability distribution given by Eq. (3.54) provides an important classification of random wanderings below.

It is essential to realize that if $H = 1/2$, $\gamma = 2$ and $F(\ldots) = \exp(\ldots)$ then we are dealing with the Brownian & Gaussian motion. Brownian and non-Gaussian wandering also occur for the scaling function above, but when $H = 1/2$ and $\gamma \neq 2$.

On the other hand, if $H \neq 1/2$ and $\gamma = 2$, then we are dealing with the normal fractal Brownian motion (nfBm), i.e., the non-Brownian and Gaussian wandering. If $H \neq 1/2$ and $\gamma \neq 2$, this wandering is non-Brownian and non-Gaussian. This type of wandering is called the fractal Brownian motion (fBm).

Now, we provide an illustrative/specific example of a stochastic process X governed by the distribution (3.54) for $\gamma < 0$. Subsequently, it is beneficial to substitute $\gamma = - \mid \gamma \mid$, which results in a scaling function F and a scaling variable ξ of the form,

$$F(X) = F(\xi) \stackrel{\text{def.}}{=} \mid \gamma \mid \xi^{(|\gamma|+1)/|\gamma|} \exp(-\xi),$$
$$\xi \stackrel{\text{def.}}{=} \mid X \mid^{-|\gamma|} t^{|\gamma| H(t)}. \tag{3.55}$$

Thus, according to Eq. (3.54), we obtain the PDF, $P(X, t)$, which shows a fat tail that asymptotically decays as $\mid X \mid^{-|\gamma|-1}$ (for a fixed $\mid t \mid$).

However, the process properties listed in this subsection do not yet lead to multifractality[12]—we need something more. For example, the Hurst exponent can be treated as a random variable with a suitable distribution. This implies introducing an appropriate weighting that combines the different PDF modes indexed by various values of the Hurst exponent. In other words, this involves transitioning to a composite statistic that accounts for the heterogeneity of the Hurst exponent. We will continue to discuss this important issue below in subsection 3.7.1.4.

In this section, we have shown only simple ways to implement anomalous stochastic processes. Next, we will offer a more complex approach closer to the real world.

3.7.1.2 Homogeneous Scaling Equation

Let us also pay attention to the homogeneous scaling equation that the Eq. (3.48) generates. Namely,

$$P(X, bt) = \frac{1}{\mid bt \mid^H} F\left(\frac{\mid X \mid}{\mid bt \mid^H}\right) = \frac{1}{b^H} P\left(\frac{X}{b^H}, t\right), \quad b > 0. \tag{3.56}$$

---

[12] Using Eqs. (3.54) and (3.55), it can be easily verified that the absolute moment of order q of the process, $\langle \mid X(t) \mid^q \rangle \propto t^{qH(t)}$, is independent of the absolute value of $\mid \gamma \mid$. The exponent that scales a time variable is a linear function of q. This conclusion also follows from the general equality (3.52).

The more general equation corresponding to the generalized PDF, as given by formula (3.54), introduces additional complexity due to the time-dependent exponent $H(t)$. We will discuss the topic of multifractality[13] later.

Let's see how the absolute moments of the distribution (3.48) scale. Namely, it is a particular case of Eq. (3.52), where instead of the Hölder function, we insert the Hurst exponent. Hence we get,

$$\langle| X(bt) |^q\rangle = b^{qH}\langle| X(t) |^q\rangle, \tag{3.57}$$

where q is an arbitrary real number. Thanks to this, it is then often said that the random variable $X(t)$ scales in a statistical sense and is written in the form

$$X(bt) = b^H X(t), \tag{3.58}$$

adding that it is equality in the sense of equality of all absolute moments, that is, equality (3.57).

We will discuss the conclusions of the scaling equations later in our textbook.

3.7.1.3 An Example of a Random Hurst Exponent

To conclude the considerations concerning the Hurst exponent and the Hölder function, we will briefly discuss the case of a random Hurst exponent [27] and its relationship with multifractality.

The concept of the random Hurst exponent, or more generally shape exponents, has a long history dating back to the seminal publication by Mantegna and Stanley from 1997 [28] (popularized three years later in their book [29]). It shows empirically (based on the long-term performance of the S&P 500 index) that the shape exponent of the Lévy distribution and the scaling factor can behave randomly. This discovery was made possible by the observation that the probability of the S&P 500 index variations scales (the non-Gaussian data collapse) at time horizons from 1 to 10^3 min [30–32]. At the same time, this multiscaling provides strong empirical evidence that a break of the CLT

[13] To understand multifractality, it's helpful to contrast it with mono-fractality. In both cases, we examine the arbitrarily absolute moments of the process $\langle| X(t) |^q\rangle$, where q is any actual number. In particular, for the PDF given by Eq. (3.54), this type of moment scales proportionally to $| t |^{qH(t)}$, where the exponent forms a linear relationship with q. If this dependence on q were non-linear, we would identify it as multifractal anomalous diffusion.

is taking place here. A similar indication was provided by Mantegna's 1991 publication for the Milan Stock Exchange [33].

The publications mentioned above supply systematically studied examples of a CLT breakdown in the context of financial markets. It was one of the key examples that sparked a flood of interest from physicists in financial markets.

So, let's assume that the Hurst exponent is a random variable with the probability density function (PDF) $f(H)$. In other words, we assume that the Hurst exponent takes different (random) values for various trajectories (or segments of them) within the same random walk. The probability density function of such a process can be expressed as a composite PDF by superimposing the decomposition (3.48) with the ones mentioned above,

$$P(X,t) = \int_0^{\infty} \frac{1}{t^{g(H)}} F\left(\frac{\mid X \mid}{t^{g(H)}}\right) f(H)dH, \tag{3.59}$$

where $g(H)$ is the differentiable function of H. Hence, the absolute moment of order q takes the form,

$$\begin{aligned} \langle \mid X(t) \mid^q \rangle &= \int_{-\infty}^{\infty} \mid X \mid^q P(X,t)dX \\ &= \int_0^{\infty} t^{qg(H)} f(H) \int_{-\infty}^{\infty} \frac{\mid X \mid^q}{t^{qg(H)}} F\left(\frac{\mid X \mid}{t^{g(H)}}\right) \frac{dX}{t^{g(H)}} dH \\ &= \langle \mid Y \mid^q \rangle \int_0^{\infty} t^{qg(H)} f(H)dH, \end{aligned} \tag{3.60}$$

where $Y = X/t^{g(H)}$ is a random variable which has a universal $PDF(Y) = F(Y)$. Furthermore, we used the commutative property of integration over H with integration over X, resulting in a time-independent pre-factor (standing before integral).

Let us note that the PDF given by Eq. (3.59) corresponds to the following generic equation of stochastic dynamics,[14]

$$X_t = Y_t \, t^{g(H_t)}, \tag{3.61}$$

where Y_t and H_t denote random variables drawn at time t from the distributions $F(Y)$ and $f(H)$, respectively. The above equation corresponds to the following stochastic differential equation (SDE):

$$d\hat{X}_t = d\hat{Y}_t + \hat{t}g'(H_t)dH_t + g(H_t)d\hat{t}, \tag{3.62}$$

[14] As we show below, the multifractal random walk of Bacry-Delour-Muzy [34] is a special case of Eq. (3.61), where the various random variables Y_t and H_t are Gaussian and the function $g(\ldots)$ is identity.

where $g'(H_t) \stackrel{\text{def.}}{=} \frac{dg(H_t)}{dH_t}$ and for quantity Z (equals X_t, Y_t, or t) we defined $\hat{Z} = \ln Z$. If the variable Z is a random variable and follows the distribution $\mathcal{F}(Z)$, then the random variable $\hat{Z}$ follows the distribution $\hat{\mathcal{F}}(\hat{Z}) = \mathcal{F}(Z(\hat{Z}))\left|\frac{d\hat{Z}(Z)}{dZ}\right|^{-1} = \mathcal{F}(Z(\hat{Z}))\exp(\hat{Z})$.

The Eq. (3.62) can be viewed as a superposition of stochastic processes $\hat{Y}_t$ and H_t with a random drift H_t (i.e., the H_t process plays a dual role). Naturally, the solution of Eq. (3.62) fundamentally depends on the specific form of the function $g(H)$, and the distributions $F(Y)$ & $f(H)$ mentioned above with the PDF $f(H)$ playing a particularly important role. This is the case even though the distributions F and f do not appear explicitly in Eq. (3.62).

The formal solution to such an SDE (more general than Eq. (3.61)) is a stochastic process $\{\hat{X}_t,\ t \geq 0\}$ (with non-vanishing initial conditions), expressed in the following integral representation,

$$\hat{X}_t - \hat{X}_{t=0} = \hat{Y}_t - \hat{Y}_{t=0} + \int_{g(H_{t=0})}^{g(H_t)} \ln t'\, dg(H_{t'}) + \int_0^t \frac{g(H_{t'})}{t'} dt', \tag{3.63}$$

where the integrals are the Itô stochastic integrals [35].

Indeed, Eqs. (3.62), hence (3.63), raises an intriguing question about the corresponding multifractal Fokker-Planck equation. Such an equation should be based on a multifractal variant of Itô's lemma (or calculus), which, to our knowledge, has not yet been considered [36] . Unfortunately, considerations in this direction (although very interesting) are beyond the scope of this textbook.

3.7.1.4 Multifractality Generated by Random Hurst Exponent

We will now present an example of an identity function $g(H) = H$ and the function $f(H)$, where a random Hurst exponent gives rise to a multifractal wandering character. This form is analogous to the one introduced in Sect. 8.3, expressed by formula (8.31), i.e., a stretched exponential of the form,

$$f(H) = \frac{1}{2^{1+1/\alpha}\sigma\Gamma(1+1/\alpha)} \exp\left(-\frac{1}{2}\left|\frac{H-\overline{H}}{\sigma}\right|^{\alpha}\right), \tag{3.64}$$

where $\alpha > 0, \sigma > 0$, and the random variable ε has been substituted with the random variable H, while the parameter μ has been replaced by $\overline{H}$. As observed, Eq. (3.64) transforms into the Gaussian probability density function (PDF) when the exponent $\alpha = 2$.

Substituting Eqs. (3.64) into (3.60), we get, after a simple change of variables, the q moment in the intermediate form,

$$\langle \mid X \mid^q \rangle = \frac{\langle \mid Y \mid^q \rangle}{2^{1+1/\alpha}\mathbf{\Gamma}(1+1/\alpha)} l^q \, I(q), \tag{3.65}$$

where $l = \exp(\overline{H} \ln t)$, i.e. β in the formula (8.30) has been replaced by $\ln t$, and furthermore in the formula (3.65) the Euler gamma function $\mathbf{\Gamma}(1+q)$ has been replaced by a more general factor $\langle \mid Y \mid^q \rangle$ independent of $\overline{H}$, and the quantity Δt replaced by $\mid X \mid$,

$$I(q) = \int_{-\infty}^{\infty} \exp\left(-\frac{1}{2}|y|^\alpha + q\sigma y \ln t\right) dy. \tag{3.66}$$

Looking at this integral, we see at once that the convergence is assured as long as $\alpha > 1$. In other words, the moment $\langle \mid X \mid^q \rangle$ exists if $\alpha > 1$.

Still, based on the integrated form $I(q)$ given by (8.37) in Sect. 8.3, we obtain

$$\langle \mid X \mid^q \rangle \simeq \frac{\langle \mid Y \mid^q \rangle}{2^{1+1/\alpha}\mathbf{\Gamma}(1+1/\alpha)} A(q) \, t^{q\overline{H}+\Phi(q)}, \tag{3.67}$$

where the nonlinear component of the exponent is,

$$\Phi(q) = 2^{1/(\alpha-1)}(\alpha-1)\,(\mid q \mid \sigma/\alpha)^{\alpha/(\alpha-1)} \mid \ln t \mid^{\alpha/(\alpha-1)}, \tag{3.68}$$

while, the amplitude $A(q)$ is provided by (A.172) in Appendix A.18. The exponent (3.68) exhibits a slow dependence on time t, where multifractality, represented by a formula (3.67), is present in a limited range of order q (for all values of Hölder exponent α). For example, close to $\mid q \mid = 0$ the q-moment of the process X, i.e., $\langle \mid X \mid^q \rangle$ is governed by the factor $t^{q\overline{H}}$ (see Appendix A.18 for detail). We emphasize that the amplitude $A(q)$ slowly varies with $\mid q \mid$. This form of dependency is permissible within the realm of multifractality.

Let us also add that the dependence $\langle \mid X \mid^q \rangle$ on q is linear for small values of q, which we proved in Appendix A.18 in the context of the dependence of the q-moment of a random time between events, $\langle \Delta t^q \rangle$, versus q. We showed that it is possible to determine the multifractality range vs. q, indicating its lower limit by monofractality.

Note that the situation discussed here is, for $\mid q \mid = 2$, an example of generalized anomalous superdiffusion because (in particular) for $t \gg 1$ the exponent $2\overline{H} + \Phi(q=2)$ is a super-linear slowly increasing function of time. We could discuss ordinary anomalous diffusion with exponent $2\overline{H}$ if the non-linear component were absent.

It can be shown in a way identical to the above that the extension of the PDF given by the formula (3.64) to the non-stationary distribution $f(H,t)$, where we allow time-dependent t in a slowly varying manner both $\sigma(t)$ and $\overline{H}(t)$, leads to results (formally) identical to those given by Eqs. (3.67) and (3.68). In other words, this leads to multifractality based on a random Hölder exponent/function[15].

The expression (3.67) is a representative example of generalized anomalous diffusion, termed multifractal anomalous diffusion. The formulation (3.67) serves as a valuable starting point for the (advanced) analysis of the multifractal behavior of the partition function.

These considerations are a prolog to Christian Beck's concept [37, 38], enriched by Beck and Cohen [39], talking about realizing by the dynamical or Langevin equation of superstatistics and hence a multifractality [34]. We present this concept at the end of this chapter. Here, let us only note that in the (double) Gaussian case, i.e., by putting $\alpha = 2$ in the above example and assuming the PDF of the variable Y as Gaussian, we reduce the equation of stochastic dynamics (3.61) to the form describing, in fact, the Bacry-Delour-Muzy multifractal random walk.

3.7.1.5 Direct Determination of the Hölder Function

Using Eq. (3.50) or equivalent to Eq. (3.54), the Hölder function can be determined if we have a set of histograms of the $P(X, t)$ indexed by time t, based on empirical data in the form of several time series (we show this in more detail in Sect. 10 using the example of the empirical time series of the S&P 500 index). Namely, let us note that

$$\begin{aligned} P(0, t) &= \frac{F(0)}{t^{H(t)}} \\ P(0, 1) &= F(0), \end{aligned} \tag{3.69}$$

where $P(X = 0, t)$ is the probability density returning the stochastic process to the origin[16] after time t. Hence, the Hölder function

$$H(t) = \ln \frac{P(0, 1)}{P(0, t)} / \ln t. \tag{3.70}$$

Determining the ratio $P(0, 1)/P(0, t)$ using an empirical time series is necessary here. Only on this basis can we build the histogram of the $P(X, t)$ distribution and read the above ratio. Let us add that we can determine this fraction with a relatively small statistical error because this error is more minor the smaller the value of $\mid X \mid$.

---

[15] In this textbook, we do not consider the more general situation of dealing with a time-dependent exponent $\alpha(t)$.

[16] This probability density also includes (as a special event) the waiting time of the process at the origin.

3.7.2 Construction of the Trajectory of the Normal Fractional Brownian Motion

The Random Midpoint Displacement Method/Algorithm (RMDM/A) is a popular approach that enables the construction of the normal fractional Brownian motion (nfBm) trajectories. This method is based, in fact, on generating a random variable that meets the definition of anomalous diffusion (3.47). It allows us to introduce, for example, the nfBm, as well as properties such as (3.102) and (3.107) below in Sect. 3.7.7, i.e., regarding, respectively, the autocorrelation of the process and the autocorrelation of its velocity. The RMDM/A algorithm enables the generation of the nfBm in the time interval $[0, t]$.[17]

Initially, in the first step, we assume that the one-dimensional walk starts at time zero from the origin of the coordinate system. However, the location of the test particle at time t, i.e., $X(t)$, is drawn from a Gaussian distribution with mean zero and variance $\langle X(t)^2\rangle = t^{2H}\sigma^2$, where H is a given Hurst exponent, established once and for all for the entire procedure.

In the next, second step, we divide the output time interval into half (bisection procedure). Our task is to find the location of the test particle, $X(t/2)$, at time $t/2$; we assume that it is the sum of two components,

$$X(t/2) = \frac{1}{2} X(t) + \gamma_1 G_1(t), \tag{3.71}$$

where the Gaussian variable $G_1(t)$ is drawn from the same Gaussian distribution as the initial variable $X(t)$ at time t, while the scaling factor γ_1 must be determined in such a way that the variance $\langle X(t/2)^2\rangle$ scaled as for fractional Brownian motion with the Hurst exponent H introduced above. Therefore, the equality must be satisfied:

$$\begin{aligned} \langle X(t/2)^2\rangle &= \frac{1}{2^2}\langle X(t)^2\rangle + \gamma_1^2 \langle G_1(t)^2\rangle \\ &= \frac{1}{2^2} t^{2H}\sigma^2 + \gamma_1^2 t^{2H}\sigma^2 = \left(\frac{t}{2}\right)^{2H}\sigma^2, \end{aligned} \tag{3.72}$$

hence, we obtain immediately that

$$\gamma_1 = \pm\frac{1}{2^H}\sqrt{1 - 2^{2H-2}}, \tag{3.73}$$

i.e., it can take both positive and negative values (e.g., a random sign change can be allowed, i.e., dichotomous noise can be imposed); of course, this may be reflected in the sign of the random variable. We emphasize the fact that the variables $X(t)$ and $G_1(t)$ were drawn from a Gaussian distribution does not mean that the ran-

[17] The literature usually provides an algorithm defining the normal fractional Brownian motion in a unit interval—the algorithm given here is more convenient in practical applications.

dom variable $X(t/2)$ will have a Gaussian distribution (since it concerns a different moment).

In the next, 3 step we divide both halves of the segment $[0, t]$ again into two equal parts—we will, for example, deal with the first half of the segment, constructing a random variable $X(t/4)$ at time $t/4$ (for the moment $3t/4$ the construction is analogous). Let,

$$X(t/4) = \frac{1}{2}X(t/2) + \gamma_2 G_2(t), \tag{3.74}$$

where $G_2(t)$ is a Gaussian variable having the same distribution as $G_1(t)$. Similar to the previous step, this new variable should have a variance that scales according to the definition of the nfBm Hurst exponent established at the beginning:

$$\begin{aligned}\langle X(t/4)^2\rangle &= \frac{1}{2^2}\langle X(t/2)^2\rangle + \gamma_2^2\,\langle G_2(t)^2\rangle \\ &= \frac{1}{2^2}\left(\frac{t}{2}\right)^{2H}\sigma^2 + \gamma_2^2\,t^{2H}\sigma^2 = \left(\frac{t}{4}\right)^{2H}\sigma^2,\end{aligned} \tag{3.75}$$

where to derive the second equality, we used the expression (3.72). Solving the equation in (3.75) for the unknown γ_2 we get:

$$\gamma_2 = \pm\frac{1}{2^2H}\sqrt{1-2^{2H-2}}. \tag{3.76}$$

Proceeding similarly to the above, in the jth step of the procedure ($j = 1, 2, 3, \ldots$), we obtain a random variable of the form,

$$X(t/2^j) = \frac{1}{2}X(t/2^{j-1}) + \gamma_j\,G_j; \tag{3.77}$$

where the formula gives its variance,

$$\begin{aligned}\langle X(t/2^j)^2\rangle &= \frac{1}{2^2}\langle X(t/2^{j-1})^2\rangle + \gamma_j^2\,\langle G_j(t)^2\rangle \\ &= \frac{1}{2^2}\left(\frac{t}{2^{j-1}}\right)^{2H}\sigma^2 + \gamma_j^2\,t^{2H}\sigma^2 = \left(\frac{t}{2^j}\right)^{2H}\sigma^2.\end{aligned} \tag{3.78}$$

Hence, we obtain the general expression for the desired scaling factor:

$$\gamma_j = \pm\frac{1}{2^{jH}}\sqrt{1-2^{2H-2}}. \tag{3.79}$$

Still, the described algorithm can be easily generalized, for example, to processes in which the random variable G_j has a non-Gaussian distribution, and this distribution (like the Gaussian one) has finite variance. Another direction of generalization is to change the exponent H to make its value dependent on the procedure step j, i.e., the transition from H to $H(j)$—we devote Sect. 3.7.3 to this last direction.

3.7.3 Random Walk Simulation with a Given Hölder Function

The Brownian motion algorithm with a given Hölder function is obtained by a direct generalization of the equality (3.78) from Sect. 3.7.2 regarding the variance of a random variable

$$X(t/2^j) = \frac{1}{2}X(t/2^{j-1}) + \gamma_j G_j, \tag{3.80}$$

formally having the same form as Brownian fractal motion. In the case of multifractal Brownian motion, the variance of this variable takes the form:

$$\begin{aligned}\langle X(t/2^j)^2\rangle &= \frac{1}{2^2}\langle X(t/2^{j-1})^2\rangle + \gamma_j^2\,\langle G_j^2\rangle \\ &= \frac{1}{2^2}\left(\frac{t}{2^{j-1}}\right)^{2H(t/2^{j-1})}\sigma^2 + \gamma_j^2\,t^{2H(t)}\sigma^2 = \left(\frac{t}{2^j}\right)^{2H(t/2^j)}\sigma^2,\end{aligned} \tag{3.81}$$

where the exponent H is dependent on time (i.e., on the procedure step). Solving the above equation for the unknown γ_j^2 we get that

$$\begin{aligned}\gamma_j^2(t) = &\frac{1}{2^{2H(t/2^j)j}}\left[1 - 2^{2[H(t/2^j)-H(t/2^{j-1})]\,j+2H(t/2^{j-1})-2}\,t^{2[H(t/2^{j-1})-H(t/2^j)]}\right]^{1/2} \\ &\times t^{2[H(t/2^j)-H(t)]}.\end{aligned} \tag{3.82}$$

As one can see, the scale factor is now not local, depending on the differences in Hölder function in subsequent and current moments. Of course, for time $t = 1$ (which is generally assumed in the literature that presents this algorithm), the expression (3.82) is simplified to a form in which the dependence on t is no longer visible.

The expression (3.82) generalizes the corresponding one (3.79) for the fractal Brownian motion with a time-independent Hurst exponent H.

3.7.4 The Rescaled R/S Range Method

Many methods exist to determine such an essential exponent as the Hurst exponent from empirical data. Here, we present one of the most popular methods: the rescaled R/S range method.

This method is based on the relative spread of the random walk scales as the dispersion of the total displacement. The R/S method is the key and most straightforward method for determining the Hurst exponent H controlling the variability of the time series. Therefore, this exponent can be associated with investment risk - the greater the volatility, the greater the risk.

So suppose we have a discrete time series $\{y_l\}_{l=1}^{L}$ $L \gg 1$, which is, for example, the increase in the price of a security or the rate of return (ordinary or logarithmic) on it in subsequent, discrete moments $l = 1, 2, \ldots, L$.

(1) In the first step, we build a statistical ensemble of equinumerous disjoint subsequences of the initial time series:

$$\{y_l\}_{l=1}^{n},\ \{y_l\}_{l=n+1}^{2n},\ \ldots,\ \{y_l\}_{l=(j-1)n+1}^{jn},\ \ldots,\ \{y_l\}_{l=L-n+1}^{L}; \tag{3.83}$$

assuming that there are J of these subsequences, i.e., $j = 1, 2, \ldots, J,\ L = n\,J$.[18]

(2) Then, we determine the average value

$$\bar{y}^j = \frac{1}{n}\sum_{l=1}^{n} y_l^j, \tag{3.84}$$

and the "local" standard deviation

$$S^j = \left[\frac{1}{n}\sum_{l=1}^{n}(y_l^j - \bar{y}^j)^2\right]^{1/2}, \tag{3.85}$$

[18] In other words, we must choose like this lengths of the series L and n so that n is a divisor of L.

for each subsequence $j(= 1, 2, \ldots, J)$ separately, where we have now introduced double labels for the elements of the output time series for convenience; then we enter the cumulative deviation defined as follows:

$$Y_k^j \stackrel{\text{def.}}{=} \sum_{l=1}^{k} (y_l^j - \bar{y}^j), \ 1 \le k \le n. \tag{3.86}$$

(3) In the next step, we enter the span (range) for each subsequence

$$R^j \stackrel{\text{def.}}{=} max_{1 \le k \le n}\{Y_k^j\} - min_{1 \le k \le n}\{Y_k^j\}, \ j = 1, 2, \ldots, J, \tag{3.87}$$

and hence the average (after the statistical ensemble) value of the rescaled range, which is an estimate of its expected value $E(R/S)(n)$, we express as follows:

$$\overline{(R/S)}(n) = \frac{1}{J} \sum_{j=1}^{J} R^j / S^j \tag{3.88}$$

(4) Finally, we calculate the average value of the rescaled range for subsequent, discrete moments n (i.e., the lengths of the previously mentioned subsequences), remembering that the expected value of the rescaled range scales power-wise with time, for asymptotically long times:

$$E(R/S)(n) = const \, n^{E(H)}, \tag{3.89}$$

where the exponent $E(H) \ge 0$ is the expected (true) value of the Hurst exponent we are looking for.[19] We present the data obtained in this way on a $\log - \log$ scale by fitting a straight line to them (Linear Regression):

$$\log(\overline{R/S})(n) = H \log n + \log const, \tag{3.90}$$

where H is the empirical value of the Hurst exponent, of course, we obtain the slope of the line from this fitting, i.e., the exponent H with some fitting error ΔH.

The applicability of the R/S method is limited

- only for long series so that they can be divided into a sufficiently large number of satisfactorily long subsequences,
- to previously detrended and seasonless series.

As we have already discussed in Sect. 3.7.1 above, the Hurst exponent governs random walk and, therefore, its volatility and (indirectly) autocorrelations, thus

[19] It is customary not to provide a separate notation for the true value of the Hurst exponent, leaving the reader to distinguish between it and his empirical value or estimate. We, too, continue to give up on such a distinction.

constituting a tool for analysis and classification of series. Moreover, as a quantity controlling volatility, it can be (according to the concept of risk introduced by Markowitz) a measure of risk.[20]

A richer, global measure of risk is the fractal box dimension $D_f = 2 - H$ of the time series, i.e., the graph of the function Y_k^j depending on the discrete-time k, where $1 \leq k \leq n$, for any selected subsequence j.[21] The algorithm for determining this quantity is discussed in Sect. 3.7.5.

There are several better methods for determining the Hurst exponent, of which Detrended Fluctuation Analysis created by physicists [40] is very popular. We will discuss it in the next chapter.

3.7.5 *Direct Reading of the Fractal Dimension*

The direct reading of the box fractal dimension is based on:

(a) box method (hence the full name of this dimension is *box fractal dimension*), and
(b) commands that read the screen state (e.g., pixel color).

3.7.5.1 Box Method

Recall that the Hurst exponent determines the response of process $Y(t)$ to the affine rescaling of its argument as follows:

$$Y(b\,t) = b^H\,Y(t) \iff Y(t) = \frac{Y(b\,t)}{b^H}. \tag{3.91}$$

Therefore, the process $Y(t)$ remains unchanged if a corresponding rescaling of the process accompanies the time rescaling. In other words, scaling the graph's horizontal axis by b and the vertical axis by $1/b^H$ leaves the plot unchanged. Of course, these equalities are only satisfied in a statistical/probabilistic sense.

One can now perform the following procedure:

(1) We initially cover the graph of the function $Y(t)$ vs. t squares (two-dimensional boxes) with side length $\varepsilon(0) = r$ in the number $L(0) = N$. Then, we introduce a descending cover with smaller boxes with sides half the length. We calculate,

---

[20] A better global measure of risk may be the quantity $1 - H$.

[21] We can also consider several substrings or even a whole series, but then we would have to redefine the meaning of the variable k and the index j accordingly.

using Eq. (3.91), the number of smaller boxes that cover the graph; one can easily check that it is $L(1) = 2^{2-H} N$. The same applies to covering with boxes of side $\varepsilon(2) = r/2^2$ - the number of these boxes is $2^{2(2-H)} N$. Generalizing to cover the boxes in the kth generation, i.e., with side $\varepsilon(k) = r/2^k$, we get:

- the number of boxes counted horizontally $L_{\|}(k) = \sqrt{N}\, 2^k$ and
- coverage cardinality of $L(k) = 2^{k(2-H)} N$.

(2) The two above quantities can be combined using the box dimension D_f (which is special case of fractal dimension) knowing that the number of boxes counted horizontally raised to the power D_f gives the total coverage number. Therefore,

$$D_f = \lim_{k\to\infty} \frac{L(k)}{L_{\|}(k)} = 2 - H \geq 0. \tag{3.92}$$

This algorithm allows a computer to read the fractal dimension of a given series fully automatically and, thus, determine investment risk (e.g., in a security whose price dynamics are represented by a given time series).

3.7.6 *Estimation of the Continuous Hölder Function*

The next task is to present a method that allows us to estimate the Hölder function, having an empirical time series. Let us immediately add that this function will be calculated within the approximation framework, treating it as a slowly changing function or as a constant interval. Then, in each of these intervals, *de facto* deals with a locally constant Hurst exponent (generally different in each interval). Although we have already determined the Hurst exponent earlier (see Sect. 3.7.4) using the R/S method, here (for a change), we will use the relation between the variance of the Gaussian process and its average absolute value, valid for an asymptotically long time t,

$$\langle \delta X(t)^2 \rangle = \frac{\pi}{2} \langle \mid \delta X(t) \mid \rangle^2, \tag{3.93}$$

where $\delta X(t) = X(t + \delta t) - X(t)$. This relation 3.93 is a simple variant of the fluctuation-dissipation theorem for nfBm.

So, let's assume that we are looking for the value of the Hölder exponent at some distant moment in time t. In the first stage the environment of the moment t should be discretized - let the discretization step be $\delta t \ll t$, assuming that in this environment the Hölder exponent is constant (to a good approximation). This environment is as follows time interval: $[t - n\delta t/2,\ t + n\delta t/2]$, where $T = n\,\delta t,\ n \gg 1$, determines its size. It should be chosen so that the variability of the Hölder function is small, that is, the Hölder function is more slowly variable, that is, $Span(H)/H \ll 1$ (where $Span(H)$ determines the span of the exponent Hurst in the range $[t - n\delta t/2,\ t + n\delta t/2]$) and $n \gg 1$.

In the second stage, we determine the estimate of the average absolute value of noise in the mentioned environment of the moment t

$$\overline{\mid \delta X(t) \mid} = \frac{1}{n} \sum_{j=-n/2}^{n/2-1} \mid \delta X(t + j\delta t) \mid, \tag{3.94}$$

where n here and below is an even number, which for simplicity we omitted from the left side of the equation.

Then, using (3.93) and remembering that for stationary anomalous processes

$$\langle \delta X(t)^2 \rangle = const\, (\delta t)^{2H(t)} \tag{3.95}$$

(see the general expression (3.47) in Sect. 3.7.1) we get that

$$2H(t)\ln(\delta t) + \ln const = 2\ln \overline{\mid \delta X(t) \mid} + \ln(\pi/2). \tag{3.96}$$

Let us also note the fact that Eq. (3.95) allows for a precise (theoretical) definition of the Hölder function/exponent in the form of the following limit,

$$H(t) = \lim_{\delta t \to 0} \frac{\langle \delta X(t)^2 \rangle}{\ln \delta t^2}. \tag{3.97}$$

As one can see, this definition is local in time and no longer depends on the nature of $H(t)$ variability. However, the estimate of $H(t)$ requires taking into account the (time) environment of the moment t (see considerations in the third stage below). Unfortunately, this distorts the genuine value of $H(t)$.

In the third stage, we will remove the unknown constant $\ln const$, i.e., we will derive the final formula from which we determine the local exponent $H(t)$. To do this, both stages should be repeated for the same size of the environment T but a different value n, e.g., it may be m and be a factor smaller than the previous n. From Eqs. (3.94) and (3.96) we finally get that

$$H(t) = \frac{\ln\left(\overline{\mid \delta_n X(t) \mid} / \overline{\mid \delta_m X(t) \mid}\right)}{\ln(m/n)}, \tag{3.98}$$

where δ_n means the discretization of the environment t with step T/n (similarly δ_m). As one can see, the Hölder exponent $H(t)$ has, as expected, a fractal structure.

Let us add that the Hurst exponent can also be calculated analogously, using the (unbiased) estimate of dispersion directly

$$\left(\overline{[\delta X(t)]^2}\right)^{1/2} = \left(\frac{1}{n-1}\sum_{j=-n/2}^{n/2-1}[\delta X(t+j\delta t)]^2\right)^{1/2}, \tag{3.99}$$

which should be substituted into the formula (3.98) instead of both absolute value estimates found there. Then, the modified formula (3.98) will apply to anomalous diffusion in general and not only to the nfBm.

Extension of the above algorithm to the discontinuous Hölder function, that is, to the generalized multifractal Brownian motion, is immediate - consider the environments between the discontinuity points and apply the above algorithm to them.

3.7.7 *Autocorrelations*

We continue to consider the autocorrelation of the process $X(t)$ and its velocity $V(t) = dX(t)/dt$. We continue to consider a family of stochastic processes for which $\eta \neq 1$ (the marginal case $\eta = 1$ is our reference Brownian motion).

3.7.7.1 Relation Between Variance and Autocorrelations

The starting point for stationary processes in the sense of variance, i.e., weak stationarity, is the following relationship between the process variance and the autocorrelation function,[22]

$$\begin{aligned}\langle [X(t_2) - X(t_1)]^2\rangle &= \langle [X(t_1)]^2\rangle + \langle [X(t_2)]^2\rangle - 2\langle X(t_1)X(t_2)\rangle \\ &= \mid t_1 \mid^{\eta} + \mid t_2 \mid^{\eta} - 2\langle X(t_1)X(t_2)\rangle,\ 0 \le t_1 \le t_2.\end{aligned} \tag{3.100}$$

We already talked about weak stationarity in Sects. 2.3.4 and 2.4.2 above, while stationarity in the strong sense (strong stationarity) was defined in Sect. 2.3.2.

On the other hand, using the Eq. (3.47), we can write (for a process without drift),

$$\langle [X(t_2) - X(t_1)]^2\rangle = \mid t_2 - t_1 \mid^{\eta}. \tag{3.101}$$

Comparing the two above equations, we get the position autocorrelation function,

---

[22] For simplicity, we assume that the (fractal) diffusion coefficient here is equal to 1. Furthermore, the 'process/position autocorrelation function' often replaces the 'position-position autocorrelation function.' For simplicity, in this subsection, we have not included in our calculations the correction to the scaling component.

$$\langle X(t_1)X(t_2)\rangle = \frac{1}{2}\left[\mid t_1 \mid^{\eta} + \mid t_2 \mid^{\eta} - \mid t_2 - t_1 \mid^{\eta}\right], \tag{3.102}$$

that is the dependence on two moments of time of the process/position autocorrelation function. As can be seen, the process is non-stationary at the level of the position autocorrelation function, as it depends on two moments and not on their difference, as is the case for the variance. When $t_1 = t_2 = t$ then expression (3.102) becomes (3.47). We are still dealing with an autocorrelated non-stationary process for $t_1 \neq 0$ even for Brownian motion (i.e., for $\eta = 1$). However, the situation is different for noise, as shown in the following section.

To conclude this section, let us consider, for example, the asymptotic situation when $t_2 \gg t_1 \geq 0$. Then, Eq. (3.102) takes the following form,

$$\langle X(t_1)X(t_2)\rangle = \frac{1}{2}t_1 t_2^{\eta-1}\left[\left(\frac{t_1}{t_2}\right)^{\eta-1} + \eta\right] = \frac{1}{2}t_1 \begin{cases} t_1^{\eta-1}, & \eta < 1 \\ 2, & \eta = 1 \\ \eta\, t_2^{\eta-1}, & \eta > 1. \end{cases} \tag{3.103}$$

As one can see, in the case of persistent random walk ($\eta > 1$), the process behaves similarly to the random walk of a stiffened polymer chain, where the autocorrelation of the process increases with the final time (t_2) in a power-law fashion (with an exponent $\eta - 1 > 0$). For an antipersistent process ($\eta < 1$), the power-law increase (with exponent η) is for the initial time. Also, for Brownian motion, the autocorrelation of the process does not disappear when the initial time is different from zero.

3.7.7.2 Velocity Autocorrelation Function

Now, we consider the autocorrelation of process noise given by the formula,

$$\langle \Delta X(t_1)\Delta X(t_1 + t)\rangle = \langle [X(t_1 + \Delta t) - X(t_1)][X(t_1 + t + \Delta t) - X(t_1 + t)]\rangle. \tag{3.104}$$

Hence, after simple transformations, we get

$$\langle \Delta X(t_1)\Delta X(t_1 + t)\rangle = \mid t - \Delta t \mid^{\eta} + \mid t + \Delta t \mid^{\eta} - 2 \mid t \mid^{\eta}. \tag{3.105}$$

Assume further that $\Delta t \ll t$. Then,

$$\langle \Delta X(t_1)\Delta X(t_1 + t)\rangle \propto H(2H - 1)\frac{\Delta t^2}{\mid t \mid^{2-2H}} \tag{3.106}$$

or in differential form,

$$\langle V(t_1)V(t_1+t)\rangle \propto H(2H-1)\frac{1}{\mid t\mid^{2-2H}}, \tag{3.107}$$

where (we remind) the process velocity $V(t) = \frac{dX(t)}{dt}$. We are dealing here with a weak stationary process at the velocity autocorrelation function level. As one can see, the equations (3.107) and (3.103) describe two different characteristics of the same process—this is a result worth remembering.

Furthermore, for the persistent process ($H > 0.5$), we deal with long-term positive autocorrelations, and for the antipersistent ($H < 0.5$), with negative ones. For the Brownian process ($H = 0.5$), the autocorrelations of the process velocity disappear (as expected), a fundamental difference from the autocorrelation of the process itself.

The above considerations/calculations were made without the correction to scaling, $CS(t)$, term (for the definition, see Sect. 3.7.1). If it is present, even velocity long-term autocorrelations for the Brownian motion are possible [26].

3.7.7.3 Relationship Between Bilinear and Nonlinear Autocorrelations

We now formulate an important issue about the relationship between bilinear and non-linear long-term autocorrelations in both the process and its velocity.[23] However, it should be noted that the presence of bilinear autocorrelations does not necessarily imply the presence of nonlinear autocorrelations and vice versa.[24] That is, the non-linear autocorrelations can manifest independently of bilinear ones. For this issue, we present a bilinear autocorrelation function in the form

$$\langle Y(t_0)\cdot Y(t_0+t)\rangle = \langle sgn\,[Y(t_0)]\cdot sgn\,[Y(t_0+t)]\cdot \mid Y(t_0)\cdot Y(t_0+t)\mid\rangle, \tag{3.108}$$

where $\langle Y(t')\rangle = \langle sgn\,[Y(t')]\rangle = 0$ and $\langle\mid Y(t')\mid\rangle > 0$, where t' means any moment, the variable Y can mean the position X of the process or its speed V and $sgn\,[Y(t')]$ is the sign of the variable Y at the moment t'. As one can see, nonlinear autocorrelations mean here autocorrelations of the absolute values of a random variable at different times.

---

[23] We recall that a function $L(x, y)$ is bilinear if it is linear for each of the arguments x and y separately.

[24] We are talking in this textbook about the nonlinear autocorrelations of random variables $x(t_1)$ and $x(t_2)$ if the absolute values of these variables are autocorrelated. This is a special but useful case of the nonlinear autocorrelation function.

We continue to assume the interesting case of the statistical independence of the product of signs, $sgn\,[Y(t_0)] \cdot sgn\,[Y(t_0+t)]$, and the absolute value of the product $\mid Y(t_0) \cdot Y(t_0+t) \mid$. Therefore,

$$\begin{aligned} &\langle sgn\,[Y(t_0)] \cdot sgn\,Y[(t_0+t)] \cdot \mid Y(t_0) \cdot Y(t_0+t) \mid \rangle \\ &= \langle sgn\,Y[(t_0)] \cdot sgn\,Y[(t_0+t)] \rangle \cdot \langle \mid Y(t_0) \mid \cdot \mid Y(t_0+t) \mid \rangle, \end{aligned} \tag{3.109}$$

where we assume the existence of autocorrelation of signs, i.e., $\langle sgn\,Y[(t_0)] \cdot sgn\,Y[(t_0+t)] \rangle \neq 0$. The lack of such autocorrelation, i.e., factorization of this average, leads to its disappearance due to the assumption of isotropic space (no drift/floating). For the existence of bilinear autocorrelations of the random variable Y, the autocorrelation of signs is necessary. Still, autocorrelation of the absolute values of this variable is not necessary. The Eq. (3.109) (supported by Eq. (3.108)) is the sought relationship between both types of autocorrelation, leading to interesting conclusions.

So, let us consider a characteristic case. Suppose that the moduli of the random variable $\mid Y(t_0) \mid$ and $\mid Y(t_0+t) \mid$ are uncorrelated. Then the right side of Eq. (3.108) is factorized, leading to the equation,

$$\langle Y(t_0) \cdot Y(t_0+t) \rangle = \langle sgn\,Y[(t_0)] \cdot sgn\,Y[(t_0+t)] \rangle \cdot \langle \mid Y(t_0) \mid \rangle \cdot \langle \mid Y(t_0+t) \mid \rangle. \tag{3.110}$$

This equation shows that autocorrelations of the absolute values of the Y variable are not needed for autocorrelation of the Y variable (including the long-term type) - autocorrelation of signs is sufficient.

It is now established that anomalous diffusion results in the emergence of bilinear long-term autocorrelations in both the process and its velocity. Let us also note that the considerations in this section are general, i.e., they do not concern only anomalous diffusion. For instance, short-term bilinear autocorrelations do not exclude the existence of long-term nonlinear autocorrelations because the autocorrelation of signs also influences it.

It is crucial to highlight our interest in multifractal anomalous diffusion, focusing on investigating the existence of nonlinear autocorrelations within multifractality. This inquiry aims to discern whether multifractality primarily arises from fluctuation cascades, nonlinear autocorrelation cascades, or a combination of both phenomena (we discuss it in detail in Sect. 5.3).

3.7.7.4 Power Spectra: Generic Formulas

We recall the definition of the velocity power spectrum for the discussed family of stochastic processes,

$$S_V(f) = \langle V^{\mathcal{F}}(f) \cdot V^{\mathcal{F}}(-f) \rangle = \langle | V^{\mathcal{F}}(f) |^2 \rangle, \tag{3.111}$$

where the Fourier transform is

$$V^{\mathcal{F}}(f) = \int_{-\infty}^{\infty} dt V(t) \exp(-2\pi i f t), \tag{3.112}$$

and averaging is performed for a fixed frequency f. Moreover, we have a time-symmetric process velocity, i.e., $V(t) = V(-t)$ and hence $V^{\mathcal{F}}(-f) = V^{\mathcal{F}}(f)^*$, where $\ldots^*$ denotes a complex conjugate.

Similarly, we can express the power spectrum for the process noise (its increments), which is the basis of the process speed,

$$S_{\Delta X}(f) = \langle \Delta X^{\mathcal{F}}(f) \, \Delta X^{\mathcal{F}}(-f) \rangle = \langle \left| \Delta X^{\mathcal{F}}(f) \right|^2 \rangle, \tag{3.113}$$

where $\Delta X^{\mathcal{F}}(f)$ is the Fourier transform of the noise $\Delta X(t)$.

On the other hand, by the well-known Wiener-Khintchin theorem [41] for stationary processes, the autocorrelation function, we can express the power spectrum as follows,

$$S_V(f) = \int_{-\infty}^{\infty} dt C(t) \exp(-2\pi i f t) = 2\Re \int_{0}^{\infty} dt C(t) \exp(-2\pi i f t), \tag{3.114}$$

where we used the abbreviation for velocity autocorrelation function

$$C(t) = \langle V(t_1) \cdot V(t_1 + t) \rangle$$

independent on t_1. Moreover, we have a time-symmetric process autocorrelation function, i.e., $C(t) = C(-t)$. Therefore, the velocity power spectrum, $S_V(f)$, can be calculated for stationary processes in two ways, either from Eq. (3.111) or from Eq. (3.114).

We have a similar situation for the noise power spectrum $S_{\Delta X}(f)$. Namely,

$$S_{\Delta X}(f) = \int_{-\infty}^{\infty} dt K(t) \exp(-2\pi i f t) = 2\Re \int_{0}^{\infty} dt K(t) \exp(-2\pi i f t), \tag{3.115}$$

where we used the abbreviation for increment/displacement autocorrelation function

$$K(t) = \langle \Delta X(t_1) \cdot \Delta X(t_1 + t) \rangle.$$

It can be seen that the equality $K(t) = C(t)\Delta t^2$ holds.

Let us emphasize that we are considering here the power spectra of three types of dynamic variables characterizing the stochastic process: its instantaneous velocity, increment/displacement, and its instantaneous position/location (hereinafter referred to as process).

3.7.7.5 Power Spectra for the Brownian Motion

For the arithmetic (standard) Brownian motion,[25] which we deal with not only in physics, the velocity autocorrelation function is stationary, and decays exponentially[26] what was derived in the textbook by Kubo et al. [41],

$$C(t) = \langle [V(0)]^2 \rangle \exp(- \mid t \mid / \tau), \tag{3.116}$$

where τ is its relaxation time. The independence of C from t_1 implies that the variance,

$$\langle [V(t_1)]^2 \rangle = \langle [V(0)]^2 \rangle, \tag{3.117}$$

i.e., it is independent of time t_1. The power spectrum here takes the form:

$$S_V(f) = 2\Re \int_0^{\infty} dt C(t) \exp(-2\pi i f t) = \frac{\langle [V(0)]^2 \rangle}{1 + (2\pi f \tau)^2}. \tag{3.118}$$

As one can see, for low frequencies ($f \ll 1/(2\pi\tau)$), i.e., for times much longer than τ, the power spectrum is (to a good approximation) constant. Therefore, the stochastic process velocity is white noise that can easily be simulated on a computer.

The power spectrum for the random variable $X(t)$ itself, i.e., for the Wiener process itself, is given by the following expression:

$$S_X(f) \propto \frac{1}{f^2 (f^2 + (1/2\pi\tau)^2)}. \tag{3.119}$$

---

[25] It should be realized that Brownian motion is a stationary stochastic process in terms of increment/displacement, $\Delta X(t) (\rightarrow dX(t))$, of the random variable and velocity of the Brownian particle but non-stationary in terms of the random variable itself (or its position) $X(t)$ (or the process of the Brownian particle).

[26] The results we present here are based on the Langevin equation with $\delta-$correlated random force. From this equation, the formulas (3.118) and (3.119) came.

For the long time intervals, we are interested in (much longer than τ), we are dealing with $1/f^2$ noise, i.e., a completely different form of the power spectrum. Hence, the question arises about the form of the autocorrelation function of the Wiener stochastic process $X(t)$.

The above question is even more justified because the Wiener process is nonstationary, and the Wiener-Khintchin theorem, which allows the application of the inverse Fourier transform to the power spectrum given by the expression (3.119), cannot be used. The sought correlation function can be determined directly, keeping in mind the stationarity of the variance of the $X(t)$ variable, namely:

$$\langle X(t_1)X(t_2)\rangle \propto t_1 \propto t_2 + t_1 - (t_2 - t_1),\ 0 < t_1 < t_2. \tag{3.120}$$

This function is nonstationary, like for the arithmetic Brownian motion, which is nonstationary in the autocorrelation of positions but weak stationary in the variance of positions.

3.7.7.6 Power Spectrum of $1/f^\gamma$ noise

We will now calculate the power spectrum for the power law decay function of the velocity autocorrelation given by (3.107), which is $C(t) = \frac{1}{2}\eta(\eta - 1)1/\mid t \mid^{2-\eta}$. Then, after simple transformations based on the Eq. (3.114), we obtain

$$S_V(f) = \frac{1}{(2\pi)^\gamma}\Gamma(\gamma)\cos\left(\frac{\pi}{2}\gamma\right)\frac{\gamma(\gamma+1)}{\mid f \mid^\gamma},\ \gamma = \eta - 1 > 0. \tag{3.121}$$

The velocity autocorrelation function decaying according to the power law in time decreases the power spectrum according to the power law in frequencies. This type of power spectrum is abbreviated as $1/f^\gamma$ noise. This type of noise can be found, for example, in physical, biological, and economic problems.

The above expression for the power spectrum shows that for $H = 1/2$ (i.e., $\gamma = 0$), we are dealing with white noise (constant power spectrum), and for $H \neq 1/2$ with colored noise (frequency-dependent power spectrum). Often, noise with a power spectrum given by the formula (3.121) is called $1/f^\gamma$ noise, and for the particular case of $\gamma = 1$, $1/f$ noise. For example, this noise can be found in physical, biological, and economic issues.

3.7.8 Fourier Filtering Method

A crucial issue that can now be formulated concerns an alternative way of generating correlated random walks.

The proposed algorithm assumes that the power spectrum $S(f)$ is known random walk that we intend to simulate. In other words, we must, of course, know what kind of wandering we aim to simulate from the point of view of autocorrelation. Since we are primarily interested in the long-term correlated walk/flight, it is enough to assume only the knowledge of the power-law exponent γ of the power spectrum (see the expression (3.121) below).

It can be seen that this approach does not clearly define a stochastic process but only a class of stochastic processes with the same time dependence on the autocorrelation function. We still assume that in this way, we will simulate stationary (in a broader sense) changes (increments) $\Delta X(t)$ of the random variable $X(t)$, e.g., we will simulate stationary, fractional Gaussian noise or, more general, a fractional Brownian motion.

The algorithm, called Fourier Filtering Method [42], consists of three simple steps.

(1) In the first step, we need to build a one-dimensional sequence of uncorrelated values Gaussian,[27] the stationary, random variable $\Delta Y(t_j),\ j = 1, 2, \dots n$, and then perform their discrete Fourier transformation

$$\Delta Y^{\mathcal{F}}(f) \approx \sum_{j=1}^{n} \Delta Y(t_j) \exp(-i\, f_j),\ f_j \stackrel{\text{def.}}{=} 2\pi \frac{j}{n},\ j = 1, 2, \dots, n, \tag{3.122}$$

and there are no limits on the number of n draws. Furthermore, $\Delta Y(t_j) \stackrel{\text{def.}}{=} Y(t_j + \Delta t) - Y(t_j)$ and $\Delta t \ll t_j$. However, it should be emphasized that we achieve discrete-time wandering in this way. This is important for the process itself because, unfortunately, it makes it impossible to construct an anti-autocorrelated random variable, i.e., a process for which $\eta < 1$ (see Eq. (3.47)). The complexity of this situation is that the noise of such a process may be anti-autocorrelated.

(2) Now, we need to define the Fourier transform of the simulated noise as follows

$$\Delta X^{\mathcal{F}}(f) = [S(f)]^{1/2}\, \Delta Y^{\mathcal{F}}(f). \tag{3.123}$$

In analogy to the first equality in Eq. (3.111), we can write (with a good approximation),

$$S(f) = \langle \Delta X^{\mathcal{F}}(f)\, \Delta X^{\mathcal{F}}(-f) \rangle = S(f) S_W(f), \tag{3.124}$$

where $S_W(f) = \langle \Delta Y^{\mathcal{F}}(f)\, \Delta Y^{\mathcal{F}}(-f) \rangle$ is the power spectrum of white noise, i.e., the quantity constant (here normalized to 1) and the averaging $\langle \dots \rangle$ was

[27] One can also draw these values simply from a uniform distribution.

performed for a fixed frequency f. We now need to perform the inverse Fourier transformation (from the variable f to the variable t), thus obtaining required noise $\Delta X(t_j),\ j = 1, 2, \ldots, n$, of discrete process $X(t_j)$.

(3) In the last step, the desired process $X(t_j)$ is obtained thanks to the obvious relationship between process and its noise.

$$X(t) = \sum_{j=1}^{n} \Delta X(t_j). \tag{3.125}$$

Let us add that the above algorithm is an alternative to the one presented in Sect. 3.7.2 which is based on the Random Midpoint Displacement Method.

3.7.9 CLT Versus Continuous Time

This subsection shows that satisfying the Central Limit Theorem by a stochastic chain (so in discrete time) is not an obstacle to building non-Brown processes (so in continuous time) that can be Gaussian or non-Gaussian.

The Central Limit Theorem deals with stochastic processes in discrete time, i.e., stochastic chains. This raises an important, delicate question about the distribution of the summary random variable in continuous time (e.g., the summary displacement of the test particle) in a situation where it satisfies CLT (in discrete time).

To answer the question posed above, we consider the composite/complex statistics (superstatistics) of the form,

$$D(X, t) = \sum_{n=0}^{\infty} G(X, n) V(n, t), \tag{3.126}$$

where the Gaussian PDF,

$$G(X, n) = \frac{1}{\sqrt{2\pi\sigma^2 n}} \exp\left(-\frac{X^2}{2\sigma^2 n}\right), \tag{3.127}$$

and the quantity $\sigma^2 n$ is the n-step variance of the distribution (3.127), here σ^2 is a single-step variance, and $V(n, t)$ is the (normalized) probability of the n-step process, performing these steps in the time interval $(0, t)$. Therefore, $D(X, t)$ describes the probability density function that a particle will be found at position X at time t. The lengths of the individual time intervals needed to complete each step may fluctuate and may be statistically dependent on each other. As one can see, the general equation (3.126) allows us to go from discrete-time n to continuous-time t. It says that the test particle can perform any number of steps (or none) in the time interval mentioned.

Now, we can calculate the dependence of the variance of the distribution $D(X,t)$ on the continuous time t. Namely,

$$\langle X^2(t)\rangle = \int_{-\infty}^{\infty} X^2 D(X,t)dX = \sigma^2 \sum_{n=1}^{\infty} nV(n,t) = \sigma^2\langle n(t)\rangle. \quad (3.128)$$

In the next stage, which is the canonical one, we deal with a stochastic Brownian & Gaussian chain (i.e., in discrete time). Let's assume that the time steps mentioned above are statistically independent, namely the execution times of individual steps. Then we are dealing with a Poisson process, where the probability $V(n,t) = \frac{(t/\tau)\exp(-t/\tau)}{n!}$, with $t/\tau = \langle n(t)\rangle$ is the average number of steps the process takes in t, and τ is the average time it takes to complete a single step. Still, according to CLT and Poisson process definition $n,\ t \to \infty$ such that $t/n = \tau = const.$ Assuming $1/\sqrt{\tau} \ll 1$ we get with good approximation that $V(n,t) \approx \delta(n - t/\tau)$, where $\delta(\ldots)$ is the Kronecker delta. Hence,

$$D(X,t) \approx G(X,t) = \frac{1}{2\pi\sigma^2(t)} \exp\left(-\frac{X^2}{2\sigma^2(t)}\right), \quad (3.129)$$

where $\sigma^2(t) = \langle X^2\rangle = \sigma^2\sigma^2_{Poiss}t$ is the multi-step variance of the X process in continuous time t and $\sigma^2_{Poiss} = 1/\tau$ is the variance of the Poisson process mentioned above. Note also that the linear dependence of $\sigma^2(t)$ on time t can be obtained without assuming that the standard deviation of the Poisson process $\sigma_{Poiss} \ll 1$. Just assume that the average time $\tau < \infty$.

The second extreme case concerns the time steps, i.e., the times needed to perform elementary/single displacements of the process are statistically dependent. We will use this assumption in a non-Brown and non-Gaussian process. Note that this process is anomalous when the $V(n,t)$ distribution is in the scaling form $V(n,t) = V(\zeta)$, where $\zeta = \frac{n}{(t/\tau_0)^\eta}$, and $\eta > 0$ while τ_0 is a time unit. Then, from the Eq. (3.128), we get

$$\langle X^2\rangle = \sigma^2\,\langle\zeta\rangle(t/\tau_0)^\eta, \quad (3.130)$$

where the mean of the scaling variable,

$$\langle\zeta\rangle = \sum_{n=0}^{\infty} \zeta(n)V(\zeta(n)) \quad (3.131)$$

for a fixed, asymptotically long time t and $0 < \langle\zeta\rangle < \infty$. In addition, we can put $\eta = 2H$, where H is the Hurst exponent.

To sum up, in this subsection, we started with a stochastic chain in Eq. (3.126) that satisfies CLT in discrete time and showed that depending on the choice of $V(n,t)$, the resulting stochastic process (in continuous time) can be both Gaussian or non-Gaussian, and Brownian or non-Brownian.

Generally, the distribution given by (3.126) is not a Gaussian distribution but a distribution of the non-Gaussian fractal walk. Normal (Gaussian) fractal wandering becomes for a particular case $V(\zeta) = \delta(\zeta - 1)$. Then, the variance $\langle X^2 \rangle = 2\sigma^2 (t/\tau_0)^\eta$ is a nonlinear function of time t, i.e., it does not meet CLT.

3.8 Dynamical Realization of Mono and Multifractality

In this section, we will show (starting from the level of stochastic dynamics) the conditions in which we deal with monofractality or multifractality. This section is significant for relating multifractality to the *ab initio* level of statistical physics. Therefore, the background of our considerations is the overdamped Langevin equation of a single macromolecule located at X at time t,

$$0 = -\frac{dX(t)}{dt} + F(X) + \sigma L(t), \tag{3.132}$$

where the first term on the right is the drag force of the viscous medium (with the coefficient of viscous resistance assumed to be equal to 1), the second term is the external force/drift force acting on the macromolecule, and the third term is the white Gaussian noise,[28] while σ is its strength. We continue to assume that $F(x)$ is the elastic force with potential $V(X) = \frac{1}{2}\kappa X^2$, i.e., force $F(X) = -\frac{dV(X)}{dX} = -\kappa X$, where κ is Hook's spring coefficient.

3.8.1 *Monofractlity*

The overdamped Langevin equation and its consequences were already considered in the context of constant drift force in Sect. 2.5.2. Our considerations here are more general because we use a drift force that depends on a spatial variable. Nevertheless, the static/stationary/steady conditional solution of the Fokker-Planck equation, equivalent to the Eq. (3.132), takes an analogous but more general form here,

$$P(X \mid \beta) = \mathcal{N}(\beta) \exp\left(\frac{2}{\sigma^2} V(X)\right) = \mathcal{N}(\beta) \exp\left(-\frac{\beta}{2} X^2\right), \tag{3.133}$$

[28] We could take a broader look at Eq. (3.132) by introducing white noise instead of white Gaussian noise. However, according to the Lévy-Itô white noise decomposition, this would imply dealing with the sum of white Gaussian noise and white Poisson noise, potentially complicating the presentation. Interested readers can refer to the work of Kanazawa-Sornette [43] (and references therein), containing the original extension for history-dependent processes.

where $\mathcal{N}(\beta) = \sqrt{\beta/2\pi}$ is a normalization constant and $\beta = 2\kappa/\sigma^2$. It is the local equilibrium distribution restricted to a Gaussian form.

We assume that the parameter β is stochastic, i.e., it can fluctuate according to $f(\beta)$ probability density. Beck et al. reported that empirical data roughly fall into three χ^2 classes [44, 45]:

(i) χ^2 or gamma superstatistics;
(ii) inverse χ^2 superstattistics;
(iii) log-normal superstatistics.

In the case (i), parameter β is χ^2 distributed with degree n:

$$f(\beta) = \frac{1}{\boldsymbol{\Gamma}(n/2)} \left(\frac{n}{2\bar{\beta}}\right)^{n/2} \beta^{n/2-1} \exp\left(-\frac{n}{2\bar{\beta}}\beta\right), \tag{3.134}$$

where $\boldsymbol{\Gamma}(y)$ is the gamma-Euler function of y variable and $\bar{\beta}$ is the mean of β. We will use this expression, for example, to determine $\langle| X |^q\rangle$, using the marginal probability distribution $p(X)$ defined in the form of superstatistics [46],

$$P(X) = \int_0^\infty P(X \mid \beta) f(\beta) d\beta, \tag{3.135}$$

Therefore, the sought absolute q-moment takes the form,

$$\begin{aligned} \langle| X |^q\rangle &= \int_{-\infty}^{\infty} | X |^q \, P(X) dX \\ &= \frac{1}{\boldsymbol{\Gamma}(n/2)} \left(\frac{n}{2\bar{\beta}}\right)^{n/2} \langle| Y |^q\rangle \int_0^\infty \beta^{(n-q)/2-1} \exp\left(-\frac{n}{2\bar{\beta}}\beta\right) d\beta \end{aligned} \tag{3.136}$$

where along the way, we used the explicit form $P(X \mid \beta)$ given by Eq. (3.133), and thanks to this, we could introduce a random variable Y having a probability distribution function in the standardized form $PDF(Y) = 1/\sqrt{\pi} \exp(-Y^2)$. Hence, $\langle| Y |^q\rangle = \int_{-\infty}^{\infty} | Y |^q \, PDF(F) dY$ takes the invariant form, independent of any parameter, e.g., β or scale parameter. Therefore, this form does not concern multiscale and only includes the amplitude (pre-exponential factor) of the q-moment $\langle| X |^q\rangle$. We will calculate below the q-moment $\langle| Y |^q\rangle$ and the integral in the second equality in Eq. (3.136) to obtain their explicit dependency on q. Therefore,

$$\langle| Y |^q\rangle = \frac{2}{\sqrt{\pi}} \int_0^\infty \exp(-F(Y) dY, \;\; F(Y) \overset{\text{def.}}{=} Y^2 - q \ln Y. \tag{3.137}$$

We calculate this integral approximately using Saddle Point Approximation. For this purpose, we find the minimum of the function $F(Y)$,

$$F'(Y) = 2Y - \frac{q}{Y} = 0 \Leftrightarrow Y = Y_{min} = \sqrt{\frac{q}{2}},$$
$$F''(Y) \mid_{Y=Y_{min}} = \left(2 + \frac{q}{Y^2}\right)_{Y=Y_{min}} = 4 > 0. \tag{3.138}$$

As one can see, the minimum of the function $F(Y)$ is at the point $Y_{min} = \sqrt{q/2}$.

Expanding this function into a Taylor series around this minimum with precision to the quadratic term (parabolic approximation) we obtain for $q > 0$,

$$\begin{aligned}
\langle \mid Y \mid^q \rangle &= \frac{2}{\sqrt{\pi}} \exp(q/2 - q \ln q/2) \int_0^\infty \exp(-2(Y - Y_{min})^2) dY \\
&= \frac{2}{\sqrt{2\pi}} \exp(q/2 - q \ln q/2) \int_{-Z_{min}}^\infty \exp(-(Z - Z_{min})^2) dZ \\
&= \frac{2}{\sqrt{2\pi}} \exp(q/2 - q \ln q/2) \operatorname{erfc}(-Z_{min}) \\
&= \frac{2}{\sqrt{2\pi}} \exp(q/2 - q \ln q/2)[1 + \operatorname{erf}(-\sqrt{q})],
\end{aligned} \tag{3.139}$$

where $Z_{min} = \sqrt{q}$. As we expected, this q-moment is independent of any parameter and depends only on the order of q.

Now, the integral in the second equality (3.136) can be expressed for $q < n$ in the following form,

$$\int_0^\infty \beta^{(n-q)/2-1} \exp\left(-\frac{n}{2\bar{\beta}}\beta\right) d\beta = \mathbf{\Gamma}\left(\frac{n-q}{2}\right)\left(\frac{2\bar{\beta}}{n}\right)^{n/2}\left(\frac{2\bar{\beta}}{n}\right)^{-q/2} \tag{3.140}$$

Substituting Eqs. (3.139) and (3.140) to Eq. (3.136) we get the sought q-moment,

$$\langle \mid X \mid^q \rangle = A(q) S^{-q/2}, \tag{3.141}$$

where amplitude $A(q) = \langle \mid Y \mid^q \rangle \mathbf{\Gamma}\left(\frac{n-q}{2}\right)\left(\frac{2\bar{\beta}}{n}\right)^{n/2}$ and the scale $S = \frac{2\bar{\beta}}{n}$.

Equation (3.141) shows that we deal with a monofractal relationship here. Moreover, it can be shown that also in case (ii), we are dealing with this type of dependency. Therefore, the question arises about the conditions under which we will obtain a multifractal relationship (if at all).

3.8.2 Parabolic Multifractality

Consider items (iii) listed in the previous subsection to answer the above question. Therefore, we take the probability density $f(\beta)$ in the form of log-normal distribution,

$$f(\beta) = \frac{1}{\sqrt{2\pi}\sigma\beta} \exp\left(-\frac{\ln^2(\beta/\mu)}{2\sigma^2}\right), \quad \beta,\ \mu,\ \sigma > 0. \tag{3.142}$$

The above distribution enables (after simple transformations) rewrite Eq. (3.136) in the following exact form,

$$\begin{aligned}
\langle | X |^q \rangle &= \int_0^\infty \int_0^\infty | X |^q \, P(X \mid \beta) f(\beta) dX d\beta \\
&= \langle | Y |^q \rangle \int_0^\infty \frac{1}{\beta^{q/2}} f(\beta) d\beta \\
&= \frac{1}{\sqrt{\pi}} \boldsymbol{\Gamma}\left(\frac{q+1}{2}\right) \int_0^\infty \left(\frac{2}{\beta}\right)^{q/2} \frac{1}{\sqrt{2\pi}\sigma\beta} \exp\left(-\frac{\ln^2(\beta/\mu)}{2\sigma^2}\right) d\beta \\
&= \frac{1}{\sqrt{\pi}} \boldsymbol{\Gamma}\left(\frac{q+1}{2}\right) \left(\frac{\mu}{2}\right)^{-q/2} S^{q^2},
\end{aligned} \tag{3.143}$$

where

$$\langle | Y |^q \rangle = \sqrt{\frac{2}{\pi}} \int_0^\infty | Y |^q \exp\left(-\frac{Y^2}{2}\right) dY = \frac{2^{q/2}}{\sqrt{\pi}} \boldsymbol{\Gamma}\left(\frac{q+1}{2}\right), \tag{3.144}$$

i.e., in a parabolic multifractal factorized/separated representation. As one can see, the scale $S \overset{\text{def.}}{=} \exp(\sigma^2/8)$ is only generated by fluctuations, while the pre-factor is based on the expected value.

We employed the harmonic potential induced by an external force in our investigation of the multifractality emerging from the steady state at the *ab initio* level. Introducing a more intricate force, such as anharmonic effects, would not significantly alter the scenario, as multifractality depends primarily on the fluctuations of the β parameter, which are governed by its distribution. Specifically, these fluctuations arise from the stochastic component of the Langevin equation rather than its deterministic counterpart.

3.8.3 General Scaling Function

This subsection introduces a scaling function that generalizes singular processes, including anomalous diffusion and Lévy flights (considered in Sect. 3.7 and especially in Sect. 3.7.7), and ultraslow diffusion (as presented in the following) for systems without drift, convection, advection, etc., i.e., those whose expected value $\langle X(t)\rangle$ or any floating current vanishes (but $\langle \mid X(t) \mid\rangle \neq 0$).

Let the (normalized) probability density function (PDF) assume for $\mid t \mid \neq 0$ the form

$$P(X,t) = [U(t)]^{-1/2} F\left(\frac{\mid X \mid}{[U(t)]^{1/2}}\right), \tag{3.145}$$

or equivalently in the Fourier domain,

$$\tilde{P}(k,t) = \tilde{F}\left(\mid k \mid \cdot [U(t)]^{1/2}\right), \tag{3.146}$$

here F is (as previously, normalized) scaling function and $\tilde{F}$ is its Fourier representation where $\tilde{F}(0) = 1$, while k is the Fourier variable in this context; the presence of the absolute value of k is a result of the parity of scaling function $F(k)$, i.e., $F(k) = F(-k)$. In particular, we do not impose additional conditions on the function $U(t)$, requiring it to be strictly positive for now. The expression (3.145), and thus (3.146), covers an extensive class of distributions, which can be represented using a scaling function, from the Gaussian distribution through distributions based on generalized exponential functions (e.g., Tsallis q-exponential and the Mittag-Leffler function) to the Lévy distribution. We emphasize the scaling function in Eq. (3.145) may have a general integral form that cannot be expressed in a closed form using any known elementary or non-elementary function - an example would be the Lévy PDF. The class mentioned above contains two subclasses for which $U(t)$ is a linear function of t and a non-linear function of t, respectively. We show below that the former significantly differs because it does not generate process velocity autocorrelation.

Next, we proceed directly

$$\langle \mid X(t) \mid^q \rangle = [U(t)]^{q/2} \langle \mid Y \mid^q \rangle \tag{3.147}$$

where

$$\langle \mid Y \mid^q \rangle = \int_{-\infty}^{\infty} \mid Y \mid^q F(\mid Y \mid) dY, \tag{3.148}$$

represents a time-independent but q-dependent factor. The case $q = 0$ specifies the normalization condition for the scaling function $F(|Y|)$. The function $U(t)$ plays a pivotal role in the stochastic asymptotic evolution of the system.

Moreover, a multifractal random walk can be derived from Eq. (3.147) by averaging over disorder, that is, averaging over random values of the control parameters of the function $U(t)$ (here, implicitly influences this function). However, this function must be amenable to such treatment, i.e., U(t) must contain these parameters.

In other words, multifractality appears only when we average over a disordered substrate/landscape of parameters/exponents (however, not every averaging of this type leads to multifractality). Only the existence of autocorrelation is insufficient; it is a necessary condition. The above comments are significant because they indicate universal causes of multifractality, regardless of the type of system. We elaborated on this aspect extensively in Sect. 3.7.1 concerning, for example, the random behavior of the Hurst exponent.

3.8.3.1 A Useful Digression

We want to demonstrate the Fourier representation approach to multifractality as a useful digression. For this purpose, we present the absolute q-moment of the stochastic process in the form,

$$\begin{aligned}\langle | X |^q \rangle &= 2\,[U(t)]^{-1/2} \int_0^\infty | X |^q \, F\left(\frac{| X |}{[U(t)]^{1/2}}\right) dX \\ &= \frac{2}{\pi} \int_0^\infty \int_0^\infty | X |^q \cos(| k || X |) \tilde{F}\left(| k | \, [U(t)]^{1/2}\right) dk dX \\ &= \frac{2}{\pi} \int_0^\infty \tilde{F}\left(k\,[U(t)]^{1/2}\right) \lim_{a \to 0} \int_0^\infty \exp(-a\,X) X^q \cos(k\,X) dX dk,\end{aligned} \tag{3.149}$$

where the inner integral over the Fourier variable k in the second equality is related to the Fourier representation of the integrand $F\left(\frac{|X|}{[U(t)]^{1/2}}\right)$,

$$F\left(\frac{| X |}{[U(t)]^{1/2}}\right) = [U(t)]^{1/2} \int_0^\infty \cos(| k || X |) \tilde{F}\left(| k | \, [U(t)]^{1/2}\right) dk. \tag{3.150}$$

We continue to use the well-known equality,

$$\int_0^\infty \exp(-a\,X) X^q \cos(k\,X) dX = \frac{\Gamma(q+1)}{\left(a^2+k^2\right)^{(q+1)/2}} \cos\left((q+1) \arctan\left(\frac{k}{a}\right)\right) \tag{3.151}$$

which is finite for $a \to 0$. Combining Eq. (3.149) with Eq. (3.151) we get,

$$\langle | X |^q \rangle = \frac{2}{\pi}\mathbf{\Gamma}(q+1) \lim_{a\to 0} \int_0^{\infty} \frac{\tilde{F}\left(k\,[U(t)]^{1/2}\right)}{\left(a^2+k^2\right)^{(q+1)/2}} \cos\left((q+1)\arctan\left(\frac{k}{a}\right)\right) dk. \tag{3.152}$$

The integral in the above equation may converge or diverge, depending on how quickly the integrand disappears in the k variable. It mainly depends on the $\tilde{F}$ form (including $U(t)$). Therefore, the integrand behavior mentioned should be considered separately in each specific case. Finally, the transition to multifractality can be achieved by averaging $\langle | X |^q \rangle$ over the parameters controlling the Fourier transform $\tilde{F}$ (including $U(t)$), which can be considered as random. We write it symbolically as follows,

$$\begin{aligned}\langle\langle | X |^q \rangle\rangle_{param} &= \frac{2}{\pi}\mathbf{\Gamma}(q+1) \lim_{a\to 0} \\ &\langle \int_0^{\infty} \frac{\tilde{F}\left(k\,[U(t)]^{1/2}\right)}{\left(a^2+k^2\right)^{(q+1)/2}} \cos\left((q+1)\arctan\left(\frac{k}{a}\right)\right) dk\rangle_{param}\end{aligned} \tag{3.153}$$

where $\langle \ldots \rangle_{param}$ is just this averaging over the control parameters.

For example, for time-dependent Lévy PDF we have

$$\tilde{F}\left(| k |\ [U(t)]^{1/2}\right) = \exp\left(-\left(| k |\ [U(t)]^{1/2}\right)^{\alpha}\right), \tag{3.154}$$

where $0 < \alpha < 2$. The parameter (shape exponent) α could be considered a random quantity. It should be noted that multifractality can only be obtained for suitably decaying distributions of the control parameters.

3.8.3.2 A Fundamental Role of $U(t)$

We will now demonstrate the fundamental role of the function $U(t)$ in the autocorrelation of the homogeneous in-time stochastic process, that is, the stochastic process of weak stationarity or in the stationary variance (it is also said 'in the sense of variance'),[29] and in the autocorrelation of the velocity associated with this process, extending our approach in Sect. 3.7.7. Therefore, employing Eq. (3.147), we can express, for $q = 2$,

$$\langle [X(t_2 - t_1)]^2 \rangle = \langle Y^2 \rangle U(t_2 - t_1). \tag{3.155}$$

---

[29] The weak stationarity of the process is identified here with its homogeneity in time, i.e., its dependence on the difference between moments. In the following text, we omit the term 'weak' for simplicity - it will be the default term.

On the other hand,

$$\langle [X(t_2) - X(t_1)]^2 \rangle = \langle Y^2 \rangle [U(t_2) + U(t_1)] - 2\langle X(t_1) X(t_2) \rangle \tag{3.156}$$

Due to the homogeneity/stationarity (in the sense of variance) of the $X(t)$ process, we can compare both Eq. (3.155) and (3.156), which allows us to determine the autocorrelation function of the process in the form,

$$\langle X(t_1) X(t_2) \rangle = \frac{1}{2} \langle Y^2 \rangle \left[U(t_2) + U(t_1) - U(t_2 - t_1) \right]. \tag{3.157}$$

This implies that the process is not stationary in the sense of the process/position autocorrelation. To better grasp the meaning of Eq. (3.157), we will apply them to the Wiener process (Brownian and Gaussian). Then, $F(|\, Y \,|) = \exp(-Y^2/2)$, $U(t) = \sigma^2 t$, $\langle Y^2 \rangle = 1$, and autocorrelation of the process $\langle X(t_1) X(t_2) \rangle = t_1$, disappearing only when $t_1 = 0$ (we have assumed here that time evolves forward, i.e., $t_2 > t_1$). Therefore, in general, even such a simple process is autocorrelated. As we shall see below, the velocity of such a process is not autocorrelated.

We will derive the autocorrelation function of the process velocity based on the above result. Thus, after straightforward algebraic manipulations, we obtain

$$\langle \Delta X(t_1) \Delta X(t_2) \rangle = \frac{1}{2} \langle Y^2 \rangle \left[U(t_2 - t_1 + \Delta t) + U(t_2 - t_1 - \Delta t) - U(t_2 - t_1) \right], \tag{3.158}$$

where $\Delta X(t_j) = X(t_j + \Delta t) - X(t_j)$ for $j = 1, 2$. Expanding the right side of the above equation by a Taylor series around the point $t_2 - t_1$ and considering at most the quadratic terms in Δt due to its smallness, we obtain:

$$\lim_{\Delta t \to 0} \langle \frac{\Delta X(t_1)}{\Delta t} \frac{\Delta X(t_2)}{\Delta t} \rangle = \langle V(t_1) V(t_2) \rangle = \frac{1}{2} \langle Y^2 \rangle \ddot{U}(t_2 - t_1), \tag{3.159}$$

where $V(t_j) = \lim_{\Delta t \to 0} \frac{\Delta X(t_j)}{\Delta t}$, and we introduce the convenient notation $\ddot{U}(t_2 - t_1) = \frac{d^2}{dt^2} U(t) \big|_{t = t_2 - t_1}$. The process exhibits stationarity in the sense of autocorrelation of the process velocity. Furthermore, it is autocorrelated only for $U(t)$, a nonlinear function of t.

Paradoxically, even the residual nonlinear addition to the linear component in $U(t)$ (i.e., the correction to scaling vanishing asymptotically) must not be neglected when computing the velocity autocorrelation function.

3.8.3.3 Example of Ultraslow Diffusion

Utilizing the result (3.147), we present a particularly intriguing example of ultraslow diffusion [47–49]. In this case, we assume $U(t) = \ln^{\nu}(t)$. Then

$$\langle | X(t) |^q \rangle = \langle | Y |^q \rangle \ln^{\nu q/2} t. \tag{3.160}$$

It is worth noting that the absolute q-moment defined by Eq. (3.160) exhibits slower growth over time compared to any power-law growth characteristic of subdiffusion. The significant case of subdiffusion discussed below, namely Sinai ultraslow diffusion, corresponds to $\nu = 4$.

The category of ultraslow diffusion also includes the process for which $U(t) = \ln \ln t$.

By the way, it is worth adding information about the possibility of intermediate diffusion between the anomalous and ultraslow ones. Then we should put $U(t) = t^{\mu} \ln^{\nu} t$.

Physically, as a specific instance of models with site-dependent jumps, the Sinai model [50] characterizes a particle's one-dimensional thermal random motion in a random potential, particularly one constructed from a Brownian path. Assuming an infinite spatial domain, as commonly done, the potential can feature arbitrarily high barriers and profoundly deep wells. Nevertheless, owing to the recurrence properties of Brownian paths in one dimension, there will be potential zero-crossings at arbitrarily distant locations from the origin. In the context of continuous space and time, the long-term behavior of the Sinai diffusion probability density function (PDF) is as follows [51–54],

$$\begin{aligned} P(X,t) &= \frac{4}{\pi} \frac{1}{\lambda \ln^2(t/\tau)} \sum_{n=0}^{\infty} \frac{(-1)^n}{2n+1} \exp\left(-\frac{\pi^2(2n+1)^2}{4} \frac{| X |}{\lambda \ln^2(t/\tau)} \right) \\ &= \frac{1}{\lambda \ln^2(t/\tau)} \sum_{n=0}^{\infty} F_n \left(\frac{| X |}{\lambda \ln^2(t/\tau)} \right), \end{aligned} \tag{3.161}$$

with a partial scaling function

$$F_n(Y) = \frac{(-1)^n}{2n+1} \exp\left(-\frac{\pi^2(2n+1)^2}{4} | Y | \right), \quad Y = \frac{| X |}{\lambda \ln^2(t/\tau)}, \tag{3.162}$$

and hence the (sum) scaling function

$$F(Y) = \sum_{n=0}^{\infty} F_n(Y) \tag{3.163}$$

consists of partial scaling functions, and time-dependent q-moment takes the form,

$$\langle | X(t) |^q \rangle = \int_{-\infty}^{\infty} | X |^q \, P(X,t)dX = \lambda^q \langle | Y |^q \rangle \ln^{2q}(t/\tau), \tag{3.164}$$

where dimensionless q-moment

$$\langle | Y |^q \rangle = \sum_{n=0}^{\infty} \langle | Y |^q \rangle_n, \quad \langle | Y |^q \rangle_n = \int_{-\infty}^{\infty} | Y |^q \, F_n(Y)dY; \tag{3.165}$$

The parameters λ and τ, which govern the Sinai subdiffusion, define the space's physical length and time scales, respectively.

3.9 Monofractal Subdiffusion Equation

Analogously to the approach in Sect. 3.5, the fractal diffusion equation can be obtained purely formally [55, 56]. The stochastic process leading to this equation is called the **fractal time random walk**.

So, let's consider the one-dimensional Fick diffusion equation (with no drift for simplicity)

$$\frac{\partial f(x,t)}{\partial t} = D \frac{\partial^2 f(x,t)}{\partial t^2}; \tag{3.166}$$

we continue to assume for simplicity that the diffusion coefficient D is constant.

Integrating this equation side by side over time, we get to the intermediate stage,

$$f(x,t) - f(x,t=0) = \frac{\partial^{-1}}{\partial t^{-1}} D \frac{\partial^2 f(x,t)}{\partial x^2} \tag{3.167}$$

and a generalization of the above equation for $0 \leq \alpha < 1$; hence the name **subdiffusion**,

$$f(x,t) - f(x,t=0) = \frac{\partial^{-\alpha}}{\partial t^{-\alpha}} D_\alpha \frac{\partial^2 f(x,t)}{\partial x^2}, \tag{3.168}$$

where D_α is the generalized diffusion coefficient (here constant), whose dimension is length^2/time$^\alpha$, which we will discuss in detail later in the textbook.

Partially differentiating with respect to time (3.168), we obtain the sought fractal equation of subdiffusion,

$$\frac{\partial f(x,t)}{\partial t} = \frac{\partial^{1-\alpha}}{\partial t^{1-\alpha}} D_\alpha \frac{\partial^2 f(x,t)}{\partial x^2}. \tag{3.169}$$

The exact, closed-form solution to this equation can be expressed in terms of the Fox H-function,

$$f(|x|,t) = \frac{1}{\sqrt{4D_\alpha t^\alpha}} H_{11}^{10}\left[\frac{|x|}{\sqrt{4D_\alpha t^\alpha}}\middle| \begin{matrix}(1-\frac{\alpha}{2},\frac{\alpha}{2})\\(0,1)\end{matrix}\right]. \tag{3.170}$$

The index "0" that appears in Fox H-function in Eq. (3.170) requires the use of a simplified Fox H-function [56] (see Chapt. 6 and Appendix A.1 for details).

It can be seen that for both alternative forms of the solution of the fractal equation of subdiffusion, we obtain the following common asymptotic form for the absolute moment of the order of q,

$$\langle |x(t)|^q\rangle = \int_{-\infty}^{\infty} |x|^q\, f(|x|,t)dx = \langle |y|^q\rangle \left(\sqrt{4D_\alpha}\right)^q t^{\alpha q/2},\quad |y| \overset{\text{def.}}{=} \frac{|x|}{\sqrt{4D_\alpha t^\alpha}}, \tag{3.171}$$

where $\langle |y|^q\rangle = \int_{-\infty}^{\infty} |y|^q\, f_s(|y|)dy$ is an amplitude independent of time, and scaling/distribution function $f_s(|y|)$ we get from $f(|x|,t)$ after the following replacing,

$$f_s(|y|) = H_{11}^{10}\left[|y|\middle| \begin{matrix}(1-\frac{\alpha}{2},\frac{\alpha}{2})\\(0,1)\end{matrix}\right]. \tag{3.172}$$

Equation (3.171) shows that we deal with a monofractal anomalous diffusion.

We would only deal with multifractal diffusion here if the exponent α were a random variable subject to an appropriate distribution (e.g., of the stretched exponential or stretched Gaussian forms).

Control and Review Questions and Exercises

3.1 Exercise

Show that the Pareto distribution is fat-tailed.

3.2 Exercise

Find all moments of the Pareto distribution.

3.3 Exercise and Problem

Assume that in some countries, the distribution of wealth can be described by a Pareto distribution. For $x_m = 1$ calculate:

(a) The parameter α assuming that 20% of the most wealthy citizens own 80% of the country's wealth.
(b) What the top 1% owns % of the total country's wealth?
(c) What is the difference between the mean wealth of that country and the median wealth?

3.4 Exercise
The distribution of returns from some investments comes from a distribution with zero mean and variance equals 2:
(a) normal
(b) t-student.
If a return on someday is less than -5, the firm is bankrupt. Find the probability of bankruptcy in cases (i) and (ii). Assuming independence of returns, find the probability that the firm will survive for the next year (365 days).

3.5 Exercise
Show that the Pareto distribution satisfies the singular scaling equation for positive values.

3.6 Exercise
Fit Eq. (3.12) to a chosen stock price index. Use your favorite software or programming language. You can search and download daily stock price data from the internet for free.

3.7 Problem
The continuous probability distribution is stable when this distribution also describes any linear combination of variables from this distribution. Check the necessary condition of stability of the Gauss PDF - find a PDF of a sum of n variables with distribution $N(0, \sigma^2)$. Check the sufficient condition of the stability of the Gauss distribution—find the distribution of the linear combination $Z = aX + bY$ of variables with distributions $N(\mu_1, \sigma_1^2)$ and $N(\mu_2, \sigma_2^2)$.

3.8 Exercise
The variables X and Y are described by the Poisson distribution. Check the stability of the sum $X + Y$ with parameters λ_1 i λ_2. Check the stability of the linear combination $aX + bY$ with parameter λ.

3.9 Exercise
Check the necessary condition of the stability of the exponential distribution—find the distribution of the sum of n variables with identical parameter λ.

3.10 Exercise
Check the stability of the Cauchy distribution - find the linear combination of variables $Z = aX + bY$ with Cauchy distribution with γ_1, x_1 and γ_2, x_2.

3.11 Problem
The characteristic function of the Cauchy distribution is $\phi(k) = \exp(ikx_0 - \gamma|k|)$. What is the distribution of a variable with the following characteristic function? $G(k) = \exp(-\gamma|k|)$? Does the sum of those variables follow the Gaussian distribution? Is the Central Limit Theorem fulfilled?

3.12 Problem

Check the stability of the Lévy distribution - find the distribution of $Z = aX + bY$ with different parameters c and μ. The PDF of the Lévy distribution is:

$$p(x) = \sqrt{\frac{c}{2\pi}}(x-\mu)^{-3/2}\exp\left(-\frac{c}{2(x-\mu)}\right),$$

and its characteristic function:

$$\phi(t) = \exp[it\mu - \sqrt{c|t|}(1 + i\ \mathrm{sign}(t))].$$

3.13 Problem

An α-stable distribution has the property that a sum of independent random variables, each with the same stability index α, is also an α-stable variable. It can be defined through its characteristic function:

$$G(t) = \exp\Big[\ it\mu - |c\,t|^{\alpha}\,(1 - i\beta\ \mathrm{sign}(t)\Phi)\ \Big], \tag{3.173}$$

where

$$\Phi = \begin{cases} \tan\frac{\pi\alpha}{2} & \text{if } \alpha \neq 1 \\ -\frac{2}{\pi}\log|ct| & \text{if } \alpha = 1 \end{cases} \quad \text{and } 0 < \alpha \leq 2, -1 \leq \beta \leq 1, c > 0, \mu \in \mathbb{R}.$$

(i) Shows this distribution is stable.
(ii) Show that for $\alpha = 2$ we get a normal distribution.
(iii) Show that for $\alpha = 1$, $\beta = 0$ we get a Cauchy distribution.
(iv) Show that for $\alpha = 0.5$, $\beta = 1$ we get a Levy distribution.

3.14 Question

Check the necessary condition of the stability of the exponential distribution - find the distribution of the sum of n variables with identical parameter λ.

3.15 Exercise

Show analytically which relaxation functions, Debye, KWW, or Nutting, decay the fastest and which decays the slowest. Which of them has a fat tail? Under what conditions will they become heavy-tail?

3.16 Exercise

Using the definition of a fractional differentiation in Eq. (A.6), find what the 1/2-th derivative of a linear function of t is. Assume that the lower integration limit is $a = 0$.

3.17 Exercise

Derive the formula (3.37) from Eq. (3.36).

3.18 Exercise

Prove that Eqs. (3.38) and (3.45) leads to Eq. (3.46).

3.19 Exercise
Prove that the Poisson PDF $\lambda \gg 1$ and $k \approx \lambda$ (so $k = \lambda(1+\delta)$, $\delta \ll 1$) can be described by the normal distribution $N(\lambda, \lambda)$.

3.20 Questions
If a process is random in discrete time and satisfies CLT, can it be non-Gaussian if the length of the time steps is decreased so it becomes continuous? Why? Why not? And under what conditions?

3.21 Questions
Under what conditions is the velocity autocorrelation function positive, and under what conditions is it negative? Make plots corresponding to these two cases and interpret the obtained results.

3.22 Question
Are formulas (3.102) and (3.107) correct or not for a non-stationary process? Justify your answer.

3.23 Exercise
Prove that the fractal subdiffusion Eq. (3.169) leads to monofractal anomalous diffusion.

3.24 Exercise
Choose the forms of the scaling function F given by the Eq. (3.145) or (3.146) so that they reproduce the following distributions: Gaussian, the Tsallis q-Gaussian, the one based on the Mittag-Leffler function, stretch exponential, and Lévy distribution.

3.25 Exercise
Assume that the function $U(t)$ satisfies the scaling equation $U(u\,t) = g(u)U(t)$. Using Eq. (3.146), derive the scaling equation for the Fourier transform $\tilde{P}(k,t)$ of PDF $P(X,t)$ and for its Laplace transform $\tilde{P}(k,s)$.

3.26 Exercise & Question
Derive Eq. (3.157) from Eqs. (3.155) and (3.156). Why is the process not stationary in the sense of process autocorrelation?

3.27 Exercise & Question
Derive Eq. (3.158) from Eqs. (3.157) and (3.159) from Eq. (3.158). Why is the process stationary in the sense of velocity autocorrelation?

References

1. Tsinober, A.: Variability of anomalous transport exponents versus different physical situations in geophysical and laboratory turbulence. In: Shlesinger, M.F., Zaslavsky, G.M., Frisch, U. (eds.) Lévy Flights and Related Topics in Physics. Lecture Notes in Physics, vol. 450, pp. 3–33. Springer, Berlin (1995)

2. Dragulescu, A., Yakovenko, V.M.: Exponential and power-law probability distributions of wealth and income in the United Kingdom and the United States. Physica A **299**, 213–221 (2001). https://doi.org/10.1016/S0378-4371(01)00298-9
3. Okuyama, K., Takayasu, M., Takayasu, H.: Zipf's law in the income distribution of companies. Physica A **269**, 125–131 (1999). https://doi.org/10.1016/S0378-4371(99)00086-2
4. Toda, M., Kubo, R., Saito, N.: Statistical Physics I. Statistical mechanics of equilibrium states. Springer, Berlin (1978)
5. Huang, K.: Statistical Mechanics. Wiley, New York (1963)
6. Sornette, D.: Discrete scaling invariance and complex dimensions. Phys. Rep. **297**, 239–270 (1998). https://doi.org/10.1016/S0370-1573(97)00076-8
7. Sornette, D., Johansen, A., Bouchaud, J.-P.: Stock market crashes, precursors and replicas. J. Phys. I France **6**, 167–175 (1996). https://doi.org/10.1051/jp1:1996135
8. Sornette, D., Johansen, A.: Large financial crashes. Physica A **245**, 411–422 (1997). https://doi.org/10.1016/S0378-4371(97)00318-X
9. Kwapień, J., Drożdż, S.: Physical approach to complex systems. Phys. Rep. **515**, 115–226 (2012). https://doi.org/10.1016/j.physrep.2012.01.007
10. Blanchard, O.J.: Speculative bubbles, crashes, and rational expectations. Econ. Lett. **3**, 387–389 (1979)
11. Blanchard, O.J., Watson, M.W.: Bubbles, rational expectations, and speculative markets. In: Wachtel, P. (ed.) Crisis in Economic and Financial Structure: Bubbles, Bursts, and Shocks. Lexington Books, Lexington MA (1982)
12. Johansen, A., Ledoit, O., Sornette, D.: Crashes as Critical Points. Int. J. Theor. Appl. Finance **03**(02), 219–255 (2000)
13. Gnaciński, P., Makowiec, D.: Another type of log-periodic oscillations on Polish stock market. Physica A **344**(1–2), 322–325 (2004). https://doi.org/10.1016/j.physa.2004.06.143
14. Jacobs, K.: Stochastic Processes for Physicists. Understanding Noisy Systems. Cambridge University Press, Cambridge (2010)
15. Paul, W., Baschnagel, J.: Stochastic Processes. From Physics to Finance, 2nd edn. Springer, Berlin (2013). https://doi.org/10.2307/3085855
16. Bouchaud, J.-P., Potters, M.: Theory of Financial Risks. From Statistical Physics to Risk Management, 2nd reprint. Cambridge University Press, Cambridge (2001)
17. Bouchaud, J.-P., Georges, A.: Anomalous diffusion in disordered media: statistical mechanics, models, and physical applications. Phys. Rep. **195**(4&5), 127–293 (1990). https://doi.org/10.1016/0370-1573(90)90099-N
18. Ghosh, M., Chakrabarti, B.K.: Relaxation in disordered systems. Indian J. Phys. **65**(A), 1–24 (1991)
19. Montroll, E.W., Shlesinger, M.F.: On the wonderful world of random walks. In: Lebowitz, J.L., Montroll, E.W. (eds.): Nonequilibrium Phenomena II. From Stochastics to Hydrodynamics. Studies in Statistical Mechanics, vol. XI, pp. 1–121. North-Holland, Amsterdam (1984)
20. Hempelmann, R.: Hydrogen diffusion in proton conducting oxides in nanocrystalline metals. In: Pękalski, A., Sznajd-Weron, K. (eds.) Anomalous Diffusion. From Basics to Applications. Lecture Notes in Physics, vol. 519, pp. 247–252. Springer, Berlin (1999)
21. Bardou, F., Bouchaud, J.-P., Aspect, A., Cohen-Tannoudji, C.: Lévy Statistics and Laser cooling. How Rare Events Bring Atoms to Rest. Cambridge University Press, Cambridge (2002)
22. Glöckle, W.G., Nonnenmacher, Th.F.: Fox function representation of non-debye relaxation processes. J. Stat. Phys. **71**, 741–757 (1993). https://doi.org/10.1007/BF01058445
23. Nonnenmacher, Th. F., Metzler, R.: Applications of fractional calculus techniques to problems in biophysics, Chap. VIII. In: Hilfer, R., (ed.) Applications of Fractional Calculus in Physics, pp. 377–428. World Scientific, Singapore (2000). https://doi.org/10.1142/9789812817747-0008
24. Metzler, R., Klafter, J.: The restaurant at the end of the random walk: recent developments in the description of anomalous transport by fractional dynamics. J. Phys. A: Math. Gen. **37**, R161–R208 (2004). https://doi.org/10.1088/0305-4470/37/31/R01
25. Zwanzig, R.: Memory effects in irreversible thermodynamics. Phys. Rev. **124**, 983–992 (1961). https://doi.org/10.1103/PhysRev.124.983

26. Van Beijeren, H., Kutner, R.: Mean square displacement of a tracer particle in a hard-core lattice gas. Phys. Rev. Lett. **55**, 238–241 (1985). https://doi.org/10.1103/PhysRevLett.55.238
27. Balcerek, M., Burnecki, K., Thapa, S., Wyłomańska, A.: Fractional Brownian motion with random Hurst exponent: accelerating diffusion and persistence transitions. Chaos **32**(9), 093114 (2022). https://doi.org/10.1063/5.0101913. Erratum: Chaos **34**(4), 049901 (2024). https://doi.org/10.1063/5.0210418
28. Mantegna, R.N., Stanley, H.E.: Physics investigation of financial markets. In: Mallamace, F., Stanley, H.E. (eds.) The Physics of Complex Systems. Proceedings of the International School of Physics "Enrico Fermi," Course CXXXIV. IOS Press, Amsterdam (1997)
29. Mantegna, R.N., Stanley, H.E.: An Introduction to Econophysics. Correlations and Complexity in Finance. Cambridge University Press, Cambridge UK (2000). https://doi.org/10.1063/1.1341926
30. Mantegna, R.N., Stanley, H.E.: Scaling behavior in the dynamics of an economic index. Nature **376**(6535), 46–49 (1995). https://doi.org/10.1038/376046a0
31. Wang, B.H., Hui, P.M.: The distribution and scaling of fluctuations for Hang Seng index in Hong Kong stock market. Eur. Phys. J. B **20**(4), 573–579 (2001). https://doi.org/10.1007/PL00022987
32. Mantegna, R.N., Stanley, H.E.: Turbulence and financial markets. Nature **383**(6601), 587–588 (1996). https://doi.org/10.1038/383587a0
33. Mantegna, R.N.: Lévy walks and enhanced diffusion in Milan stock exchange. Physica A **179**(2), 232–242 (1991). https://doi.org/10.1016/0378-4371(91)90061-G
34. Bacry, E., Delour, J., Muzy, J.F.: A multifractal random walk. Phys. Rev. E **64**, 026103 (2001). https://doi.org/10.1103/PhysRevE.64.026103
35. Shoutens, W.: Lévy Processes in Finance. Pricing Financial Derivatives. Wiley Series in Probability and Statistics. Wiley, Chichester (2003). https://doi.org/10.1002/0470870230
36. Kutner, R., Masoliver, J.: The continuous time random walk, still trendy: fifty-year history, state of the art and outlook. Eur. Phys. J. B **90**, 50 (2017). https://doi.org/10.1140/epjb/e2016-70578-3
37. Beck, C.: Dynamical foundations of nonextensive statistical mechanics. Phys. Rev. Lett. **87**, 180601 (2001). https://doi.org/10.110/PhysRevLett.87.180601
38. Kiyono, K., Hidetoshi Konno, H.: Log-amplitude statistics for Beck-Cohen superstatistics. Phys. Rev. E **87**, 052104 (2013). https://doi.org/10.1103/PhysRevE.87.052104
39. Beck, C., Cohen, E.G.D.: Superstatistics. Physica A **322**, 267–275 (2003). https://doi.org/10.1016/S0378-4371(03)00019-0
40. Kantelhardt, J.W., Kościelny-Bunde, E., Rego, H.H.A., Havlin, S., Bunde, A.: Detecting long-range correlations with detrended fluctuation analysis. Physica A **295**, 441–454 (2001). https://doi.org/10.1016/S0378-4371(01)00144-3
41. Kubo, R., Toda, M., Hashitsume, N.: Statistical Physics II. Nonequilibrium Statistical Mechanics. Springer, Berlin (1985)
42. Makse, H.A., Havlin, S., Schwartz, M., Stanley, H.E.: Method for generating long-range correlations for large systems. Phys. Rev. E **53**(5), 5445–5449 (1996). https://doi.org/10.1103/PhysRevE.53.5445
43. Kanazawa, K., Sornette, D.: Standard form of master equations for general non-Markovian jump processes: the Laplace-space embedding framework and asymptotic solution. Phys. Rev. Res. **6**, 023270 (2024). https://doi.org/10.1103/PhysRevResearch.6.023270
44. Beck, C., Cohen, E.G.D., Swinney, H.L.: From time series to superstatistics. Phys. Rev. E **72**, 056133 (2005). https://doi.org/10.1103/PhysRevE.72.056133 Erratum: Phys. Rev. E **73**(049905), 10 (2006). https://doi.org/10.1103/PhysRevE.73.049905
45. Kiyono, K., Konno, H.: Log-amplitude statistics for Beck-Cohen superstatistics. Phys. Rev. E **87**, 052104 (2013). https://doi.org/10.1103/PhysRevE.87.052104
46. Kiyono, K., Hayano, J., Watanabe, E., Struzik, Z.R., Yamamoto, Y.: Non-Gaussian heart rate as an independent predictor of mortality in patients with chronic heart failure. Heart Rhythm **5**(2), 261–268 (2008). https://doi.org/10.1016/j.hrthm.2007.10.030

47. Chechkin, A.V., Klafter, J., Sokolov, I.M.: Fractional Fokker-Planck equation for ultraslow kinetics. Europhys. Lett. **63**(3), 326–332 (2003). https://doi.org/10.1209/epl/i2003-00539-0
48. Padash, A., Aghion, E., Schulz, A., Barkai, E., Chechkin, A.V., Metzler, R., Kantz, H.: Local equilibrium properties of ultraslow diffusion in the Sinai model. New J. Phys. **24**, 073026 (2022). https://doi.org/10.1088/1367-2630/ac7df8
49. Murthy, K.P.N., Giacomettib, A., Kehr, K.W.: Multifractal scaling in Sinai diffusion. Physica A **224**, 232–238 (1996). https://doi.org/10.1016/0378-4371(95)00323-1
50. Sinai, Y.G.: The limiting behavior of a one-dimensional random walk in a random medium. Theory Probab. Appl. **27**(2), 256–268 (1982). https://doi.org/10.1137/1127028
51. Nauenberg, M.: Random walk in a random medium in one dimension. J. Stat. Phys. **41**(5–6), 803–810 (1985). https://doi.org/10.1007/BF01010004
52. Kesten, H.: The limit distribution of Sinai's random walk in a random environment. Physica A **138**, 299–309 (1986). https://doi.org/10.1016/0378-4371(86)90186-X
53. Bunde, A., Havlin, S., Roman, H.E., Schildt, G., Stanley, H.E.: On the field dependence of random walks in the presence of random fields. J. Stat. Phys. **50**(5–6), 1271–1276 (1988). https://doi.org/10.1007/BF01019166
54. Comtet, A., Dean, D.S.: Exact results on Sinai's diffusion. J. Phys. A: Math. Gen. 31. https://doi.org/10.1088/0305-4470/31/43/004
55. Metzler, R., Klafter, J.: The random walk's guide to anomalous diffusion: a fractional dynamics approach. Phys. Rep. **339**(1), 1–77 (2000). https://doi.org/10.1016/S0370-1573(00)00070-3
56. Mainardi, F., Pagnini, G., Saxena, R.K.: Fox H functions in fractional diffusion. J. Comput. Appl. Math. **178**, 321–331 (2005). https://doi.org/10.1016/j.cam.2004.08.006

Chapter 4
Non-deterministic Fractals

Abstract Before delving into the primary focus of the upcoming chapter on multifractal structures, it is essential to examine a spectrum of fractals to establish a solid foundational understanding. It is crucial to emphasize the intrinsic stochasticity in the fractal structures of random-walk trajectories, as this significantly influences and defines their fractal dimensions. Without structural noise, deterministic fractals serve merely as preliminary reference points. The visualizations presented in this chapter prepare us for the transition to multifractality, which will be further explored in the subsequent chapter.

4.1 Definition of Stochastic Fractals

Stochastic fractals are formulated by incorporation of noise (i.e., statistical perturbations) into deterministic (conventional/canonical) fractal architectures, including self-similar or (more general) self-affine fractals. Stochastic fractals diverge from their deterministic counterparts by introducing probabilistic elements capable of modifying the inherent deterministic fractal architecture with the potential for numerous subsequent alterations. These stochastically induced fractal variants typically segregate into universality classes distinct from those associated with deterministic fractals. Inherently, *stochastic fractals more accurately mirror natural phenomena*, such as porous substances or, more expansively, percolating systems or interfaces, compared to deterministic fractals, which act as idealized benchmarks for theoretical exploration.

Our methodology for characterizing stochastic fractals is bifurcated into two distinct phases. Initially, we explore the domain of *random fractals*, subsequently transitioning to a more expansive category termed *statistical fractals*. The inception of the former is marked by elementary, stochastic modifications applied to deterministic fractals (see Sect. 4.3), which do not modify the fractal dimension of the

M. Chorowski et al., *Anomalous Stochastics*, Understanding Complex Systems,
https://doi.org/10.1007/978-3-031-80392-5_4

deterministic progenitors. Conversely, statistical fractals are generally characterized by fractal dimensions diverging from their deterministic antecedents (see Sect. 4.4). Such a two-step approach allows for a better understanding of the role of confounding noise.

4.2 Mathematical Fractals Versus Physical Fractals

When we discuss fractals without qualifiers, we specifically denote deterministic mathematical fractals. However, these idealized or mathematical fractals are not observable in natural phenomena. This raises the question: Why do physicists invest in their study? Such an engagement mirrors the utilization of differential and integral calculus within physics. We inherently interact with finite increments instead of infinitesimally small limiting values in the natural world. For example, a velocity measurement quantifies the finite distance traversed by an entity within a bounded temporal interval and nothing beyond. The theoretical advancement in defining velocity as a limit constitutes a mathematical abstraction; its practicality is derived from the consistent observation of progressively regular patterns in diminishing spatial increments over increasingly brief durations. This is identified as a convergent process. This methodology permits the cessation of measurements at a requisite level of fineness since it becomes unfeasible to ascertain the velocity of a swiftly moving vehicle by examining millimetric segments of the trajectory within millisecond intervals.

The same principle applies to self-similar/self-related objects, fractals, or multifractals. We are constrained to consider merely a finite series of generations of a self-similar or self-related structure or, expressed in more straightforward terms, a restricted assortment of space-time scales. Fundamentally, *physical fractals are prefractals or stochastic premultifractals*. However, through the methodology of fractal analysis, we are capable of forecasting the behavior of a natural system on an extensive range of scales relevant to our study, thus illustrating multiscale behavior with meticulous precision.[1]

4.3 Random Fractals

In Sects. 4.3 and 4.4, we delve into two categories of fractal structures: constrained/bounded fractals, where the total linear size L remains constant regardless of granularity (scale) and unconstrained/unlimited/infinite fractals, where the incremental linear size increases as one moves to a coarser (larger scale) image. Section 4.3 focuses on random fractals, while Sect. 4.4 delves into statistical fractals.

---

[1] It is imperative to acknowledge that, for instance, critical behavior manifesting near the critical point of a continuous phase transition, mandates attaining the limit of the number of generations.

In summary, *bounded fractals are generated through the deliberate incorporation of imperfections into a uniform matrix at multiple scales*, while *infinite fractals arise from systematically managed defective proliferation.* This section methodically investigates the self-similar dimension and the box-counting dimension across both fractal categories, highlighting their pragmatic relevance.

4.3.1 Constrained Self-similar Fractals

We commence our discourse by examining entities engendered through the incorporation of a rudimentary stochastic disturbance into self-similar[2] (deterministic) fractal constructs.

4.3.1.1 Generator of Random Fractal

Figure 4.1 shows two successive generations[3] of a randomly defective Sierpiński carpet—its one-dimensional equivalent is the Cantor set and the three-dimensional is Sierpiński sponge. For the Sierpiński carpet, the dimension of the Euclidean space d in which the carpet is immersed/embodied is $d = 2$, for the Cantor set $d = 1$, and for the Sierpiński sponge $d = 3$.

In Fig. 4.1, we delineate the initial ($k = 1$) and subsequent ($k = 2$) iterations generated from a principal square with a constant side length $L(= 1)$. Observably, this dimension is partitioned into $n(= 3)$ equivalent segments; conventionally, $1/n$ is designated as the reduction factor and m/n^d as the defective factor, where $1 \leq m < n^d$ represents the number of diminutive squares excised at each level from every resultant square for the respective generation (further, by m we still denote not only the number of elements removed but also the number of elements added/introduced in each generation of a given structure, i.e., the scale). The generation under discussion was methodically engineered by the excision of a singular (i.e., $m = 1$) smaller square. Subsequently, progressively smaller squares, not necessarily centrally located (unlike

[2] Self-similar and self-affine structures are termed scale-free owing to their consistent or analogous appearance across all scales—no specific scale is discernible.

[3] In this chapter, and consequently throughout the entire textbook, we exclusively employ the technique of covering a given object with boxes, referred to as box covering. This approach differs from other coverings, such as those using balls. Various dimensions are derived independently, and within the dimension's zoo, some exhibit simple or linear relationships with each other. This textbook focuses primarily on the self-similar dimension d_s. Using this dimension as an example, we effectively demonstrate the impact of noise (probabilistic disturbance) on the dimension. Thus, our exploration is centered on dimensions that lack topological properties—see, for detail, [1, 2].

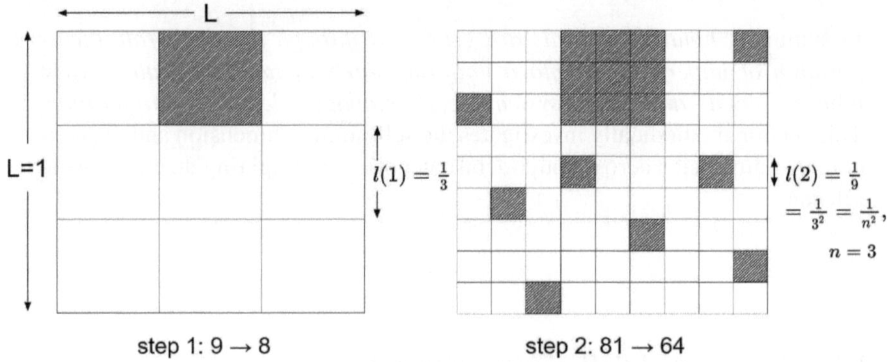

Fig. 4.1 The initial two iterations, or scales, of the bounded, descending, and accidentally defective Sierpiński carpet are depicted. The progression from level/generation k to $k + 1$, where $k = 1, 2, \ldots$, involves the strategic excision of elements (defects) as delineated in the primary discourse. Observably, the fractal dimension d_s remains consistent with the deterministic archetype, given that at each hierarchical level, a single square ($m = 1$) is selectively eliminated from the total of $n^d (= 3^2 = 9)$ squares. Constructing this fractal involves systematically removing progressively smaller squares at each successive scale

the deterministic Sierpiński carpet), are removed from the emergent configuration. The removal of a solitary, smaller square is consistently executed in each generation k. However, the specific square excised remains indeterminate and inconsequential, as the selection is randomized. Consequently, the count of squares removed in each generation remains constant and mirrors that of its deterministic analog; this figure is not prone to variations. Hence, the process is essentially a shuffling operation.

Thus, *generator of a random fractal structure* (i.e., the structure defecting on any descending scale) was defined, which allows the construction of a restricted/limited (but consisting of an infinite number of generations) self-similar fractal object in the probabilistic (statistical) sense. This definition gives the self-similarity of a random fractal structure. The immediate conclusion is that the *self-similar dimension* is identical to the fractal dimension of the deterministic Sierpiński carpet. We still call it fractal, because it is a fractional dimension and not an integer like the Euclidean (topological) dimension of space. We denote it by d_s.[4] The self-similarity dimension is (here) insensitive to which square we removed. In each generation/scale, we have one elementary (smallest) square we drew from the larger one, i.e., the one that was elementary for the one generation earlier.

[4] See the useful and inspiring books by Peitgen-Jürgens-Saupe [1] and by Vicsek [3].

4.3.1.2 Self-similar Fractal Dimension d_s

On the one hand, for the considered randomly defective structures, we can write the multiplicative homogeneous recursive equation in the form,

$$N(k) = N(k-1)(n^d - m),\ k = 1, 2, \ldots, \text{ and } N(0) = 1, \tag{4.1}$$

where $N(k)$ is the number of squares remaining in the k generation after the (above) random defect procedure. Equation (4.1) is also valid for a deterministic Sierpiński carpet.

The solution of Eq. (4.1) takes, for each generation $k(= 1, 2, \ldots)$, the form,

$$N(k) = (n^d - m)^k = (9-1)^k = 8^k, \tag{4.2}$$

(In the second equality in the above equation, we substituted the numbers characterizing the example shown in Fig. 4.1. We are dealing with a similar situation in the last equality in (4.6) below.)

On the other hand, thanks to the property of self-similarity (here in a statistical sense because it is additionally burdened with an element of randomness), a key relationship between the number of squares $N(k)$ and the number $n(k)$, can be written.

$$n(k) = n^k = \frac{L}{l(k)} \tag{4.3}$$

into which the side of the original square of length L was divided in the k-th generation. Note that $l(k)(= L/n^k)$ is the side length of the small square in the k generation. The question of the value of the self-similarity exponent, d_s, of the fractal structure is answered by the following constitutive equation:

$$N(k) = n(k)^{d_s} = \left(\frac{l(k)}{L}\right)^{-d_s},\ k = 1, 2, \ldots. \tag{4.4}$$

Often, the expression (4.4) is written in abbreviated form,

$$N(k) \sim (l(k))^{-d_s}, \tag{4.5}$$

saying that the number of covering squares that exist in the kth generation scales with the linear size of the covering squares using a power law with a self-similarity exponent d_s. This notation allows us to extend the analysis to the case of fractals that are not self-similar (i.e., subject to an appropriate scaling only asymptotically in the limit of a large number of generations k, see Sect. 4.3.2 below for details).

From (4.2), taking into account (4.3) and (4.4), we get

$$d_s = \frac{\ln N(k)}{\ln(L/l(k))} = \frac{\ln(n^d - m)}{\ln n} = 3\,\frac{\ln 2}{\ln 3} = 1.892789261\ldots, \quad m = 1, 2, \ldots, n^d - 1, \tag{4.6}$$

where, of course, the exponent

$$0 \le d_s < d, \tag{4.7}$$

which results from the fact that the discussed structure is a defective square in a sense shown in Fig. 4.1, where $m > 0$; if our squares have no holes ($m = 0$), then, of course, $d_s = d$. As can be seen, the dimension of self-similarity d_s is independent of the generation number k, which results from the self-similar nature of the structure.

In Sect. 4.3.3, we demonstrate that the expression delineated in Eq. (4.6) is amenable to extension to encompass negative values of m. This extension is contingent upon the specific delineation of the fractal structure generator, namely, the judicious elimination or incorporation of elements.

In the subsequent discourse, we shall address the pivotal inquiry: under which specific conditions does randomness directly influence the self-similarity dimension? To elucidate this, we intend to generalize the formula (4.6), our fundamental point of departure.

4.3.1.3 Minkowski Dimension

Let's consider the meaning of the self-similar dimension d_s, which is a special case of the Hausdorff dimension. For this purpose, note that $d(= 2)$-dimensional volume $V(k)$ (here area or surface) of the object in generation k is,

$$V(k) = N(k)(l(k))^d, \tag{4.8}$$

where $[l(k)]^d$ is the area/surface of the small square in the k generation. The volume $V(k)$ measures a given fractal structure.

On the other hand, by the Eqs. (4.8) and (4.4) we get,

$$V(k) = L^{d_s}(l(k))^{d-d_s} = L^d\left(\frac{l(k)}{L}\right)^{d-d_s} \tag{4.9}$$

or

$$d_s = d - \frac{\ln(V(k)/L^d)}{\ln(l(k)/L)}. \tag{4.10}$$

Sometimes the exponent $\Delta_s = d - d_s$ is called the Minkowski dimension or self-similar dimension deficit.

Often, the expression (4.9) is written shorter,

$$\frac{V(k)}{L^d} \sim (l(k))^{\Delta_s}, \tag{4.11}$$

the (relative) measure of a given fractal structure scales with the linear size $l(k)$ of the covering squares using the Minkowski dimension. This type of notation makes it possible to extend the analysis to the case of fractal structures that are not self-similar (see Sect. 4.3.2 for details).

One has to be aware that the expressions (4.6) and (4.10) are valid for each generation of k only because we are dealing with self-similarity (regardless of whether it is self-similar in a deterministic sense or the probabilistic/random sense defined above).

4.3.1.4 Limit Paradox the Structure Empty Almost Everywhere

Based on the formulas (4.8), (4.4), (4.2) and (4.3), the following border crossing can be made,

$$\lim_{k\to\infty} \frac{V(k)}{L^d} = \lim_{k\to\infty} \frac{N(k)[l(k)]^d}{[n(k)]^d\,[l(k)]^d} = \lim_{k\to\infty} \frac{N(k)}{n(k)^d} = \lim_{k\to\infty} \left(1 - \frac{m}{n^d}\right)^k = 0. \tag{4.12}$$

The above result is paradoxical because, as can be seen, in the limit $k \to \infty$, the volume of the structure and the relative number of its elements (elementary squares) are negligibly small concerning the volume of the non-defected structure, and the number of squares forming it. In other words, the number of holes equals the number of (almost) all the squares that make up a homogeneous (non-defective) object. This result is valid only in the limit $k \to \infty$, indicating that the self-similar boundary structure is maximally defective, i.e., the size of the hole area is (almost) equal to the size of the entire area of the structure. Although at each level of granularity, only the minimum number ($m = 1$) of squares needed to be removed to ensure the (nontrivial) self-similarity of a bounded random Sierpiński carpet.

A fundamental limitation must be considered when creating fractal structures: for the structure to exist, all its elements must be related, that is, touch each other. This means that any two structure elements can be connected by a line that belongs entirely to that structure. This is the definition of a connected set.

4.3.2 Box Dimension of Bounded Fractal Structures

The self-similar dimension is just a special case of the Hausdorff fractal dimension. For structures that are not self-similar, the expressions (4.6) and (4.10) should be written in a more general form,

$$d_b \overset{\text{def.}}{=} \lim_{k\to\infty} \frac{\ln N(k)}{\ln(L/l(k))} \tag{4.13}$$

or

$$d_b = d - \lim_{k\to\infty} \frac{\ln(V(k)/L^d)}{\ln(l(k)/L)}. \tag{4.14}$$

The formulas (4.13) and (4.14) were obtained by covering a given fractal structure with such boxes, whose linear size $l(k)$ decreases according to power-law with the asymptotic value of exponent k, similarly as for self-similar structures (i.e., $l(k) = L/n^k$). The quantity d_b obtained is called the asymptotic box dimension (or, shortly, the box dimension).

The created, thanks to the above, definition of the box dimension is if and only if for vanishing granularity l (that is, the linear size of the box, which in general is the independent variable),

$$d_b \overset{\text{def.}}{=} \lim_{l\to 0} \frac{\ln N(l)}{\ln(L/l)} \tag{4.15}$$

or

$$d_b = d - \lim_{l\to 0} \frac{\ln(V(l)/L^d)}{\ln(l/L)}. \tag{4.16}$$

As one can see, the minimum volume $V(l)$ in which it gives to close a given fractal structure scales up (with disappearing granularity l) like

$$V(l) \propto l^{d-d_b} \tag{4.17}$$

while the minimum number of the non-overlapping boxes $N(l)$ scales as

$$N(l) \propto l^{-d_b}. \tag{4.18}$$

Note that it is often more convenient to operate with quantities dimensionless—for this purpose, let us introduce dimensionless granularity defined as a fraction,

$$\epsilon = \frac{l}{L}. \tag{4.19}$$

Using definitions (4.19) we can expressions (4.4) and (4.9) represent as

$$N(\epsilon) = \epsilon^{-d_b} \tag{4.20}$$

and

$$\frac{V(\epsilon)}{L^d} = \epsilon^{d-d_b}. \tag{4.21}$$

Hence, the box dimension d_b can be written as

$$d_b = \lim_{\epsilon \to 0} \frac{\ln N(\epsilon)}{\ln(1/\epsilon)} \tag{4.22}$$

or

$$d_b = d - \lim_{\epsilon \to 0} \frac{\ln(V(\epsilon)/L^d)}{\ln(\epsilon)}, \tag{4.23}$$

which facilitates further generalizations (e.g., to covers other than boxes), allowing the introduction of the concept of dimension Hausdorff. Apparently, the self-similarity dimension d_s and box-counting dimension d_b are not identical (for self-similar structures).

The definition of d_s is sufficient to explicitly show the influence of probabilistics on the fractal dimension (see Sect. 4.4 below regarding statistical fractals). However, d_b is particularly useful from the operational/practical point of view (see the paragraph below).

4.3.2.1 Instructive Example

As a typical example, we determine the box dimension, d_b, of the curves that are the coastlines of Great Britain and Ireland.[5] In Fig. 4.2, we present two generations of square coverage of the areas of the countries mentioned above shown on the map.

The left panel is for $\epsilon = 1/24$, and the right one is for $\epsilon = 1/32$ (for simplicity, the bottom edge of both panels is set to 1). We use the formula (4.22), saving it for each panel separately,

$$\begin{aligned} d_b \ln(1/\epsilon_1) &\approx \ln N(\epsilon_1) \\ d_b \ln(1/\epsilon_2) &\approx \ln N(\epsilon_2), \end{aligned} \tag{4.24}$$

hence

$$d_b \approx \frac{\ln N(\epsilon_2) - \ln N(\epsilon_1)}{\ln(1/\epsilon_2) - \ln(1/\epsilon_1)}, \tag{4.25}$$

where $\epsilon = \epsilon_1 = 1/24$ and $\epsilon = \epsilon_2 = 1/32$. The expressions above are approximations since the values of $1/\epsilon_1$ and $1/\epsilon_2$ are finite and not infinitesimal. However, these

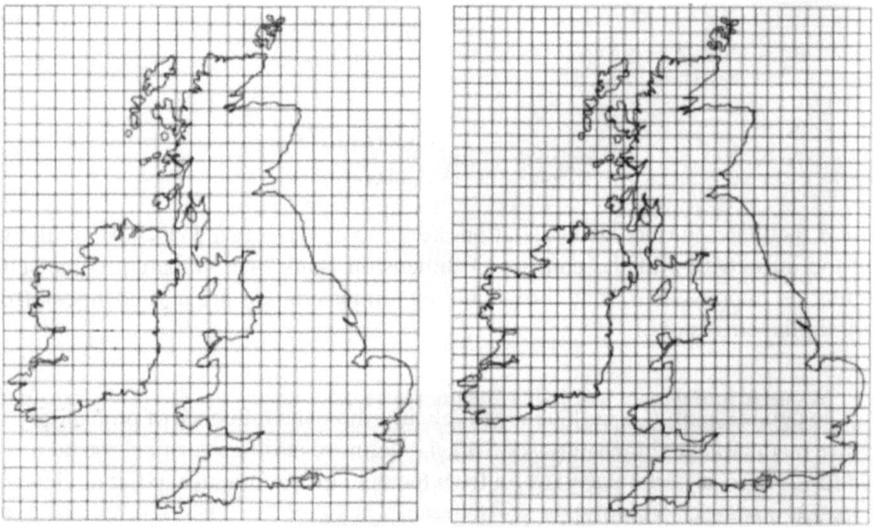

Fig. 4.2 Two panels showing two generations of box coverage, i.e., two generations of surface coverage (left panel: for $\epsilon_1 = 1/24$ and right panel: for $\epsilon_1 = 1/32$) of the same rectangle containing a schematic map of Great Britain and Ireland. The coastline of both countries deserves special attention—finding their box fractal dimensions is the goal of our considerations here. Reprinted with permission from [4]. ©2000, Springer Nature Limited. All rights reserved

[5] This canonical example is taken from the Peitegen-Jürgens-Saupe book [1].

approximations suffice for illustrative purposes, facilitating the manual enumeration of grid squares that intersect or are adjacent to the curve. From this method, the counts $N(\epsilon_1) = 194$ and $N(\epsilon_2) = 283$ were derived. Using these data, an approximate value of $d_b \approx 1.31$ was calculated.

The formula (4.25) is extensively employed for numerical and empirical ascertainment of the fractal's box dimension. This formula is particularly effective in computing the box dimensions of fractal random-walk trajectories, including those associated with Brownian motion, Weierstrass processes, and Lévy flights. The ability to determine the fractal dimension of random walk trajectories is a pivotal motivation for the integration of fractal theory elements into our textbook.

This methodology facilitates the categorization of curves into universality classes, delineating their intrinsic properties. Moreover, it facilitates a coherent progression to the discussion of multifractality in the subsequent chapter.

4.3.2.2 Remark on the Structure Density

Note that the number density of the $\rho(k)$ structure scales under these conditions identically to a non-defective structure, namely,

$$\rho(k) = \frac{N(k)}{V(k)} = (l(k))^{-d}, \tag{4.26}$$

meaning it is insensitive to the system's defective operation. Thus, the number density $\rho(k)$ remains invariant and unsuitable for describing fractal structures. A similar situation applies to the vanishing granularity l, where the number density $\rho(l) \propto l^{-d}$. This is a straightforward result/property that one should know.

An interesting property we point out in this context is the answer to the question of how empty/defective a fractal structure is. We have included the answer in the Appendix A.2.

4.3.3 Unlimited Self-similar Fractals

Similarly, as in Sect. 4.3.1, we consider a self-similar random fractal, i.e., a self-similar structure in the statistical sense, created by appropriate duplication of the output object (as shown in Fig. 4.3).

The structure depicted in Fig. 4.3 represents an infinitely expanding, stochastic (non-deterministic) Sierpiński carpet, composed of elementary squares with a linear dimension (side length) $L(0) = l(= 1)$. The dimensionless (linear) magnification factor of the carpet is, for instance, $n = L(1)/L(0) = 3$ (the grain factor $\epsilon = 1/n$ is often used interchangeably). Each larger square comprises $(n^d - m)^k$ smaller squares, where $d(= 2)$ denotes the dimension of the Euclidean space in which the Sierpiński carpet is embedded, and $m(= 1)$ represents the number of smaller squares excised from each duplicated square of the preceding generation. Currently, k signifies the current generation (scale or level of the structure's granularity) at which calculations are performed. Evidently, at each level of granularity, the pattern of defects can vary; specifically, one of the duplicated squares (larger) is removed from the square at the level of granularity k, and randomly, from any location within these large duplicated squares, an elementary square at the previous scale $k - 1$ is removed.

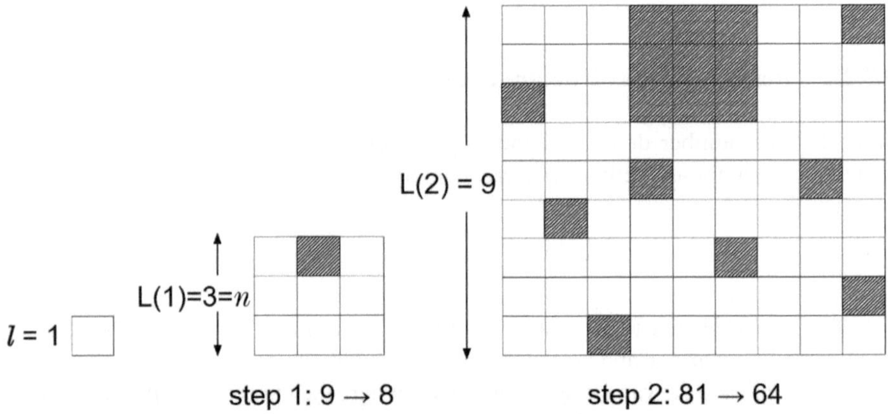

Fig. 4.3 The first two generations of an unconstrained (ascending) randomly defective Sierpiński carpet. The structure in step $k+1$, $k = 0, 1, 2, \ldots$, where $L(0) = l = 1$ is obtained from the structure in step k by enlarging it n times (here, for example $n = 3$) and appropriate defects. Namely, we first remove one smaller square with a fixed (non-random) location (e.g., always top center) of linear size $L(k)$. Then, in all remaining squares of the same linear size $L(k)$, we randomly remove m (here, for example, $m = 1$) even smaller squares of linear size $L(k - 1)$. The number of carpet elements in each generation is the same as shown in Fig. 4.1 for the descending carpet. We offer that the formulas describing these numbers are identical in both cases

4.3.3.1 Initiating Questions

We are now positioned to formulate the foundational inquiry: what is the quantity of $N(k)$ elementary squares (each with side length l) encompassed within the larger square at the kth generation (or, equivalently, at the kth level of granularity)?

On the one hand, the configuration illustrated in Fig. 4.3 demonstrates through direct enumeration that a straightforward formula can articulate the desired quantity,

$$N(k) = N(k-1)(n^d - m) = (n^d - m)^k = (3^2 - 1)^k = 8^k, \ k = 1, 2, \ldots, \tag{4.27}$$

i.e., it is analogous to the bounded random Sierpiński carpet, see equations (4.1) and (4.2)); furthermore, it corresponds to the respective values for deterministic Sierpiński carpets, whether bounded or unbounded.

On the other hand, analogous to the scenario of the bounded Sierpiński carpet, we inquire into the existence of a number d_s that remains invariant regarding the generation number k and which fulfills the constitutive equation:

$$N(k) = n(k)^{d_s} = \epsilon(k)^{-d_s} = \left(\frac{L(k)}{l}\right)^{d_s}, \ k = 1, 2, \ldots, \tag{4.28}$$

where we entered the size,

$$L(k) = n(k)l, \ k = 1, 2, \ldots, \tag{4.29}$$

being the linear size of the square in the kth generation, the same as for bounded structures

$$n(k) = n^k, \ k = 1, 2, \ldots, \tag{4.30}$$

moreover, we have,

$$\epsilon(k) = \epsilon^k, \ k = 1, 2, \ldots. \tag{4.31}$$

The constitutive equation (4.28) is written often as,

$$N(k) \sim (L(k))^{d_s}, \ k = 1, 2, \ldots, \tag{4.32}$$

saying that the "mass" of the $N(k)$ structure visible on the kth granularity scales is the power-law with its linear size $L(k)$; the exponent appearing in the Eq. (4.32) is called the self-similar dimension—later we will see that it is a special case of the fractal dimension (or more precisely, the fractal of the Hausdorff dimension).

Also, based on an obvious soft inequality

$$n^{d_s} \leq n^d \tag{4.33}$$

(resulting from the construction of the carpet) we get,

$$d_s \leq d. \tag{4.34}$$

Naturally, if the structure were homogeneous/uniform (i.e., not lacking certain elements in each generation), we would observe $d_s = d$. Identifying the exponent d_s as consistent across all generations, levels, and scales is critical for these analyses and is attributed to the self-similar nature of the constructed objects. However, the distinction between statistical and deterministic self-similarity is inconsequential.

4.3.4 Fractals Growing with Segments/Squares/Cubes Added

Moreover, if we consider analogous structures formed by appropriately adding elements at each level of granularity k—a good example would be the Koch curve, or Koch snowflakes [1], then we have,

$$N(k) = N(k-1)(n^d + m) = (n^d + m)^k = (3^2 + 1)^k = 10^k, \; k = 1, 2, \ldots,$$
$$N(0) = 1, \tag{4.35}$$

where $m \geq 0$. Using the above and proceeding similarly to the previous subsection, we obtain:

$$N(k) = n(k)^{d_s} \tag{4.36}$$

which gives the expression for the self-similar fractal dimension in the form,

$$d_s = d + \frac{\ln\left(1 + \frac{m}{n^d}\right)}{\ln n}. \tag{4.37}$$

analogous to (A.18). Basing on expression (4.37), we get

$$n(k)^{d_s} > n(k)^d, \; k = 1, 2, \ldots, \tag{4.38}$$

and to the fractal dimension

$$d_s > d, \tag{4.39}$$

exceeding the dimension of the Euclidean space in which the original, non-defected object is embedded.

All previously examined fractal dimensions lacked probabilistic influences. The subsequent section delves into more intricate and realistic scenarios wherein such dependencies are explicitly manifested. This phenomenon arises when the quantity of elements excised from a given foundational structure or substrate at each scale exhibits variability, a condition not present in the prior analyses.

4.4 Statistical Fractals

This section examines statistical structures called statistical fractals, which may exhibit bounded or unbounded characteristics.

The elucidation of this concept is facilitated through the utilization of the statistical Cantor set.[6] Specifically, it involves a multiplicative probabilistic cascade of the Cantor set, focusing on the abundance of generational average elements.

4.4.1 *Limited Statistical Fractals*

Figure 4.4 schematically illustrates a statistical ensemble comprising $\mathcal{N}^{k(=1)}$ segments/intervals divided into three equal parts/fragments. As depicted, some of the central sections have been removed.[7] We assume this occurred with a probability of $0 \leq p \leq 1$.

One can now pose a question regarding the undeleted intervals, specifically, what is the average number of them, $\langle N(k(=1))\rangle$, in a given generation/scale k (in this case, the first because $k = 1$)? The answer is straightforward—it is the limit of the following weighted average, referred to as the *average over a statistical ensemble*,

[6] For a discussion on the deterministic Cantor set, refer to work such as the book by Peitgen-Jürgens-Saupe [1].

[7] We could just as easily remove any segments, which would not change the self-similar dimension. This results from previously conducted considerations. We always remove the middle fragment here to simplify our considerations.

k=1: A statistical ensemble of a mixture of defective and non-defective segments

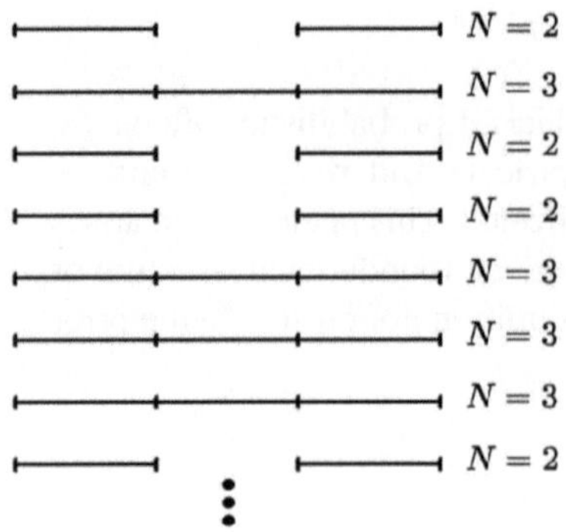

N : Number of leftover segments

$\mathcal{N}^1$: Size of the statistical ensemble (for k=1)

$\mathcal{N}_j^1, j = 2, 3$: Size of the statistical ensemble (for k=1) with the middle segment removed (j=2) and not removed (j=3)

Note: Central segment is removed with probability p

Average number of segments of the statistical Cantor set in generation (step) k=1:

$$\langle N(k=1)\rangle \approx \frac{1}{\mathcal{N}^1}(2\mathcal{N}_2^1 + 3\mathcal{N}_3^1) = 2\frac{\mathcal{N}_2^1}{\mathcal{N}^1} + 3\frac{\mathcal{N}_3^1}{\mathcal{N}^1} \approx$$

$$2p + 3(1-p) = (3-p)^{k(=1)},\ p = \lim_{\mathcal{N}^1\to\infty} \frac{\mathcal{N}_2^1}{\mathcal{N}^1}$$

Fig. 4.4 The statistical ensemble illustrating the initial generation of the randomly defected Cantor set (for $k = 1$) is presented herein. This ensemble includes both defective and non-defective segments. The central portion of the interval is systematically excised with a probability of $0 \le p \le 1$. We introduce a distinctive implementation of our scenario or statistical ensemble. Our objective is to determine the average number of segments, $\langle N(k)\rangle$, $k = 1, 2, \ldots$, that persist following a defect in each generation, constituting a multiplicative cascade of generations—a matter of considerable importance

$$\begin{aligned}\langle N(k(=1))\rangle &= \lim_{\mathcal{N}^{k(=1)}\to\infty}\left(2\,\frac{\mathcal{N}_2^{k(=1)}}{\mathcal{N}^{k(=1)}} + 3\,\frac{\mathcal{N}_3^{k(=1)}}{\mathcal{N}^{k(=1)}}\right) = 2p + 3(1-p)\\ &= 3 - p,\end{aligned} \tag{4.40}$$

where $\mathcal{N}_2^{k(=1)}$ represents the total number of segments without a central part (i.e., composed only of two fragments), and $\mathcal{N}_3^{k(=1)}$ represents segments that include a central part. Meanwhile, $\mathcal{N}^{k(=1)}\left(= \mathcal{N}_2^{k(=1)} + \mathcal{N}_3^{k(=1)}\right)$ denotes the total number of segments on the level/scale $k = 1$. Note that individual segments are statistically independent, and p is independent of k. As can be seen, we are dealing here with a probabilistic (weighted) mixture of defective and non-defective sections.

In the subsequent generation ($k = 2$), where each defected segment comprises two shorter pieces separated from each other or a non-defected piece consists of three shorter segments, we attribute distinct statistical ensembles (refer to Fig. 4.5).

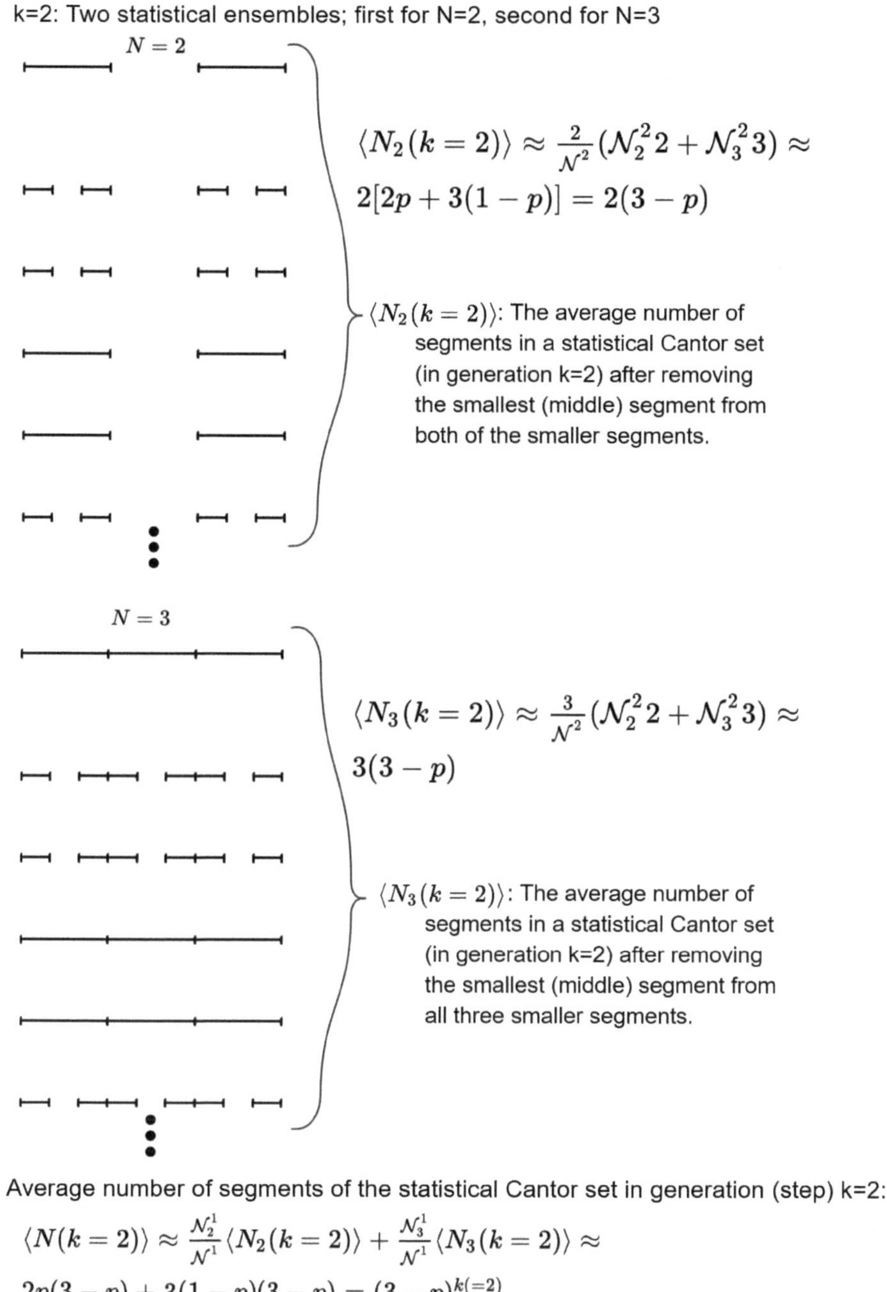

Fig. 4.5 The statistical ensemble, consisting of two subordinate ensembles derived from the preceding generation, corresponds to defective and non-defective intervals in the second generation ($k = 2$). This configuration inadvertently perturbs the Cantor set. These intervals, originating from defects at the $k = 1$ level, possess the potential to progress to the $k = 2$ level. Consequently, this framework facilitates the determination of the desired average value, $\langle N(k = 2)\rangle$

Similarly to the first generation, each shorter segment is also statistically defective. To convey the essence more efficiently, we introduce synchronization so that, for instance, a defect event co-occurs for all shorter pieces of a given generation (number k), constituting one more extended segment of the previous generation (number $k-1$). Following the defect procedure in the second generation, we then calculate the remaining average number of segments, $\langle N(k(=2))\rangle$:

$$\begin{aligned}\langle N(k(=2))\rangle &= \lim_{\mathcal{N}^{k(=1)}\to\infty}\lim_{\mathcal{N}^{k(=2)}\to\infty}\left\{2\,\frac{\mathcal{N}_2^{k(=1)}}{\mathcal{N}^{k(=1)}}\times\left(2\,\frac{\mathcal{N}_2^{k(=2)}}{\mathcal{N}^{k(=2)}}+3\,\frac{\mathcal{N}_3^{k(=2)}}{\mathcal{N}^{k(=2)}}\right)\right.\\ &\quad\left.+3\,\frac{\mathcal{N}_3^{k(=1)}}{\mathcal{N}^{k(=1)}}\times\left(2\,\frac{\mathcal{N}_2^{k(=2)}}{\mathcal{N}^{k(=2)}}+3\,\frac{\mathcal{N}_3^{k(=2)}}{\mathcal{N}^{k(=2)}}\right)\right\}\\ &= (2\,p+3\,(1-p))^{k(=2)} = (3-p)^{k(=2)},\end{aligned} \tag{4.41}$$

where $\mathcal{N}_2^{k(=2)}$ and $\mathcal{N}_3^{k(=2)}$ are defined analogously to $\mathcal{N}_2^{k(=1)}$ and $\mathcal{N}_3^{k(=1)}$ but for generation $k=2$.

Proceeding similarly for the subsequent generations, we derive a general and concise expression for the average number of segments remaining from the k-generational statistical defect procedure,

$$\langle N(k)\rangle = (2p+3(1-p))^k = (3-p)^k. \tag{4.42}$$

The expression (4.42) enables us to determine the fractal dimension d_s of a statistical fractal structure. Specifically, by definition,

$$\langle N(k)\rangle = \left(3^k\right)^{d_s} \tag{4.43}$$

and the expression (4.42) immediately follows,

$$d_s = \frac{\ln\left(3^{d(=1)}-p\right)}{\ln 3}. \tag{4.44}$$

The equality (4.43) is often interpreted as the relationship between the (dimensionless) "volume" of a stochastic fractal structure (left side of the equation) and the (dimensionless) "mass" contained in it in the generation kth (right side of this equation); this is more clearly seen below. Of course, a statistical structure fractal becomes deterministic only when $p=1$; then its fractal dimension $d_s=\ln 2/\ln 3$. When $p=0$, we are dealing with the second extreme case concerning the non-defected segment—then $d_s=d(=1)$, i.e., the self-similar dimension equals the space dimension.

4.4.2 Different Ways of Defecting Structures

A more generalized form of the formula (4.42) and consequently (4.44) is linked to the method of defecting, i.e., the procedure by which a given fractal structure/substrate is derived from a uniform structure and the minimum dimension of the space in which it is embedded. For instance, if instead of removing the middle segment, we replace it with a "roof" of two segments, then in the following formula, $m = +1$ instead of $m = -1$ as in the case of the Cantor set (compare it with the expressions (4.42) and (4.44)). Therefore, we can easily extend formula (4.42) and consequently (4.44) to the form,

$$\langle N(k)\rangle = \left(\left(n^d + m\right) p + n^d(1-p)\right)^k = \left(n^d + mp\right)^k = \left(n^k\right)^{d_s}, \tag{4.45}$$

where $-1 \le m/n^d \le 1$, with n representing the output linear measure of the initial defected uniform structure and d denoting the minimum Euclidean dimension of the embedding space. The upper limit in the double inequality mentioned above is introduced solely to render m/n^d symmetrical about zero for aesthetic purposes; utilizing the full range of m/n^d is unnecessary. In particular, there is no upper bound on the fraction m/n^d.

From the third equality in (4.45), we obtain the following:

$$d_s = \frac{\ln(n^d + mp)}{\ln n} = d + \frac{\ln\left(1 + p\frac{m}{n^d}\right)}{\ln n}, \tag{4.46}$$

which directly generalizes the formula (4.44). The defecting method reveals that we can address both a fractal with increasing mass (where $m > 0$) and decreasing mass (where $m < 0$). For $m = 0$, the object's mass, no longer a fractal structure, remains unchanged from the outset. Hence, we can broaden the applicability of the formula (4.46) by assuming $0 \le p \le 1$.

In Fig. 4.6 we show the surface plot of d_s versus m/n^d and p determined by Eq. (4.46). Only the points of this surface for which m is an integer and n a natural number have physical meaning (where $-n^d \le m \le n^d \le$). This surface has a structure, which results in each of its edges having a different shape, and furthermore its nonphysical part (for $d_s < 0$) has singularity at the point ($m/n^d = -1$, $p = 1$), which results in a specific bending of the physical part of this surface (i.e. for $d_s \ge 0$). The surface d_s versus (m/n^d, p) shown in Fig. 4.6 indicates the wealth of information in the self-similar dimension d_s.

This chapter presents the profound impact of fluctuations, or probabilistic behavior, on fractals. Specifically, we explore their influence through fractal dimensions, fundamental characteristics that define the universality classes of fractals. This is even more important because significant examples of statistical fractals are random-walk trajectories.

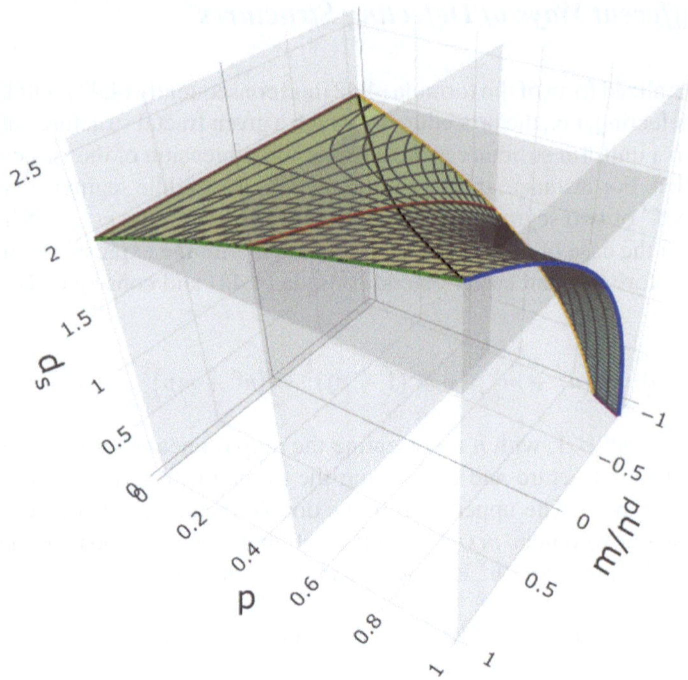

Fig. 4.6 The plot of the surface in three dimensions depicting $d_s (\geq 0)$ as a function of the variables m/n^d and p, with n and d fixed at 3 and 2, respectively. An extended version of the formula (4.46) has been employed within the interval $-1 \leq m/n^d \leq 1$, encompassing the real values of m. The relationship (4.46) is a continuous surface to facilitate a more comprehensive visualization. The structure of this surface is visible, resembling a gutter (drain) in the direction $p \to 1$ and $\frac{m}{n^d} \to -1$ of the disappearing self-similar dimensionm d_s

Essentially, we offered a quantitative validation of Smoluchowski's hypothesis, an increasingly pivotal idea, by emphasizing the critical role of fluctuations in complexity. As an illustration, we analyzed fluctuation cascades leading to multifractal behavior. Consequently, this chapter is an authentic introduction to our textbook, highlighting the profound significance of investigating the role of fluctuations. In this chapter, we mainly analyze expected values, while the higher moments that constitute multifractality are considered in the next chapter.

Control and Review Questions and Exercises

4.1 Questions
What distinguishes random fractals from statistical fractals? Why does the former leave the self-similar fractal dimension unchanged while the latter introduces alterations? Which one better mimics a deterministic fractal? What does it mean self-similar fractals? Give the range of self-similar fractal dimensions.

4.2 Question
How do bounded and unbounded self-similar fractals differ?

4.3 Questions
What does "a self-similar fractal structure is empty almost everywhere" imply? Under what conditions does this occur for self-similar fractals?

4.4 Question
How does a box fractal dimension differ from a self-similar one?

4.5 Question
Is the self-similar fractal dimension a topological property?

4.6 Question
Under what conditions is the self-similar fractal dimension greater, and under what conditions is it smaller than the dimension of the Euclidean space in which a given fractal structure is embedded?

4.7 Questions
What quantity is unsuitable for analyzing a self-similar fractal structure? In other words, what physical quantity does not exhibit self-similar scaling?

4.8 Questions
What assumptions should be met for the self-similar fractal dimension to depend on the probability of fractal structure defects?

4.9 Exercise and Question
Calculate the fractal box dimension: (i) individually for Ireland and Great Britain, and (ii) for the eastern and western coasts of both countries. What conclusions can be drawn from this regarding the geological history of both islands?

4.10 Questions
Is it feasible to identify at least one statistical fractal within the same universality class as, for instance, the coast of Great Britain? What characteristics would such a fractal exhibit?

4.11 Question
What are limited statistical fractals? Give examples.

4.12 Questions
Discuss the formula for the self-similar fractal dimension for statistical fractals. Under what conditions can we expect this dimension to increase, and under what conditions to decrease? When is this dimension equal to the Euclidean dimension?

References

1. Peitgen, H.-O., Jürgens, H., Saupe, D.: Fractals for the Classroom: Part One Introduction to Fractals and Chaos. Springer, New Jersey (1991)
2. Peitgen, H.-O., Jürgens, H., Saupe, D.: Fractals for the Classroom: Part Two: Complex Systems and Mandelbrot Set, 2nd edn. Springer, Berlin (1992)
3. Vicsek, T.: Fractal Growth Phenomena, 2nd edn. World Scientific, Singapore (1992). https://doi.org/10.1142/1407
4. Peitgen, H.-O., Jürgens, H., Saupe, D.: Chaos and Fractals. New Frontiers of Science, 2nd edn. Springer, New York (2004)

Chapter 5
Signatures and Causes of Multifractality

Abstract This chapter is systematically divided into three distinct stages. The initial stage elucidates how bi-fractality and multifractality emerge by applying a unipolar invariant measure. We scrutinize the implications of these findings, extending the discussion to a broader context through the widely utilized Saddle Point Approximation and clarifying the interconnection between multifractality and thermodynamic formalism. The second stage delves into multiplicative binomial cascades, presenting an alternative and prevalent methodology for generating multifractality. In the final stage, we introduce and juxtapose fluctuation multifractality with autocorrelation multifractality. The insights derived from this chapter represent a significant advancement in examining diverse multifractal random walks.

5.1 Bi-Fractality and Multifractality

The discourse will commence with an exposition on the concept of multifractality. Multifractality transcends mere fractality by presenting a scenario wherein a diverse spectrum of fractal exponents/dimensions/singularities exists, collectively termed the multifractal spectrum (or a spectrum of exponents/dimensions/singularities). The singularity spectrum is recognized as the quintessential indicator of multifractality. This concept will be elucidated through a convincing and elucidative example that investigates the singular invariant measure, facilitating the necessary generalization.

We note that there are at least the following causes of multifractality:

(a) The distribution of fluctuations, distinguished by a fat-tail, exhibits positive excess kurtosis, indicative of its leptokurtic characteristics.
(b) Parametric scaling of the probability density function characterized by a stochastic Hurst exponent.

M. Chorowski et al., *Anomalous Stochastics*, Understanding Complex Systems,
https://doi.org/10.1007/978-3-031-80392-5_5

(c) Long-range/long-term correlations and autocorrelations, particularly those that exhibit nonlinear characteristics.
(d) Binomial and polynomial cascades, whether in the absence or presence of laws that conserve physical quantities (referred to as constants of motion; see Sect. 5.2 for further elucidation).
(e) Thermally activated over barrier charge hopping on the amorphous matrix, exemplified by the Gaussian substrate within a valley model.[1] This procedure engenders characteristic intricate statistical distributions of interevent durations (waiting/pausing intervals), culminating in multifractal phenomena.
(f) Singular invariant density/measure.[2] It discriminates between distinct zones within the domain, specifically proximal to the singularity, and those remotely situated. This differentiation fosters a multifractal representation of the domain (see Sects. 5.1.1 and 5.1.3 for further elucidation).

The ramifications of these causative factors are analogously evidenced, mainly via the generalized Hurst exponent. This discourse explores the multiscale exponent and the singularity spectrum.

We intend to deliberate on the notably intriguing phenomenon of multifractal anomalous diffusion, specifically emphasizing the role of nonlinear long-term autocorrelations concerning the origins of multifractality.

5.1.1 Singular Invariant Measure

Take, for instance, the inspiring example of a unique yet integrable invariant density represented by the following power-law decreasing function,[3]

$$\rho(x) \stackrel{\text{def.}}{=} (1-\gamma)x^{-\gamma},\ 0 < \gamma < 1,\ 0 \le x < 1. \tag{5.1}$$

It can be interpreted as the stationary probability density describing the likelihood of locating a wandering particle (confined to the interval [0, 1[) at the point x—as discussed below. As one can observe, this measure shows a power-law singularity/division at $x = 0$ rather than, for instance, a logarithmic one.

[1] This entails a substrate wherein the potential minima are treated as stochastic variables conforming to, for instance, a Gaussian probability density function, while the maxima remain uniformly elevated.

[2] It pertains to invariance arising from topological transformations, such as translations and multiplicative adjustments, including stretching or compression. Generally, in invariance discussions, it is imperative to delineate the specific transformations applied to the domain of this measure.

[3] This example is drawn from the book by Schuster [1]. It is used here to show its general character.

We persist in dividing/discretizing/graining the domain of variable x into segments of small length $l \ll 1$; the discretization is formed by equidistant ticks denoted by $x_j = (j-1)l,\ j = 1, 2, \ldots, N$, i.e., $x_{j=1} = 0, x_{j=2} = l, \ldots, x_{j=N} = (N-1)l$. For each of these small segments $[x_j, x_{j+1}[,\ j = 1, 2, \ldots, , N-1$, we are sure of the associated coarse-grain probability,

$$p_j(l) = \int_{x_j}^{x_{j+1}} \rho(x)dx = \begin{cases} l^{1-\gamma} & \text{for } j = 1 \\ \rho(x_j)l^1 & \text{for } 2 \le j \le N-1, \end{cases} \tag{5.2}$$

where (in other words) l denotes a box of the linear size l. The bottom formula (for $2 \le j \le N-1$) is approximate, and the smaller the size l, the more accurate it becomes. The upper formula cannot be presented in an analogous form because, at point $x_{j=1} = 0$, the density ρ exhibits a singularity (non-analytical divergence). This aspect significantly distinguishes these two separate regions.

Generalizing the expression (5.2), we can say that we analyze situations for which coarse-grain probabilities can be expressed in the scaling form $p_j(l) \propto l^{\alpha_j}$, where we will define the general form of the exponent $\alpha_j (> 0)$ below. As can be seen, for the example discussed where $\alpha_1 = 1-\gamma$ and $\alpha_2 = 1$, we only have two different exponents, hence the name "bi-fractal."

Let us proceed to calculate the partition function/statistical sum,[4] Z_q, provided by formula (5.3), and the well-known one, forming the basis of statistical thermodynamics, will be explained later. It should be noted that the generalized statistical sum forms the basis for multifractality. It can be asserted that multifractality originates in statistical physics. More precisely, it emerges from the partition function, revealing the multiscale, multifractal nature of the systems analyzed.

We now calculate the partition function for our example,

$$\begin{aligned} Z_q(l) &\stackrel{\text{def.}}{=} \sum_{j=1}^{N} [p_j(l)]^q = \left[\int_0^l \rho(x)dx\right]^q + \sum_{j=2}^{N-1} [\rho(x_j)l]^q \\ &\approx l^{(1-\gamma)q} + l^{q-1} \int_l^1 [\rho(x)]^q dx \\ &= l^{(1-\gamma)q} + \frac{(1-\gamma)^q}{1-\gamma q} \left(1 - l^{1-\gamma q}\right) l^{q-1} \\ &= [1-a(q)]l^{(1-\gamma)q} + a(q)l^{q-1},\ 0 \le a(q) \stackrel{\text{def.}}{=} \frac{(1-\gamma)^q}{1-\gamma q} \le 1, \end{aligned} \tag{5.3}$$

where we used Eq. (5.2) to calculate the equality of each component separately by assuming in the second component $l^q = l^{q-1}dx$ and replacing the summation

[4] The term "Statistical sum" is a faithful translation of the German term "Zustandsssumme." Both terms function in the literature (roughly speaking) in equal terms.

with integration. Note that the weight $a(q)$ is independent of the size box l. The weight $1 - a(q)$ tells how often the first singularity $l^{(1-\gamma)q}$ appears in the partition function Z_q, while similarly, $a(q)$ indicates how often the singularity l^{q-1} appears in this partition function. This interpretation applies to each order of q separately (see Fig. A.1 in Appendix A.3 for the more subtle view).

The inequality in (5.3) that defines the range of allowed values for q imposes significant constraints, namely: $q \in [0, 1] \cup \infty$, regardless of the value of γ satisfying its limitation in (5.1).

The division of the statistical sum into two components arises because, in this context, we are dealing with only one singularity in the invariant density and a non-singular rest.

Based on the first equality in (5.3), Z_q can be formally interpreted as the (stationary) probability of locating q independent wandering particles (confined variable x to the segment $[0, 1[)$ in any grain/cell of size l. This interpretation proves helpful in determining correlation integrals, providing an alternative representation of the statistical sum. This alternative representation arises directly from the definition of the probability $p_j(l)$ of finding the variable x_i (an element of the time series) in the neighborhood x_j of size l given by the formula $p_j(l) = \sum_i \Theta(l- \mid x_i - x_j \mid)l$, where the function $\Theta(\ldots)$ denotes the Heaviside theta. In Sect. 5.1.7, we will return to this issue.[5]

As one can see,

(a) $Z_{q\to 0}(l) = \frac{1}{l}(= N)$, as $a(q = 0) = 1$. Hence, it determines the size of the medium/substrate (in units of length l, i.e., $Nl = 1$).
(b) Furthermore, $Z_{q\to 1}(l) = 1$, as $a(q = 1) = 1$. This means that the procedure above did not violate the normalization of probability, a crucial condition imposed upon it.
(c) At last, $Z_{q\to\infty}(l) \approx l^{(1-\gamma)q}$, as $a(q \to \infty) \to 0$.[6]

All these properties are utilized below to characterize multifractality.

We also note that the definition of $Z_q(l)$ in (5.3) is general. Based on it, we can propose an even more prevalent, axiomatic definition of $Z_q(l)$, which can also be used in the case of a continuous dependence of $f(\alpha)$ on α (this is an anticipatory formulation to which we return in Sect. 5.1.3 below). Namely, by partition function, we will understand a function that meets the following postulates:

(1) Standardization, $Z_{q\to 1}(l) = 1$.
(2) Grain, $Z_{q\to 0}(l) = N$.

---

[5] The proposed derivation is approximate and aims primarily to expand the intuitive view of multifractality.

[6] The convergence of $a(q \to \infty)$ to zero here takes place on the side of negative values of $a(q)$. However, in this case, it does not matter because all terms containing $a(q \to \infty)$ are negligibly small compared to 1.

(3) Extremes, $Z_{q\to-\infty}(l) \approx \left(p^{\min}\right)^q$ and $Z_{q\to\infty}(l) \approx (p^{\max})^q$, where $p^{\min}$ and $p^{\max}$ are the minimum and maximum values in the set of probabilities $p_j,\ 1 \le j \le N$.[7]

Nevertheless, in Sect. 5.3.2 there will be a need to extend these axioms (more precisely, axiom (2)).

5.1.2 Exact Viewpoint

In this subsection, we present the characteristics of multifractality as continuous nonlinear functions of q in the allowed range $0 \le q \le 1$. For this purpose, using the last equation in (5.3), we can express the general scaling exponent $\tau(q)$ as follows,

$$\tau(q) = \frac{\ln Z_q(l)}{\ln l} = q - 1 + \frac{\ln\{[1 - a(q)]\, l^{1-\gamma q} + a(q)\}}{\ln l}, \tag{5.4}$$

where the nonlinear component is a slowly varying function of l. In Fig. 5.1, we have presented (in the upper left panel) the scaling exponent $\tau(q)$ versus q.

It can be seen that the exponent $\tau(q)$ consists of two parts: the first one is linearly dependent on q, and the second one is non-linear. This nonlinear component is, moreover, a slowly varying function of the scale l (which is allowed in generalized multifractal analysis).

From Eq. (5.4) we get an exact formula,

$$\alpha(q) = \frac{d\tau(q)}{dq} = 1 - \frac{1}{[1 - a(q)]l^{1-\gamma q} + a(q)} \times \left\{\gamma\, [1 - a(q)]\, l^{1-\gamma q} - a(q)\frac{1 - l^{1-\gamma q}}{\ln l}\left[\ln(1-\gamma) + \frac{\gamma}{1-\gamma q}\right]\right\}, \tag{5.5}$$

allowing, together with the formula (5.4) and the Legendre transformation, to build all the remaining multifractal system characteristics. The graphs in Figs. 5.1 and 5.2 illustrate their runs.

Note that in extreme cases,

(i) $\alpha(q = 0) = 1 - \frac{1-l}{\ln l}[\ln(1-\gamma) + \gamma]$,
(ii) $\alpha(q = 1) = 1 + \frac{1-l^{1-\gamma}}{\ln l}\left[\ln(1-\gamma) + \frac{\gamma}{1-\gamma}\right]$,
(iii) $\alpha(q \to \infty) = 1 - \gamma$.

It should be noted that $\alpha(q = 0)$ and $\alpha(q = 1)$ are slowly varying functions of l and equal 1 in the limit $l \to 0$.

In the following, we exploit a limited situation, i.e., only three points of the α domain (defined by items (i)–(iii)), significantly reducing the multifractal characteristics of the invariant measure given by Eq. (5.1).

---

[7] The point here is that $q \to -\infty$ and $q \to \infty$ mean that $q \ll -1$ and $q \gg 1$, respectively.

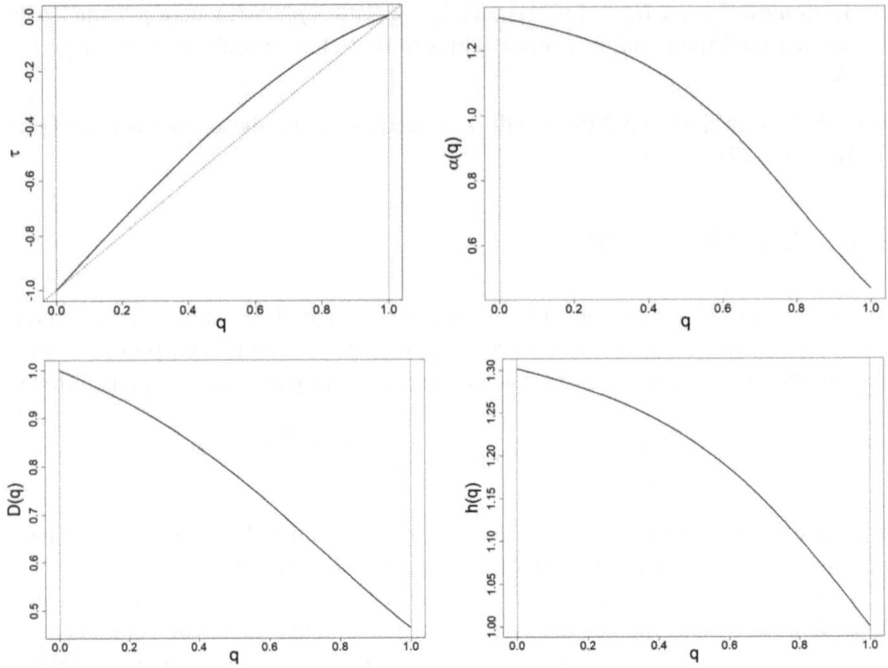

Fig. 5.1 Dependence of the basic characteristics of multifractality on variable q, for instance, for the parameters $l = 0.01$ and $\gamma = 0.9$. (i) Upper left panel: the relationship $\tau(q)$ versus q was derived from the formula (5.4). This graph represents a continuous function of q, validating that $\tau(q = 0) = -1$ and $\tau(q = 1) = 0$ are attained in extreme cases. Additionally, the non-linear characteristics of the relationship described in Eq. (5.4) are evident. (ii) Upper right panel: the relationship $\alpha(q)$ versus q was derived from the formula (5.5). This graph represents a continuous and monotonically decreasing function of the variable q, validating $\alpha(q = 0) > \alpha(q = 1)$. (iii) Lower left panel: the relationship $D(q)$ versus q was derived from the well-known formula $\tau(q) = (q - 1)D(q)$. This graph represents a continuous and monotonically decreasing function of the variable q, validating $D(q = 0) > D(q = 1)$. (iv) Lower right panel: the relationship of the Hurst exponent $h(q)$ versus q was derived from the well-known formula $\tau(q) = qh(q) - h(q = 1)$. This graph represents a continuous and monotonically decreasing function of the variable q, validating that $h(q = 0) > h(q = 1)$

5.1.3 Two-Point Multifractal Representation of Statistical Sum

We will now examine the extreme situation for which the multifractal characteristics are independent of the scale l (i.e., they are not even slowly varying functions of l). For this purpose, let us note that Eq. (5.3) can be rewritten as:

$$Z_q(l) = \int \hat{\rho}(\alpha)\, l^{\alpha q} \frac{1}{l^{f(\alpha)}} d\alpha = \int \hat{\rho}(\alpha) \exp\{-[\alpha\, q - f(\alpha)] \mid \ln l \mid\} \, d\alpha, \quad (5.6)$$

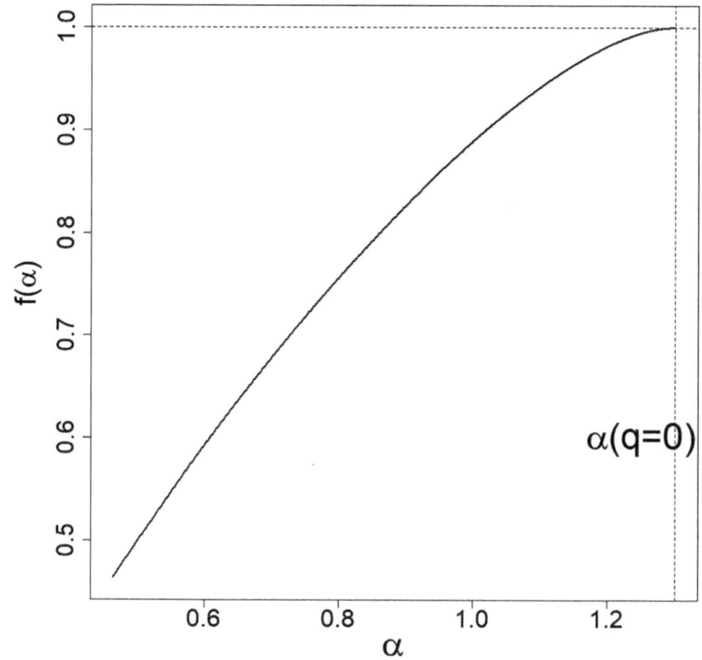

Fig. 5.2 The relationship of the spectrum of dimensions $f(\alpha)$ versus α was derived from the well-known Legendre transformation $f(\alpha) = q(\alpha)\alpha - \tau(q(\alpha))$. This graph represents the continuous left-sided spectrum of dimensions. As one can see (cf. Fig. 5.1), its maximum $f(\alpha(q = 0) = D(q = 0) = 1$, which results from the self-consistency of the procedure (as it should be) and is not imposed from the outside (as it often happens in the literature). The plot was generated using the same parameters as the previous figure

allowing for significant generalization; we introduce here a PDF

$$\hat{\rho}(\alpha) \stackrel{\text{def.}}{=} (1 - a(q))\delta(\alpha - \alpha_1) + a(q)\delta(\alpha - \alpha_2), \tag{5.7}$$

for the occurrence with only two values of singularity,

$$\alpha = \alpha(q) = \begin{cases} \alpha_1 = 1 - \gamma, & \text{for } q \to \infty, \text{ as then } a(q \to \infty) = 0 \\ \alpha_2 = 1, & \text{for } q \to 1 \text{ or } q \to 0, \text{ as then } a(q \to 0, 1) = 1. \end{cases} \tag{5.8}$$

Thus, the statistical sum Z_q has been reorganized to yield an integral over equivalence classes characterized by the index α. Each equivalence class encompasses all intervals with identical α values. In this context, we deliberately introduce a binary scenario, constraining ourselves to only two distinct points/values of α, thereby corresponding to only three values of q. To clarify, we deliberately create a discontinuous scenario, in contrast to the continuous situation outlined in Sect. 5.1.2, where we explore the exponent q continuum and the differentiable scaling exponent $\tau(q)$.

The probability density function $\hat{\rho}(\alpha)$ is associated with the multiplicity of α, revealing the number of boxes of size l that share the same singularity/exponent value α. Moreover, $l^{-f(\alpha)}$ characterizes the density of this set of singularities, where $f(\alpha)$ quantifies the fractal dimension of this density. This density satisfies an additional scaling hypothesis, supplementing the one previously presented by the formula (5.2).

In other words, the probability density function (PDF) defined by Eq. (5.7) asserts that irrespective of the value of $q \in [0, 1] \cup \infty$, we observe only two distinct values of $\alpha(q)$. Thus, we are dealing with a multifractal system, a combination of only two monofractals referred to as a bifractal.

The singularities given by Eq. (5.8) can be regarded as local box fractal dimensions, generally representing the Hausdorff dimension, defining the "volumes" l^{α_i}, $i = 1, 2$, which in turn correspond to local fractal regions defining the corresponding probabilities p_1 and p_j, $2 \leq j \leq N$. The transition from summing over j in the formula (5.3) to integrating over singularities α in the formula (5.6) also required the introduction of the number density $1/l^{f(\alpha)}$, where

$$f(\alpha) = \begin{cases} 0, & \text{for } \alpha = \alpha_1 \\ 1, & \text{for } \alpha = \alpha_2. \end{cases} \tag{5.9}$$

This density function informs us about the multiplicity of singularity α, indicating how many fractal regions share the same (uniform) singularity.[8] It should be noted that $f(\alpha)$ can be viewed as a fractal dimension (Hausdorff, associated with the singularity of α), which explains how the number density mentioned scales. Naturally, $\alpha \neq f(\alpha)$ in general. **The density function $f(\alpha)$ is, in fact, the sought singularity spectrum**—only two points here.

The expression (5.6) can be rewritten (using (5.7)–(5.9)) as

$$Z_q(l) = [1 - a(q)]l^{\alpha_1 q - f(\alpha_1)} + a(q)l^{\alpha_2 q - f(\alpha_2)}, \tag{5.10}$$

that is, the weighted sum (superposition) defining the coexistence of two monofractals. The expression (5.10) can also be rewritten in two-branch form,

$$Z_q(l) = \begin{cases} l^{\alpha_1 q - f(\alpha_1)} & \text{for } q \to \infty, \text{ as then } a(q \to \infty) = 0 \\ l^{\alpha_2 q - f(\alpha_2)} & \text{for } q \to 1 \text{ or } q \to 0, \text{ as then } a(q \to 0, 1) = 1. \end{cases} \tag{5.11}$$

It defines a bi-fractal on a three-point support q, namely for $q = 0, 1, \infty$. As one can observe, in this context, **the bi-fractal acts as a boundary object for two superposed monofractals.**

---

[8] More precisely, in the case of $\alpha = \alpha_2 = 1$, the mentioned multiplicity is $\frac{1}{l} - 1$. However, since $\frac{1}{l} \gg 1$, this justifies the adoption of the slightly simplified formula for the number density mentioned above.

The expression (5.11) can be formally rewritten as follows,

$$Z_q(l) = l^{\tau(q)}, \tag{5.12}$$

where

$$\tau(q) \overset{\text{def.}}{=} \alpha(q)q - f(\alpha(q)) \overset{\text{def.}}{=} \begin{cases} \alpha_1(q)\,q - f(\alpha_1(q)), & \text{for } q \to \infty \\ \alpha_2(q)\,q - f(\alpha_2(q)), & \text{for } q \to 1 \text{ or } q \to 0. \end{cases} \tag{5.13}$$

The first definition in (5.13) is the definition of the Legendre transform that transforms the function f into τ when going from the variable α to the variable q.

The expression (5.12) can be treated as a general power-law form (i.e., scale-free) of the partition function, which we already considered and we will use later.[9]

Note also that the Eq. (5.12) is equivalent to the following homogeneous scaling equation in variable l,

$$Z_q(\lambda l) = \lambda^{\tau(q)} Z_q(l), \tag{5.14}$$

where parameter λ is any real number and $Z_q(l = 1) = 1$ while $\tau(q)$ is generally a non-linear function of q. Proof that Eq. (5.14) follows from Eq. (5.12) is obvious. For the proof in the opposite direction, it is sufficient to substitute in Eq. (5.14) $\lambda = 1/l$.

Thanks to the interpretation of individual factors in the integrand function of the expression presented above (5.6), they can be regarded as general, independent of the specific form of the invariant density ρ. Therefore, the entire expression (5.6) can be considered a general (integral) form of the statistical sum Z_q. We deal only with a bi-fractal, as the singularity spectrum (5.9) is two-pointed as a α function. This spectrum can generally range from a single point (resulting in a monofractal, also known as a fractal) through a two- and multipoint spectrum to a continuous spectrum. In general, we can say that the combined probability $\hat{\rho}(\alpha)\, \frac{1}{l^{f(\alpha)}}\, d\alpha$ determines (by scanning) how many times the scaling relation $p_j(l) \approx l^{\alpha_j}$ is included in the total $\sum_{j=1}^{N} [p_j(l)]^q$, (then $\alpha = \alpha_j$).

We continue to examine the general situation in which we simplify the integral (5.6) using the Saddle Point Approximation/Method (SPA, outlined in Appendix A.4).

[9] More specifically, the public expression takes the form $Z_q(l) \propto l^{\tau(q)}$, where the proportionality coefficient may depend on q.

In other words, we approximate it by focusing on a dominant component associated with a specific value $\alpha = \alpha^*(q)$. This approximation is feasible when dealing with a slowly varying probability density $\hat{\rho}(\alpha)$. However, it's important to note that applying this method to our specific example is neither possible nor necessary (since in our case, quantities $\alpha_1 = \alpha^*(q = \infty)$ and $\alpha_2 = \alpha^*(q = 0) = \alpha^*(q = 1)$ are known).

In the Sect. 5.1.4 below, we take the opportunity to show in general terms how from the statistical sum $Z_q(l)$, we extract the principal component that is crucial for the behavior of $Z_q(l)$.

5.1.4 Rényi's Generalized Dimensions

5.1.4.1 Main Component of Z_q

We'll define the specific value $\alpha^*(q)$, which pertains to the general situation, as the point where the function $\alpha q - f(\alpha)$ reaches its minimum in the variable α. We are looking for the dominant component of the partition function $Z_q(l)$ (as $l \ll 1$); we have already said this above. We recall that α for which this holds is denoted by α^*. The function defined above plays the role of the function $F(x)$ in formula (A.28) (in Appendix A.4). Here, α replaces the variable x, and the parameter N is represented by $\mid \ln l \mid$. It is important to note that the Saddle Point Approximation can be applied only if a well-defined minimum of $\alpha q - f(\alpha)$ exists, allowing the approximation of this function in its minimum vicinity using a second-order polynomial (or parabolic approximation). Therefore, this subsection is complementary to the previous ones in this chapter, showing, for a general situation, the use of SPA to construct multifractality.

Using the formula (A.28) we get,

$$Z_q(l) \approx A(\alpha^*(q))\, l^{\alpha^*(q)q - f(\alpha^*(q))} = A(\alpha^*(q))\, \exp\left\{-\left[\alpha^*(q)q - f(\alpha^*(q))\right] \mid \ln l \mid\right\}, \tag{5.15}$$

where the relationship $\alpha^* = \alpha^*(q)$ is obtained from the condition of the existence of the $\alpha q - f(\alpha)$ minimum for a given q, i.e., $f'(\alpha^*(q)) = q$ (see Appendix A.4 for detail); the condition $-f''(\alpha^*(q)) > 0$ defining the minimum is needed below. In addition, the slowly-varying coefficient in the q variable,

$$A(\alpha^*(q)) = \hat{\rho}(\alpha^*(q))\sqrt{\frac{2\pi}{\mid f''(\alpha^*(q)) \mid\mid \ln l \mid}}. \tag{5.16}$$

is the product of the particular value $\hat{\rho}(\alpha^*)$ and the inverse of the normalization factor as residue after using the Saddle Point Method.

However, in our example, the connection between α^* and q is derived differently due to the nondifferentiability of our scenario of the singularity spectrum. It is essential to highlight that the method outlined above is applicable only when dealing with a slowly varying relationship $\hat{\rho}(\alpha)$ on α and its upper limits, along with $|\ln l|$ being sufficiently large for the Gaussian integral to approximate the Dirac delta adequately. For our specific case, the Saddle Point Method is unnecessary, as the integral (5.6) can be (almost) precisely computed through direct calculation.

5.1.4.2 Consequences

We proceed to introduce the profoundly significant generalized multi-scale exponent expressed as:

$$\tau(q) = \alpha^*(q)q - f(\alpha^*(q)). \tag{5.17}$$

This serves as the Legendre transform of the singularity spectrum $f(\alpha^*(q))$; the Legendre transform was discussed in Appendix A.5. Consequently, the connection between the generalized dimension of Rényi D_q and the generalized exponent $\tau(q)$ is established through the relationship

$$\tau(q) = (q-1)D_q \tag{5.18}$$

and statistical sum

$$\begin{aligned} Z_q &= A(\alpha^*(q))\exp(-\tau(q) \mid \ln l \mid) = A(\alpha^*(q))\exp(-(q-1)D_q \mid \ln l \mid) \\ &\Leftrightarrow -I_q = C_q + D_q \mid \ln l \mid, \end{aligned} \tag{5.19}$$

where the Rényi information

$$I_q = \frac{\ln Z_q}{q-1}. \tag{5.20}$$

Additionally, the reduced information of Rényi[10] is,

$$C_q = -\ln A(\alpha^*(q))/(q-1) \tag{5.21}$$

[10] As we do not delve into the reduced information in the subsequent Rényi information, we do not concern ourselves with the proper selection of the normalization factor, as it only contributes to C_q.

and is independent of l. Therefore, for a sufficiently small l we obtain from (5.18)–(5.21) a popular expression for the generalized Rényi dimension,

$$D_q \approx -\frac{I_q}{\mid \ln l \mid} = \frac{S_q}{\mid \ln l \mid}, \tag{5.22}$$

where the non-additive Rényi entropy

$$S_q \overset{\text{def.}}{=} -I_q. \tag{5.23}$$

As evident, when there's less information accessible in the system (indicating higher disorder), the system's entropy increases. This concept of entropy has paved the way for non-extensive thermodynamics. This framework characterizes systems with long-range interactions between their elements.[11]

Interestingly, it is worth mentioning that $S_{q \to 1} (= -I_{q \to 1})$ corresponds to the Shannon information entropy. To illustrate this, let's express the sum,

$$\begin{aligned} \sum_{j=1}^{N} p_j^{1+(q-1)} &= \sum_{j=1}^{N} p_j \exp((q-1)\ln p_j) \approx \sum_{j=1}^{N} p_j(1+(q-1)\ln p_j) \\ &= 1 + (q-1)\sum_{j=1}^{N} p_j \ln p_j. \end{aligned} \tag{5.24}$$

Hence, from (5.20) and from the definition of entropy $S_{q \to 1}$ we obtain the desired relationship,

$$S_{q\to 1} = -\lim_{q\to 1}\frac{1}{q-1}\ln\left(1+(q-1)\sum_{j=1}^{N} p_j \ln p_j\right) = -\sum_{j=1}^{N} p_j \ln p_j, \tag{5.25}$$

also often used in thermodynamics (and not only in information theory).

Now let's go back to our example and build a Table 5.1 of quantities characterizing considered bi-fractal.

It's evident (without resorting to the Saddle Point Method) that only three specific values of q enable the derivation of our two-point singularity spectrum. These are the desired q values. Notably, observe that the values in the third row of Table 5.1 correspond to two distinct q values—this was partially discussed in Sect. 5.1.1.

[11] In this context, the following works are recommended: by Dauxois-Ruffo-Arimondo-Wilkens [2] and by Tsallis [3].

Table 5.1 Basic quantities characterizing a bi-fractal

	D_q	$\alpha^*(q)$	$f(\alpha^*(q))$
$q \to \infty$	$(1-\gamma)\frac{q}{1-q} \to 1-\gamma$	$1-\gamma$	0
$q \to 0,\ 1$	1	1	1

5.1.5 Basic Properties of the Singularity Spectrum

We will now introduce additional properties that facilitate the construction of the singularity spectrum.

(a) It follows from the Legendre transformation (5.17) that since the generalized scaling exponent $\tau(q)$ is the negative Legendre transformation of the singularity spectrum $f(\alpha^*)$ conversely, the singularity spectrum is the negative Legendre transformation of the generalized exponent. It has consequences in the form of the following relationship: $\alpha^*(q) = \frac{d\tau(q)}{dq} \Leftrightarrow q = \frac{df(\alpha^*)}{d\alpha^*}$. We can still assume that both equalities are satisfied.

(b) Furthermore, from the Legendre transformation (5.17) and the expression (5.18) it follows[12] that $\tau(q=1)=0 \Leftrightarrow f(\alpha^*(q=1)) = \alpha^*(q=1)$ and $\frac{df(\alpha^*)}{d\alpha^*} = 0$ for $\alpha^*(q=0)$.

(c) Another property concerns the boundary values of the singularity spectrum α_{min} and α_{max}. Namely, using the first equation in (5.3) and (5.20) and (5.22) we get the necessary formula $D_q \approx \frac{1}{\ln l}\frac{1}{q-1}\ln\sum_{j=1}^{N}[p_j(l)]^q$. Hence, $D_{q\to\infty} \approx \frac{1}{\ln l}\frac{1}{q-1}\ln[p_{max}(l)]^q = \frac{q}{q-1}\alpha^*_{min} \to \alpha_{min}$, where to derive the penultimate equality, we used the relationship $p_j(l) \approx l^{\alpha_j}$, and for sufficiently large values of q, only the minimum value of the exponent $\alpha_j = \alpha^*_{min}$ remains in total. Analogous considerations can be made for $q \to -\infty$. Then, the dominant term in total is $[p_j(l)]^q \approx l^{\alpha^*_{max} q}$ which leads to $D_{q\to-\infty} \approx \alpha^*_{max}$. Therefore, the singularity spectrum is in the range $[\alpha^*_{min}, \alpha^*_{max}]$.

(d) As follows from the properties presented in point (a), the derivatives on the edge of the carrier of the singularity spectrum are unlimited: $\frac{df(\alpha^*)}{d\alpha^*}|_{\alpha^*_{min}} = \infty$ and $\frac{df(\alpha^*)}{d\alpha^*}|_{\alpha^*_{max}} = -\infty$.

Consequently, leveraging properties (a) through (d), we are equipped to delineate a schematic representation of the singularity spectrum $f(\alpha^*)$. These properties are utilized, whether directly or indirectly, comprehensively or partially, throughout the subsequent sections of this textbook. We have included additional helpful considerations regarding the Legendre Transformation in Appendix A.5.

[12] We have tacitly assumed that we are dealing here with the limited information dimension of Rényi that is, $D_{q\to 1}$.

5.1.6 Relationship Between Multifractality and Thermodynamics

The similarity of the first equality (definition) in Eq. (5.3) to the thermodynamic statistical sum/partition function is possible for both closed and open systems. This definition can be understood generally and not just in the context of this specific example. It is similar to Eq. (5.2) defining coarse-grain probability; to obtain a general definition, it is enough to replace the integral over the segment $[x_j, x_{j+1}]$ with the integral over the seed/cell, i.e., introduce

$$p_j(grain\ size) \stackrel{\text{def.}}{=} \int_{grain(j)} \rho(x)dx. \tag{5.26}$$

We continue to develop the above generalization in a one-dimensional variant.

In such a situation, it is possible to use the normalized (as it should be[13]) probabilities p_j. For example, for a canonical system (i.e., open due to the exchange of energy with the environment/reservoir/thermal bath and closed due to the number of particles in the system), we can write,

$$p_j^q \equiv \exp(-\beta E_j) = \left[\exp(-E_j)\right]^{\beta}, \; E_j \geq 0, \tag{5.27}$$

where in the multifractal description, we use the Rényi exponent q, whose thermodynamic analog is $\beta = 1/k_B T$ (see row no. 1 in the Table 5.2), and E_j is the energy of the entire system in the microscopic state with number/index j (in the thermodynamic description). Hence, p_j is the probability (independent of β) of finding a system in the microscopic state j with energy E_j. Therefore, Eq. (5.27) is the Gibbs canonical PDF (which we placed in row no. 4 in Table 5.2 and the corresponding partition functions in row no. 5). By the way, in row no. 2 we have included a multifractal equivalent of the volume of the system[14] and in the line no. 3 equivalent to the number of N cells in the system. As one can see, large values of β (i.e., q in the multifractal notation) favor, in the first equation in (5.3), events with high probability values or extreme events, i.e., low energy values (in the thermodynamic description). For small q, the situation is reversed. We can say that the first equation in (5.3) is the basis for the thermodynamic analogy of multifractality, which also considers other thermodynamic variables and potentials (in Table 5.2 we have included the starting values that allow us to build a full thermodynamic formalism). However, it should

[13] The subtlety here is that for the multifractal description, quantity p_j should be normalized and not p_j^q. Then, one can more conveniently examine the system's dependence on q, which is crucial for multifractal analysis.

[14] The volume equivalent here has a more general form than the $|\ln l|$ used in the example - this requires explanation. Namely, L is the time series length, and l is the single grain/cell size, and number of cells $N = L/l$. Therefore, the volume is $V = \ln N$. In the example, we assumed (for simplicity) $L = 1$ and l is some fraction, the volume $V = \ln(1/l) = |\ln l|$.

Table 5.2 Assignments of multifractal quantities to thermodynamic ones

No.	Multifractal	Thermodynamics
1.	q	β
2.	$\ln(L/l)$	V
3.	L/l	N
4.	p_j	$\exp\left(-\beta E_j\right)$
5.	Z_q	Z_β
6.	$\alpha^*(q)$	$\frac{U_\beta}{V}$
7.	$\tau(q)$	$\beta\frac{F_\beta}{V}$
8.	$\frac{\partial\left(\tau\ln(L/l)^{1/q}\right)}{\partial\ln(L/l)}\mid_q$	p
9.	$f(\alpha^*(q))$	$\frac{S_\beta}{V}$
10.	$c_{\ln(L/l)}(q)$	$c_V(\beta)$

be emphasized that the correspondence between multifractal and thermodynamic quantities is not complete because multifractality allows negative values of q, unlike the values of β.

We begin the analysis of the relationship between multifractality and the classical thermodynamics of equilibrium states by drawing attention to the analogy between the specific free energy, F_β/V, of a thermodynamic system and the generalized exponent $\tau(q)$. Namely, we have

$$\beta\frac{F_\beta}{V} \equiv \tau(q), \tag{5.28}$$

the equivalent of the quantity β is q (for $q \geq 0$), and the equivalent of the (macroscopic) volume of the system $V = \ln N$ (see rows 2 and 3 in Table 5.2).

From here, we can derive important correspondence,

$$\begin{aligned}\frac{U_\beta}{V} &\equiv \alpha^*(q),\\ \frac{S_\beta}{V} &\equiv f(\alpha^*(q)),\\ c_v(\beta)/k_B = \frac{1}{V}\frac{dU_\beta}{d(1/\beta)}\mid_V &\equiv c_{\ln(L/l)}(q) = \frac{d\alpha^*(q)}{d(1/q)}\mid_{\ln(L/l)}.\end{aligned} \tag{5.29}$$

where we put '$k_B = 1$' for simplicity.

All assignments are summarized in a Table 5.2.

Let us note that thanks to the Legendre transformation we can connect the specific energy of the system and its specific entropy, S_β/V, with the specific free energy of this system, i.e.,

$$\beta\frac{F_\beta}{V}=\beta\frac{U_\beta}{V}-\frac{S_\beta}{V}\equiv\tau(q)=q\alpha^*(q)-f(\alpha^*(q)). \tag{5.30}$$

The above correspondence takes place thanks to the Legendre transformation (8.50), which allowed for two further assignments (in addition to the one given above (5.28)). Furthermore, we have included a multifractal equivalent, $c_{\ln L/l}(q)$, of the specific heat at constant volume c_V, which can be obtained thanks to the prior assignment of the specific free energy to the generalized exponent.

5.1.7 *Relationship Between Multifractality and Nonlinear Dependence*

The specific objective of these considerations is to establish a connection between the statistical sum (or partition function) of order q as given by the initial equation in (5.3) and the canonical correlation integral of Grossberger and Procaccia [4] of the same order, namely, (in general) with nonlinear correlations of order q. To put it differently, the absence of nonlinear correlations eliminates the possibility of multifractality in a given random walk/time series. It is essential to highlight that the ensuing derivation is universal and not exclusively tied to the specific measure discussed in this chapter [5].

So, let's write coarse-grained probability p_i in the form,

$$p_i=\int_{grain(i)}dx\;\rho(x)=\int_{x_i}^{x_{i+1}}dx\;\rho(x)=\rho(\overline{x}_i)l,\;\;x_{i+1}-x_i=l, \tag{5.31}$$

where the index $i=1,2,\ldots,N$, numbers the grains/cells, $x_i=(i-1)l$, and $\overline{x}_i$ is the element of box i satisfying the well-known mean value theorem of integral calculus. Individual disjoint (one-dimensional) cells/grains of the same size l cover the domain where the probability is determined.

Thanks to this, the partition function can be transformed into the form containing the Grossberger-Procaccia correlation integral,

$$Z_q(l)=\sum_{i=1}^{N}p_i(l)^q=\langle p(l)^{q-1}\rangle=\sum_{i=1}^{N}[\rho(\overline{x}_i)l]^q=\sum_{i=1}^{N}\rho(\overline{x}_i)l[\rho(\overline{x}_i)l]^{q-1} \tag{5.32}$$

where $\langle f\rangle=\sum_{i=1}^{N}p_i f_i$, L is the size of the whole domain (in our bi-fractal example introduced in the Sect. 5.1.1, we assumed $L=1$), and $l\ll L$ (We haven't used this inequality here yet).

We can still, inspired by the Grossberger-Procaccia approach, introduce the following helpful equality,

$$[\rho(\overline{x}_i)l]^{q-1} = \left[\frac{1}{\mathcal{N}}\sum_{j=1}^{\mathcal{N}}\Theta(l-\mid \overline{x}_i - \tilde{x}_j \mid)\right]^{q-1}, \tag{5.33}$$

right for $\mathcal{N} \to \infty$ such that $\mathcal{N} >> N$. Recall that $\Theta(\ldots)$ is a theta Heavyside function. It is necessary to explain the origin of the points $\tilde{x}_j,\ j = 1, 2, \ldots, \mathcal{N}$. It is a long sequence of points (trajectory) drawn from the initial distribution $\rho(x)$ (e.g., using a simple method to invert the cumulative distribution function). The idea is to have many trajectory points in each l box, but remember that the distribution of points in each l box is not uniform.

Substituting Eqs. (5.33) to (5.32) we get,

$$Z_q(l) = \frac{1}{N}\sum_{i=1}^{N}\rho(\overline{x}_i)\left[\frac{1}{\mathcal{N}}\sum_{j=1}^{\mathcal{N}}\Theta(l-\mid \overline{x}_i - \tilde{x}_j \mid)\right]^{q-1} = C_q(l). \tag{5.34}$$

Meanwhile, $C_q(l)$ is a notation inspired by Grossberger-Procaccia (here, q was introduced as a subscript). Moreover, it can be seen that despite the double summation, the relationship between variables $\overline{x}_i$ and $\tilde{x}_j$ does not disappear in the sense that it affects the magnitude $Z_q(l)$.

For the particular case $q = 2$, the partition function $Z_{q=2}(l)$ (and therefore also $C_{q=2}(l)$) becomes the Grassberger-Procaccia correlation integral $C_{q=2}(l) = C(l)$, which has the meaning of the probability that the distance of two (arbitrarily selected) points of the time series/process is not greater than l.

We note that Eq. (5.34) are only correlational representations/notations (continuous and discrete, respectively) of the partition function and do not resolve what kind of correlation accompanies a given multifractality.

In the following, we present an alternative view of the statistical sum through the conditional probability distribution function. Namely,

$$\begin{aligned}
Z_q &= \sum_{grain(i)} p_i\, p_i^{q-1} = \sum_{grain(i)}\int_{grain(i)} dx\ \rho(x)\left[\sum_{grain(j)}\int_{grain(j)} dy\ \rho(x \mid y)\rho(y)\right]^{q-1} \\
&\approx \sum_{grain(i)}\int_{grain(i)} dx\ \rho(x)\left[\sum_{grain(j)}\rho(x_i \mid y_j)\int_{grain(j)} dy\ \rho(y)\right]^{q-1} \\
&= \sum_{grain(i)} p_i\left[\sum_{grain(j)} p_j\ \rho(x_i \mid y_j)\right]^{q-1} \\
&= \sum_{grain(i)} p_i\left[\sum_{grain(j)} p_j\ \rho(\mid x_i - y_j \mid)\right]^{q-1},
\end{aligned} \tag{5.35}$$

where x_i and x_j denote the left edges of the grains,[15] respectively i and j, i.e. $x_i = (i-1)l$ and $x_j = (j-1)l$. As one can see, we have introduced the assumption here of replacing the conditional $\rho(x \mid y)$ by its approximate coarse-grained form $\rho(x_i \mid y_j)$. Then, we have introduced the assumption of stationarity of this coarse-grained conditional distribution, i.e., $\rho(x_i \mid y_j) = \rho(\mid x_i - y_j \mid)$. We can still write $\rho(\mid x_i - y_j \mid) = \rho(\mid i - j \mid l)$. All this makes the last term in the above equation useful due to its single-particle, coarse-grained, and stationary nature.

At the end of this subsection, it is worth realizing that, in general, the partition function can be written in a strict form that considers fluctuations' degeneracy.

$$Z_q = \sum_{grain(i)} p_i^q = \sum_{grain(\hat{i})} w(\hat{i}) p_{\hat{i}}^q, \tag{5.36}$$

where the reduced index $\hat{i}$ numbers the equivalence classes consisting of seeds characterized by the same p_i value, while $w(\hat{i})$ is a measure of the size of the $\hat{i}$ class. This allows us to present the formula (5.35) in an approximate form reduced to classes,

$$Z_q \approx \sum_{grain(\hat{i})} w(\hat{i}) p_{\hat{i}} \left[\sum_{grain(\hat{j})} w(\hat{j}) p_{\hat{j}}\, \hat{\rho}(x_{\hat{i}} \mid y_{\hat{j}}) \right]^{q-1}, \tag{5.37}$$

where $\hat{\rho}(x_{\hat{i}} \mid y_{\hat{j}})$ denotes the distribution ρ defined on such x_i and y_j that are in closest distance from each other, i.e. for which the module $\mid x_i - y_i \mid$ is minimal.

It should be emphasized that the combination of multiplicity and q-power of the coarse-grain probabilities may lead to multifractality—the probabilities alone without the participation of multiplicity are insufficient. We showed this in Sect. 5.1.3, where we introduced multifractality by presenting multiplicity and the q-power of the coarse-grain probabilities in forms based on appropriate scaling hypotheses.

5.2 Multifractal Multiplicative Binomial Cascades

This section delineates two multiplicative binomial cascades, exemplifying statistical multifractal architectures. These instances elucidate how cascades via averaging can engender a multifractal configuration. Notably, presumably, the simplest version of the cascade model is the multiplicative deterministic binomial de Wijs cascade—

[15] In general, x_i and x_j denote characteristic points of the grains, i and j respectively, e.g. edges or centers.

see Appendix A.22. In this Appendix, we show how this cascade can extract the multifractal properties, for example, of an ideal gas—the basic model of statistical thermodynamics.

5.2.1 Statistical Cantor Mass Distribution: A Toy Model

This subsection constructs the statistical binomial cascade model for mass distribution, conceptualized here as a simplified, illustrative model, which culminates in multifractality.

We examine a bifurcated scenario as depicted in Fig. 5.3. The left branch, or experimental path, is executed with a probability p, whereas the right branch has a complementary probability of $1 - p$. Should the system initially reside on a specific branch within any experimental iteration (as part of a statistical ensemble of experimental replicates), it persistently remains on that designated branch. Within each branch, every generation undergoes a deterministic cantorization process of the sections, by

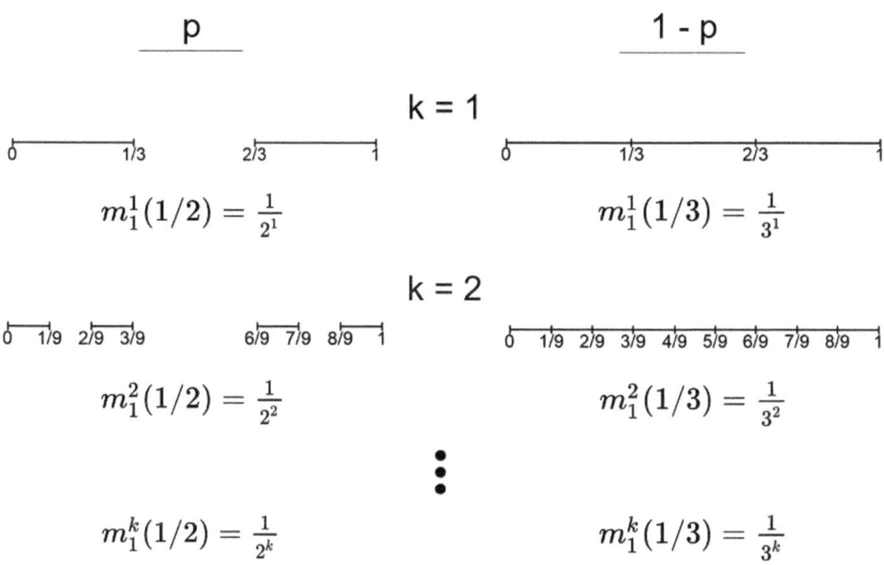

Fig. 5.3 The architecture of the statistical ensemble is delineated, comprising two distinct branches characterized by defective and non-defective segments. As one ascends to a higher ensemble level k, the ensemble manifests deterministically(or systematically) defective sections. It is a binomial cascade restricted by the principle of mass conservation in each generational/ensemble level k. That is, the total mass for a single branch (p or $1 - p$) and ensemble level k is constant and (for simplicity) equal to 1. For example, for a single branch p and level $k = 1$ the mass of a single segment $m_1^1(1/2) = 1/2^1$ and for $k = 2$ the analogous mass is $m_1^2(1/2) = 1/2^2$. An analogous situation applies to branch $1 - p$. The introduction of a probabilistic component arises from the stochastic selection of these branches according to the dichotomic statistics

which the central segment of each section is excised, and the section is subsequently partitioned into three equal segments. In addition, the mass of the excised segment is uniformly redistributed to the two remaining segments. Consequently, this process constitutes a statistically deterministic binomial cascade restricted by the principle of mass conservation in each generational stage k. Figure 5.3 illustrates each branch's initial two generational stages of sections.

Based on the comments above, we first answer the question of how the ensemble/cascade of average masses, $\langle m_1^{k-1}\rangle$, of a single segment in generation $k-1$ scales with its length, $\left(\frac{1}{3}\right)^{k-1}$. Using the definition of the ensemble average mass, we can write for $k = 1, 2, 3, \ldots$,

$$\langle m_1^{k-1}\rangle = p\left(\frac{1}{2}\right)^{k-1} + (1-p)\left(\frac{1}{3}\right)^{k-1} = \left(\frac{1}{3}\right)^{(k-1)D(k)} = \left(\frac{1}{3}\right)^{\tau(k)}. \quad (5.38)$$

Equation (5.38) is a weighted average of the masses $m_1^{k-1}(1/2)$ and $m_1^{k-1}(1/3)$ of individual sections in the $k \geq 2$ generation (Eq. (5.38) gives for $k = 1$ a normalization of dichotomic statistics). In this way, we are dealing with a multi-scale general scaling exponent $\tau(k)$ depending on the generation k. However, in the extreme cases $p = 1$ we get $D(k) = \frac{\ln 2}{\ln 3}$ regardless of k (quantity $D(k)$ in defined by Eq. (5.41) below). In another extreme case, $p = 0$, the quantity $D(k) = d = 1$, i.e., the dimension of Euclidean space, regardless of k.

The expression (5.38) allows us to define the generalized partition function for this toy model by the following assignment:

$$Z_k = \langle m_1^{k-1}\rangle, \quad k = 1, 2, 3, \ldots\ , \quad (5.39)$$

where

$$\tau(k) = (k-1)D(k), \quad (5.40)$$

with $D(k)$ extended into the form

$$D(k) = -\frac{\ln\left(p(n^d \mp m)^{-(k-1)} + (1-p)n^{-d(k-1)}\right)}{(k-1)\ln n}. \quad (5.41)$$

(Note: please do not confuse the m variable that defects segments with mass $m_1^k(1/l),\ l = 2, 3$.)

The assignment delineated in Eq. (5.39) substantiates the interpretation of the exponent $\tau(k)$ as scaling the overarching partition function. Specifically, in the scenario where $d = 1$, $n = 3$, and $\mp m = -1$, the formulation above represents an advanced generalization of the cantorization process. The multiscaling exponent $\tau(k)$, as defined by Eq. (5.40) and supported by Eq. (5.41), is depicted as a function of k in Fig. 5.4 (left panel).

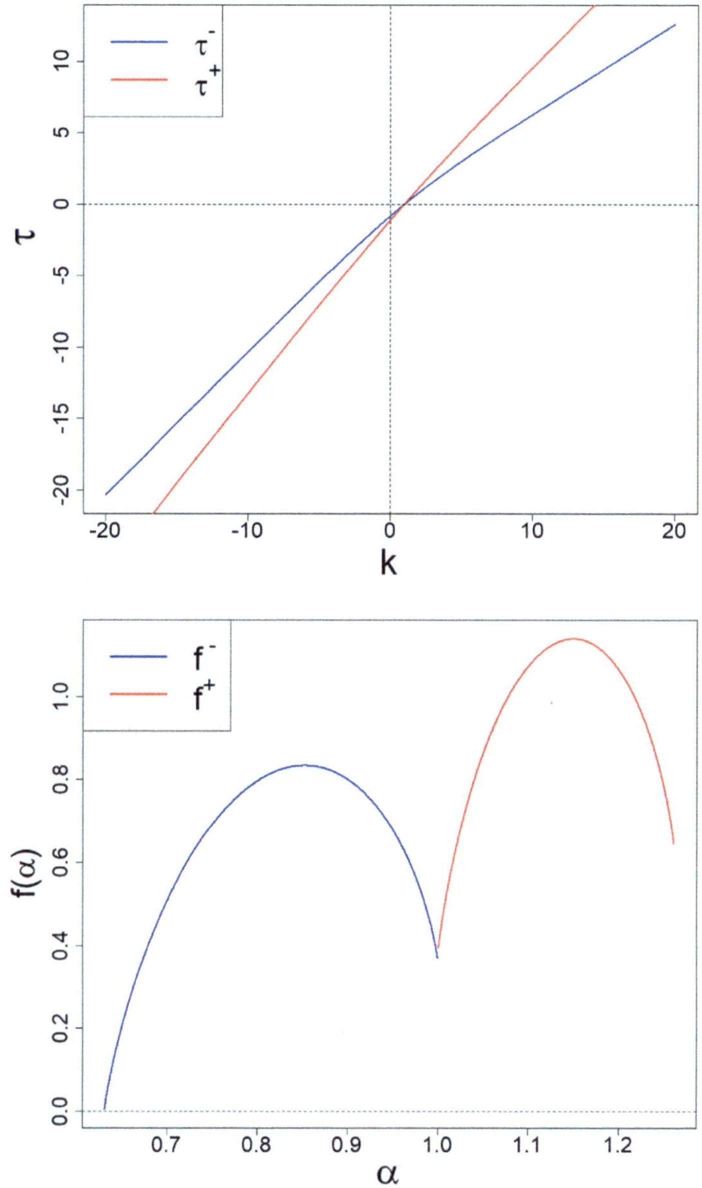

Fig. 5.4 Exploration of the generalized scaling exponent, $\tau(k)$, as a function of k (as delineated in Eq. (5.40)) and the spectrum of dimensions, $f(\alpha)$, in relation to α (articulated in Eq. (5.43)), utilizing dual methodologies of defect analysis. Herein, the symbols $+$ (red curves) and $-$ (blue curves) signify $+m$ and $-m$ within the Eqs. (5.41) and (5.43), correspondingly, with $m > 0$ (alongside other relevant equations). Each scenario reveals a non-linear interdependence, as explained in (5.40) and corroborated by Eq. (5.41). Both panels (upper and bottom) demonstrate that constructive/positive defecting (as opposed to destructive/negative) facilitates lower-order fluctuations. Collectively, the cascading and statistical attributes of the mass structure robustly validate this nonlinear linkage, thus illuminating the emergence of multifractality within this essential theoretical model

Hence singularity,

$$\alpha(k) = \frac{d\tau(k)}{dk} = \frac{1}{\ln n} \frac{p\left(n^d \mp m\right)^{-(k-1)} \ln\left(n^d \mp m\right) + (1-p)n^{-d(k-1)} \ln\left(n^d\right)}{p\left(n^d \mp m\right)^{-(k-1)} + (1-p)n^{-d(k-1)}} \tag{5.42}$$

and the singularity spectrum

$$\begin{aligned} f(\alpha(k)) &= k\alpha(k) - \tau(k) \\ &= \frac{k}{\ln n} \frac{p\left(n^d \mp m\right)^{-(k-1)} \ln\left(n^d \mp m\right) + (1-p)n^{-d(k-1)} \ln\left(n^d\right)}{p\left(n^d \mp m\right)^{-(k-1)} + (1-p)n^{-d(k-1)}} \\ &+ \frac{\ln\left(p(n^d \mp m)^{-(k-1)} + (1-p)n^{-d(k-1)}\right)}{\ln n}, \end{aligned} \tag{5.43}$$

constituting the Legendre contact transformation. We have illustrated the expression (5.43) also in Fig. 5.4 (the bottom panel). A more systematic analysis of multifractal characteristics is performed in the next subsection.

The primary origin of multifractality in this context is straightforwardly attributed to a statistical amalgamation of fractal and nonfractal (undefeated) configurations (or mixture) in mass distribution governed by the principle of mass conservation.

5.2.2 *Statistical Multiplicative Binomial Cascade*

In the preceding subsection, we delineated the multifractality cascade of segment masses. In this section, we elucidate the cascade of segment numbers within a more intricate model for comprehensiveness.

We introduce multifractal here by averaging the quantity $\langle N(k)^{q-1}\rangle$ (being an extension of Eq. (4.45) considered in Sect. 4.4.2) over generations, with some relatively simple weight. This averaging means that the individual generations of segments have been placed on the substrate according to the weight mentioned above—we explain this in more detail below.

We introduce the weight in the form of the simplest possible distribution, i.e., in the form of Kronecker delta

$$w(k) = \delta_{k,k^*} \tag{5.44}$$

which tells us with what probability an observer can draw the k generation in the statistical set of previously prepared statistically defective structures (e.g., sections from the Cantor set). We realize that such a weight should not disappear on the set of natural numbers, e.g., having the form of a discrete Gaussian distribution centered at the point k^*. The distribution (5.44) is its limiting case (with vanishing variance). Fortunately, the above simplification does not destroy multifractality, significantly simplifying the calculations.

5.2.2.1 Partition Function and Generalized Multi-scale Exponent

The partition function $Z_q(k)$ is defined below by the average number of intervals, $\langle N(k)^{q-1}\rangle$, obtained in the generation k. Namely, the expression for a moment of order $q-1$ of the average number of intervals obtained in the generation k as a result of the defecting (or cantorizing) section procedure from the previous generation $k-1$ is the following,

$$Z_q(k) \stackrel{\text{def.}}{=} \langle\langle N(k)^{q-1}\rangle\rangle = \sum_{k=1}^{\infty} w(k)\langle N(k)^{q-1}\rangle = \exp(k^*G(k^*)) = \eta(k)^{\tau(q)}, \tag{5.45}$$

where the outer bracket $\langle\ldots\rangle$ means averaging over k (giving independence from k). Furthermore, we reduce the infinite sum to a single (effective) component, which makes it easier to identify the source of multifractality. Generally speaking, such a reduction is possible if the weights have a sharp maximum, e.g., for the distributions of thermodynamic quantities. Let us add that along the way, we benefited from relationships,

$$\langle N(k)^{q-1}\rangle = \left[(n^d \mp m)^{q-1} p + n^{d(q-1)}(1-p)\right]^k, \quad n, m > 0 \text{ and } n^d - m > 0, \tag{5.46}$$

substituting $\eta(k) = n^{k^*}$ (because, by definition, we remain within self-similar structures), we get,

$$G(k^*) = G = \frac{\ln\langle N(k^*)^{q-1}\rangle}{k^*} = \ln\left(p\left(n^d \mp m\right)^{q-1} + n^{d(q-1)}(1-p)\right), \tag{5.47}$$

which is already independent of k^* and

$$\tau(q) = (q-1)D(q) = \frac{\ln\left(p\left(n^d \mp m\right)^{q-1} + (1-p)n^{d(q-1)}\right)}{\ln n}, \tag{5.48}$$

Hence,

$$\tau(q=0) = -D(q=0) = \frac{\ln\left(\frac{p}{n^d \mp m} + \frac{1-p}{n^d}\right)}{\ln n} < 0. \tag{5.49}$$

Equation (5.48) follows from the direct generalization of the expressions (4.45) and (4.46).

In Fig. 5.5, we have presented the relationship $\tau(q)$ versus q given by Eq. (5.48). It can be seen that this relationship is nonlinear—it is (to a good approximation) linear only for asymptotically small and large values of q. Namely, for $-m(<)0$ we get,

$$\tau(q) = \begin{cases} (q-1)\frac{\ln(n^d - m)}{\ln n} + \frac{\ln p}{\ln n}, & \text{for } q \to -\infty \\ (q-1)d + \frac{\ln(1-p)}{\ln n}, & \text{for } q \to \infty, \end{cases} \tag{5.50}$$

while for $+m(>0)$ we have,

$$\tau(q) = \begin{cases} (q-1)d + \frac{\ln(1-p)}{\ln n}, & \text{for } q \to -\infty \\ (q-1)\frac{\ln(n^d + m)}{\ln n} + \frac{\ln p}{\ln n}, & \text{for } q \to \infty. \end{cases} \tag{5.51}$$

The text discusses the non-linear or multifractal nature of the function $\tau(q)$, which characterizes the scaling behavior of specific structures. This function reflects the mixing, at the average level, of defective and non-defective realizations of the structure in fixed proportions. It suggests that mixing fractal structures or introducing fractal elements into nonfractal structures can lead to multifractality, a general feature.

5.2.2.2 Rényi Dimensions and Generalized Hurst Exponent

Here are Rényi's dimensions coming from Eq. (5.48),

$$D(q) = \frac{\ln\left(p\left(n^d \mp m\right)^{q-1} + (1-p)n^{d(q-1)}\right)}{(q-1)\ln n} \tag{5.52}$$

and presented in Fig. 5.5 (the upper right panel).

The generalized (multiscale) Hurst exponent, $h(q)$, connects to the generalized multi-scale exponent through the relationship,

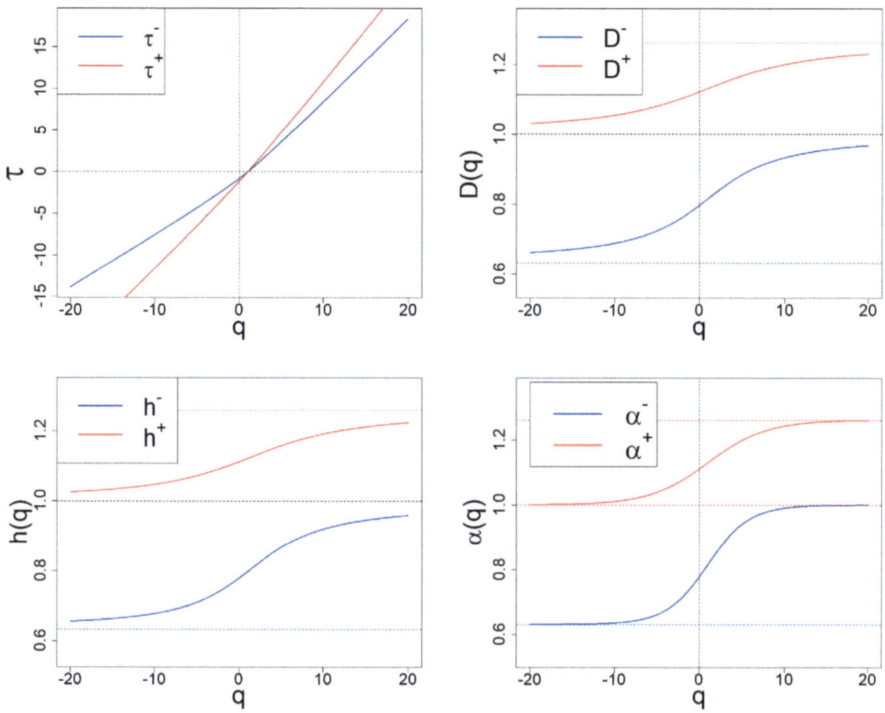

Fig. 5.5 Representations of multifractal characteristics as functions of the variable q, elucidating equations for $n = 3$, $m = \mp 1$, $p = 1/2$, $d = 1$. The positive sign (red curves) corresponds to $m = +1$, and the negative sign (blue curves) to $m = -1$. (i) In the upper left panel, the depiction of the scaling exponent $\tau(q)$, exemplifying Eq. (5.48). This panel demonstrates the convex nature of the function τ. (ii) The upper right panel displays the Rényi dimensions $D(q)$, as delineated by the Eq. (5.52). (iii) The lower left panel presents the generalized Hurst exponent $h(q)$, as described by the Eq. (5.54). (iv) The lower right panel illustrates the coarse Hölder exponent $\alpha(q)$ versus q, by Eq. (5.57). These functions exhibit a monotonically increasing relationship with q. The graphical representations suggest the presence of asymptotic behavior: a diagonal asymptote for $\tau(q)$ versus q and correspondingly horizontal asymptotes for the subsequent plots

$$\tau(q) = qh(q) - h(q = 1), \tag{5.53}$$

where,

$$h(q) = \frac{h(q = 1)}{q} + \frac{\ln\left(p\left(n^d \mp m\right)^{q-1} + (1 - p)n^{d(q-1)}\right)}{q \ln n}, \tag{5.54}$$

and from the comparison of the formulas (5.48) and (5.53) for $q = 0$ we additionally obtain that

$$h(q=1)=D(q=0)=d-\frac{1}{\ln n}\ln\left(\frac{p}{1\mp m/n^d}+1-p\right). \tag{5.55}$$

An illustration of the expression (5.54) (supported by the expression (5.55)) is shown in Fig. 5.5.

As one can see, $D(q=0)<d$ for the '–' sign and $D(q=0)>d$, for the '+' sign. In particular, if there is no defect (i.e., when $m=0$), dimension $D(q=0)=d$.

Furthermore, $D(q=2)=d^p_{\mp m}=\frac{\ln(n^d\mp pm)}{\ln(n)}$, which is the fractal dimension of the statistical structure (compare the formulas (4.45) and (4.46)).

From the definition of $D(q)$ given by the formula (5.52) it follows after simple algebraic transformations that

$$D(q\to 1)=(1-p)d+pd^{p=1}_{\mp m},\; d^{p=1}_{\mp m}=\frac{\ln\left(n^d\mp m\right)}{\ln n}, \tag{5.56}$$

where $d^{p=1}_{\mp m}$ is the dimension of the regular/deterministic fractal structure and $D(q\to 1)$ is the weighted average of the dimensions of the non-defected and defected structures, with the weights being the probabilities $1-p$ and p.

5.2.2.3 Coarse Hölder exponent

From the Eq. (5.48) we get the coarse Hölder exponent,

$$\alpha(q)=\frac{d\tau(q)}{dq}=\frac{1}{\ln n}\frac{p\left(n^d\mp m\right)^{q-1}\ln\left(n^d\mp m\right)+(1-p)n^{d(q-1)}\ln\left(n^d\right)}{p\left(n^d\mp m\right)^{q-1}+(1-p)n^{d(q-1)}}, \tag{5.57}$$

whose dependence on q we presented in Fig. 5.5 the lower right panel.

We will also determine the limit values of $\alpha(\mp\infty)$ from the formula (5.57). Namely,

$$\alpha(q=+\infty)=\alpha_{max}=\begin{cases} d & \text{for sign '–'}\\ d^{p=1}_{+m} & \text{for sign '+'},\end{cases} \tag{5.58}$$

and

$$\alpha(q=-\infty)=\alpha_{min}=\begin{cases} d^{p=1}_{-m} & \text{for sign '–'}\\ d & \text{for sign '+'},\end{cases} \tag{5.59}$$

5.2.2.4 Spectrum of Fractal Dimensions

By using the Legendre transformation, we obtain the key formula for the spectrum of fractal dimensions,

$$f(\alpha(q)) = q\alpha(q) - \tau(q) = \frac{q}{\ln n} \frac{p\left(n^d \mp m\right)^{q-1} \ln\left(n^d \mp m\right) + (1-p)n^{d(q-1)} \ln\left(n^d\right)}{p\left(n^d \mp m\right)^{q-1} + (1-p)n^{d(q-1)}}$$
$$- \frac{\ln\left(p\left(n^d \mp m\right)^{q-1} + (1-p)n^{d(q-1)}\right)}{\ln n}. \tag{5.60}$$

We have illustrated the above relationship in Fig. 5.6.

Hence,

$$f(\alpha(q=1)) = \alpha(q=1) \text{ and } \frac{df(\alpha)}{d\alpha}\Big|_{\alpha(q=1)} = 1, \tag{5.61}$$

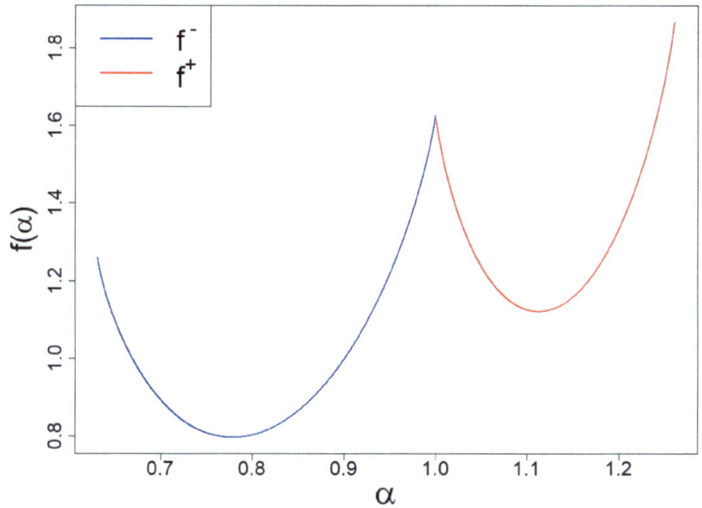

Fig. 5.6 Plot of the spectrum of dimensions f versus coarse Hölder exponent $f(\alpha)$. It illustrates the formula (5.60) for $n = 3$, $m = \mp 1$, $p = 1/2$, $d = 1$, where the notation τ^- (blue curve) refers to $m = -1$ and τ^+ (red curve) refers to $m = +1$. As one can see, we are dealing here with functions versus α, which have (absolute) minima and arms that point upward. This is a consequence of the concave shape of the function/exponent $\tau(q)$. The main text provides the recipe for transitioning to the thermodynamic representation of multifractality

which is a condition (sufficient and necessary) for the contact transformation, which is the Legendre transformation. Moreover,

$$f(\alpha(q=0)) = -\tau(q=0) = D(q=0) \text{ and } \frac{df(\alpha)}{d\alpha} |_{\alpha(q=0)}= 0. \tag{5.62}$$

Moreover,

$$\frac{df}{d\alpha} |_{\alpha_{min}}= \infty, \quad \frac{df}{d\alpha} |_{\alpha_{max}}= -\infty. \tag{5.63}$$

Although the expressions (5.48), (5.52), (5.57) and (5.60) were derived in analytical form, unfortunately, the crucial relationship $f(\alpha)$ versus α was not. We need to obtain α numerically—this can be done based on formulas (5.57) and (5.60). One can check that out here.

5.2.2.5 Linear Transformation from the Concave Into Convex Multifractality

The metamorphosis from concave to convex multifractality, as expounded in Sects. 5.2.2 and 5.2.1 respectively, constitutes a linear transformation. This transformation requires an alteration of both independent and dependent variables, specifically delineated as $(q, \tau) \rightarrow (\text{new}q = 2 - q,\ \text{new}\tau = -\tau)$. The notation "new$z$" signifies the redefined variable post-transformation z. This specified linear transformation is demonstrated in Sects. 8.2.2 and 8.2.3 elucidate the connection between parabolic multifractality and the statistical thermodynamic framework.

> From the perspective of a generalized partition function framework, it is imperative to underscore that categorizing general scaling exponents as either concave or convex is inherently relative or conventional. The transformation of these exponents is unequivocal and can be implemented without any intrinsic limitations or hesitations.

5.3 Multifractal Detrended Fluctuation and Autocorrelation Analysis

Physicists have utilized the concept of multifractal scaling to develop a significant and widely used method known as Multifractal Detrended Fluctuation Analysis (MF-DFA). This method was specifically designed to analyze various time series. Of particular interest are the time dependencies of trajectories in random walks, especially those that exhibit anomalous behavior. The detailed discussion is motivated by the promising approach to examining anomalous diffusion, where we address long-term autocorrelated stochastic processes. Multifractality extends the analysis to moments and autocorrelations of any order, enabling a comprehensive and multiscale examination of anomalous diffusion. In this section, we address MF-DFA based on the q-fluctuation function and MF-DACA (Multifractal Detrended Autocorrelation Analysis) using the q-autocorrelation function. Organizationally, we will systematically develop MF-DFA, the leading method, while integrating references to MF-DACA at appropriate stages.

5.3.1 Main Stages

The analyses described above are divided into five stages to facilitate their algorithmization and applications.

5.3.1.1 Initial Stage: Definition of a Primary Time Series

Consider a one-dimensional time series $\{x_k\}_{k=1}^{N}$ comprising $1 \leq N < \infty$ elements indexed by the natural index k. Here, only the series values are under analysis, with time discretized in a constant-step manner. This is our fundamental time series, upon which we will construct a cumulative or integrated series. An example of a basic series could involve stock market company prices or returns observed over different time intervals. We opt for the former type of quotation, as it can be directly interpreted as the random walk of a test particle. A specific empirical illustration here involves the five-minute quotes of Alphabet Inc. (Google) on the NASDAQ stock exchange (refer to Fig. 5.7 for details), demonstrating the non-Brownian and non-Gaussian nature of this quotation or wandering, attributed to significant increments or flights.

> In this subsection, we address multifractal anomalous diffusion. Figure 5.7 provides real-world examples that starkly contrast Gaussian behavior.

5.3.1.2 Stage One: Construction of the Series Profile

We will define the series profile as a cumulative/integrated centered random variable of the form:

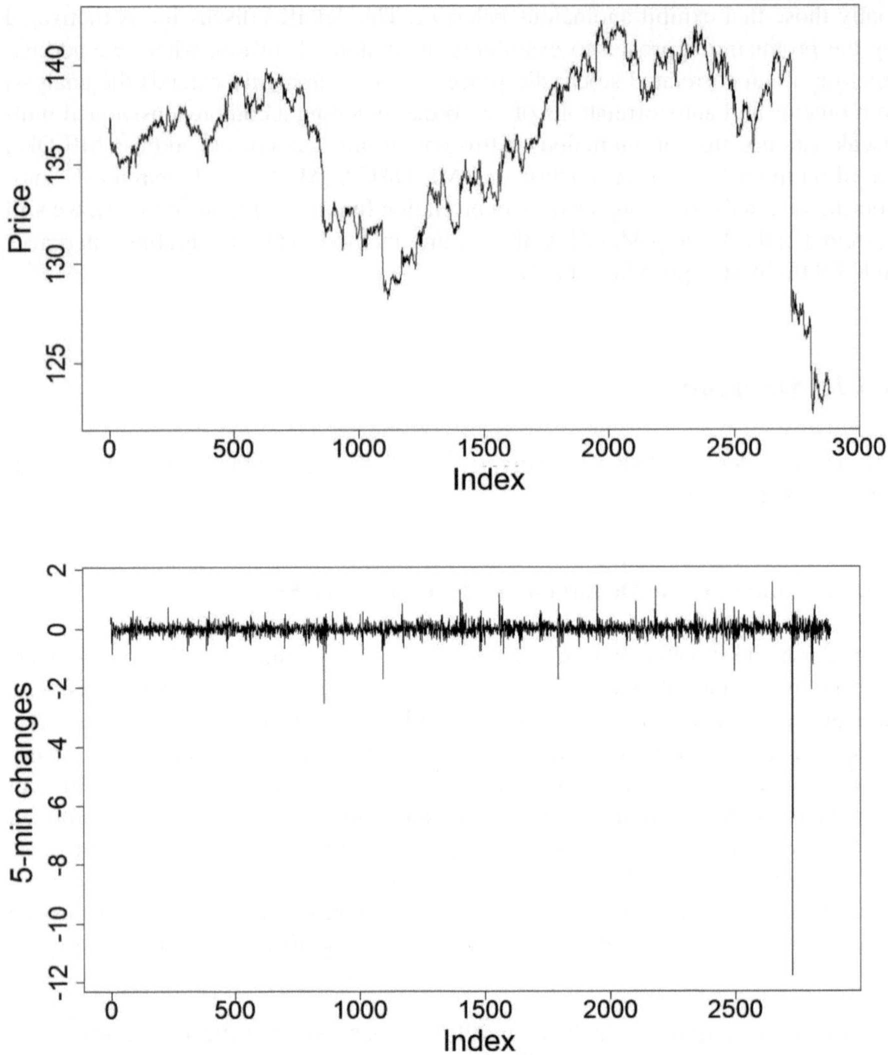

Fig. 5.7 Five-minute quotes for Alphabet Inc. (i.e., Google, GOOG) on the NASDAQ stock exchange. The quotations pertain to the period from 2023-09-06 15:35:00 CEST to 2023-10-26 21:55:00 CEST. (i) The left panel: the empirical data represents a time series comprising $N = 2885$ elements (connected here by straight-line segments). The rare large increments are well seen. (ii) The right panel: The variogram's non-Gaussian and hierarchical character is visible. Moreover, in this subsection, we show that this quote is multifractal. Moreover, it is an example of a real multifractal anomalous diffusion

$$Y(i) = \sum_{k=1}^{i} (x_k - \langle x \rangle), \ i = 1, 2, \ldots, N, \tag{5.64}$$

where $\langle x \rangle = \frac{1}{N} \sum_{k=1}^{N} x_k$ is the estimate of the expected value of the time series. It remains a constant quantity throughout a given time series. Thus, it is unnecessary to incorporate it directly into the formula (5.64). Instead, it can be incorporated below, in stage three, within the formula (5.65), as part of the polynomial detrending applied to the cumulative time series.

Subsequently, the ensemble of random variables (referred to as profiles) $Y(i)_{i=1}^{N}$ will be regarded as a fresh, integrated time series, with the variable itself representing a formal displacement over i time steps akin to that of a hypothetical financial "blind ant" (analogous to a wandering molecule or test particle). This new variable adheres to a convenient boundary condition, with $Y(i = N) = 0$.

5.3.1.3 Stage Two: Substrate Construction

We divide the time interval $[1, N]$ into $N_s = integer[N/s]$ non-overlapping segments of the same size s (see Fig. 5.8 for details).

Given that the magnitude of s frequently is not commensurate with N, the residual segment, characterized by size $N - s\ N_s$ that is less than s, remains unutilized. To avoid the waste of this segment, we construct additional N_s segments of dimension s, albeit in reverse order, transitioning from the terminal point N to the initial point 1, rather than progressing from 1 to N. Consequently, this methodology yields a substrate comprising twice the number of segments, precisely $2N_s$, each maintaining the original dimension s. This approach effectively obviates concern for the aforementioned residual fragments in both segmentation strategies.

5.3.1.4 Stage Three: Elimination of Local Trends

Fluctuation function. In each of the N_s segments counted forward, the trend is approximated by a polynomial w_ν^m of fixed degree $m(= 1, 2, \ldots)$, which remains constant across all ν segments and satisfies $m \leq s - 1$ to ensure the determinacy of all polynomial coefficients within each ν segment. These coefficients are computed independently for each ν segment using the least squares method to minimize variance during forward counting,

$$F^{far}(\nu, s) = \frac{1}{s} \sum_{i=1}^{s} [Y^{far}(i; \nu, s)]^2, \ \nu = 1, 2, \ldots, N_s, \tag{5.65}$$

where

$$Y^{far}(i; \nu, s) = Y[(\nu - 1)s + i] - w_\nu^m(i). \tag{5.66}$$

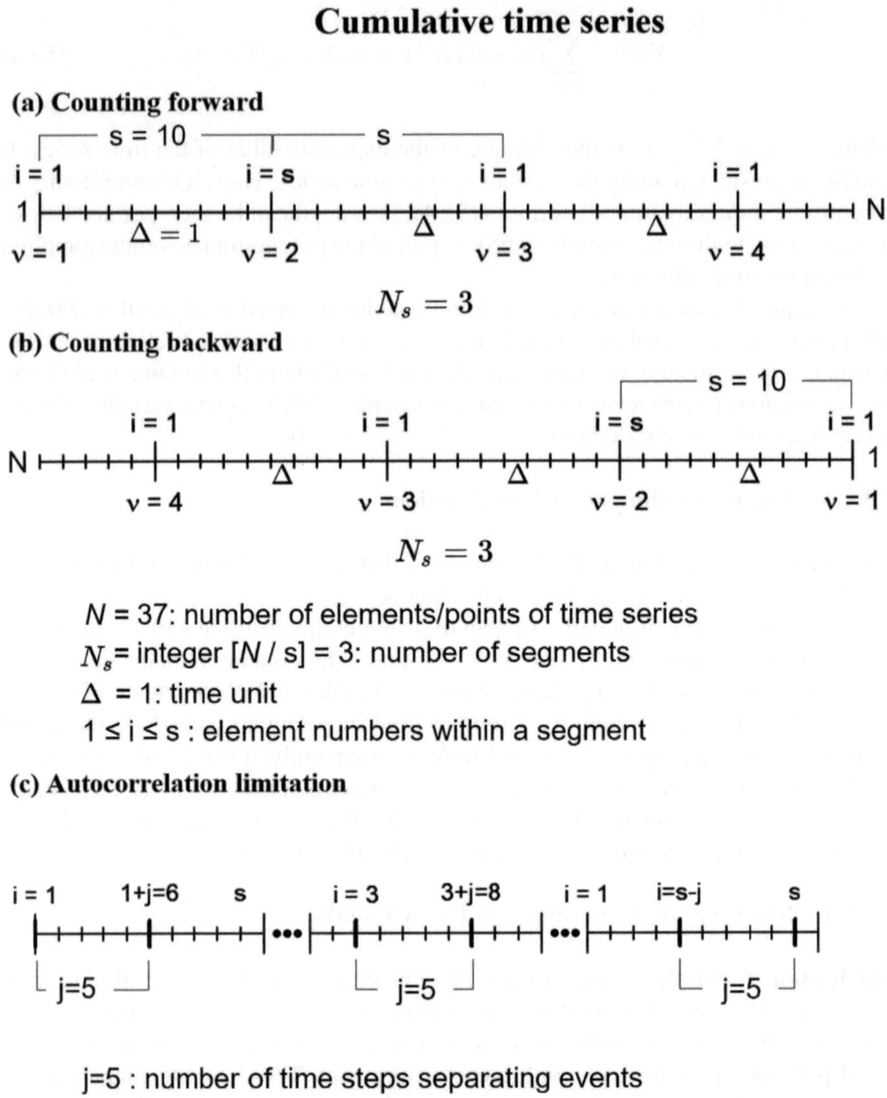

Fig. 5.8 Depiction of temporal interval segmentation: **a** progressive counting and **b** regressive counting, serving as the foundational mechanism for integrated time series within the MF-DFA framework. This segmentation methodology can be readily adapted (refer to item (c) configured for progressive enumeration) to the MF-DACA, which is discussed subsequently. This adaptation introduces a higher level of complexity, as it involves the simultaneous consideration of dual events at times i and $i + j$, thus separated by j intervals. Consequently, this imposes a general limitation on the range of i, specifically $1 \leq i \leq s - j$, with $0 \leq j \leq s - 1$

For backward counting, the polynomial coefficients are likewise determined using the least-squares method, minimizing the modified variance.

$$F^{back}(\nu, s) = \frac{1}{s}\sum_{i=1}^{s}[Y^{back}(i; \nu, s)]^2, \ \nu = N_s + 1, N_s + 2, \ldots, 2N_s, \tag{5.67}$$

where

$$Y^{back}(i; \nu, s) = Y[N - (\nu - N_s)s + i] - w_\nu^m(i), \tag{5.68}$$

also for N_s segments counting here backward; apparently, the range of i and the order of polynomial m remain unchanged.

Both single-segment variances $F^{\text{far}}(\nu, s)$ and $F^{\text{back}}(\nu, s)$, which quantify the magnitude of second-order fluctuations, serve as the foundation for analyzing multiscale fluctuations and autocorrelations. Thus, the total variance, which measures second-order fluctuations, takes the form,

$$F^2(s) = \frac{1}{2N_s}\left\{\sum_{\nu=1}^{N_s} F^{far}(\nu, s) + \sum_{\nu=N_s+1}^{2N_s} F^{back}(\nu, s)\right\}. \tag{5.69}$$

We refer to both when examining higher-order fluctuations and autocorrelations.

The polynomial degree, denoted by m, is often explicitly referenced in the method name for clarity, such as in MF-DFAm. Consequently, we can discuss the primary (linear) MD-DFA1 method, the secondary MF-DFA2 method, the tertiary MF-DFA3 method, etc. Ensuring the stability of the results obtained in different methods yields insight into the minimum order of methods required for the singularity spectrum (local exponents) to stabilize, which in our case was achieved at $m = 3$. Unfortunately, stabilizing the singularity spectrum can only be achieved numerically and necessitates separate consideration for each multifractal case.

Autocorrelation function. At the end of this step, we introduce the non-linear time-dependent process autocorrelation function (PACF) in the form,

$$C^{far}(j; \nu, s) = \frac{1}{s-j}\sum_{i=1}^{s-j} \mid Y^{far}(i; \nu, s) \mid \cdot \mid Y^{far}(i + j; \nu, s)) \mid, \tag{5.70}$$

where

$$Y^{far}(k; \nu, s) = Y[(\nu - 1)s + k] - w_\nu^m(k), \ k = i, i + j, \tag{5.71}$$

here $\nu = 1, 2, \ldots, N_s$, and j indexes the number of time steps separating the (detrended) process at time i and at time $i + j$, i.e., $1 \le i \le s - j$, within the same segment ν. This equation extends Eq. (5.65), where the time series is counted forward. We use backward counting in analogy to Eq. (5.67), get

$$C^{back}(j; \nu, s) = \frac{1}{s-j} \sum_{i=1}^{s-j} | Y^{back}(i; \nu, s) | \cdot | Y^{back}(i+j; \nu, s) |, \tag{5.72}$$

where

$$Y^{back}(k; \nu, s) = Y[N - (\nu - N_s)s + k] - w_\nu^m(k), \; k = i, i+j. \tag{5.73}$$

here $\nu = N_s + 1, N_s + 2, \ldots, 2N_s$. As can be seen, both formulas (5.72) and (5.72) transform for $j = 0$ as they should into (5.65) and (5.67), respectively. In the following, we will name the extension of MF-DFA as a Multifractal Detrended Autocorrelation Analysis (MF-DACA).

Hence, the non-linear time- and scale-dependent PACF takes the form of the following arithmetic average,

$$C(j; s) = \frac{1}{2N_s} \left\{ \sum_{\nu=1}^{N_s} C^{far}(j; \nu, s) + \sum_{\nu=N_s+1}^{2N_s} C^{back}(j; \nu, s) \right\}. \tag{5.74}$$

The autocorrelation function analogous to the above was considered in the work [6].

Equation (5.74) is an autocorrelation extension in the step time domain of variance (5.69) or the q-fluctuation function, $[F_q(s)]^q = \langle [F^2(s)]^{q/2} \rangle$ for $q = 2$, given by Eq. (5.76) below.

The autocorrelations discussed here pertain to intra-segment relationships, where moments i and $i + j$ reside within the same segment ν. While extending beyond this constraint is feasible, it substantially complicates the patterns described in Eqs. (5.72)–(5.74).

It is crucial to highlight that the extended Multifractal Detrended Fluctuation Analysis (MF-DFA), namely Multifractal Detrended Autocorrelation Analysis (MF-DACA), offers the opportunity to directly assess whether a time series or process exhibits nonlinear, time-dependent autocorrelation. This autocorrelation can be readily contrasted with the analogous bilinear function $K(j; s)$ shown in Fig. 5.9, differing only by the absence of absolute value operations in its definition.

This enables us to investigate whether nonlinear long-term autocorrelations underlie the observed multifractality. It is worth noting that the red dashed horizontal line in Fig. 5.9 signifies an uncorrelated process, resulting in a constant autocorrelation function,

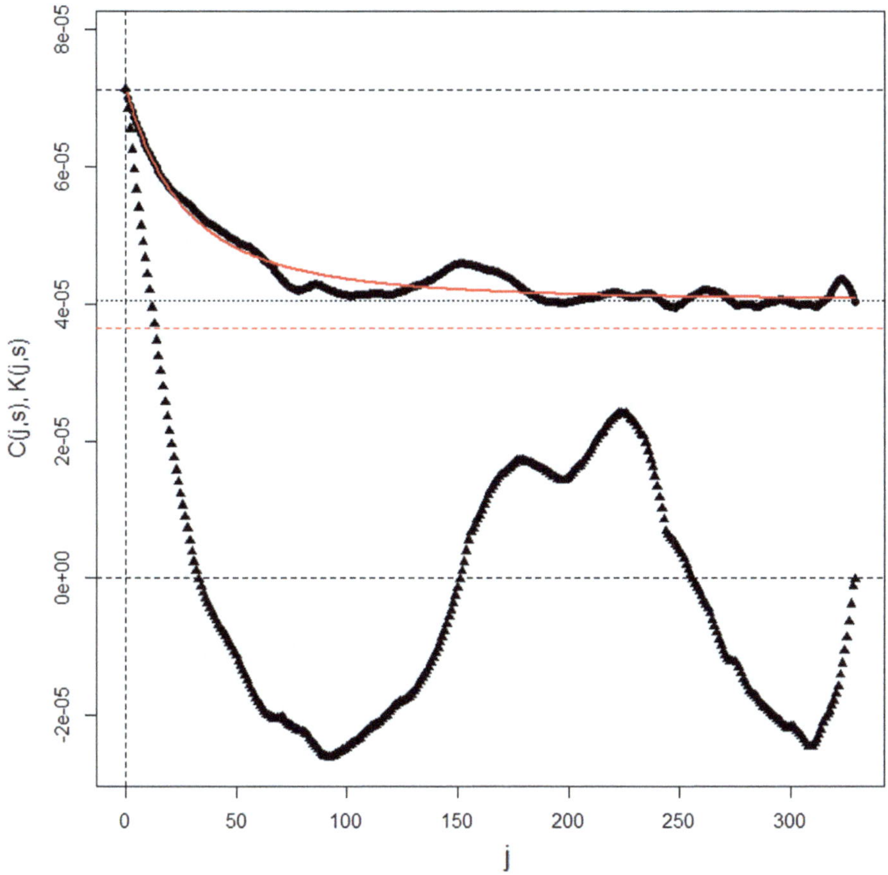

Fig. 5.9 Comparison of the non-linear autocorrelation function $C(j;s)$ defined by the formula (5.74) (the upper black empirical curve) with its bilinear counterpart $K(j;s)$ (the lower black empirical curve made of triangles explained in the main text) depending on j, for $s = 400$. There is an evident slowdown in the relaxation of $C(j;s)$ compared to $K(j;s)$. Furthermore, distinct stationary oscillations of $K(j;s)$ versus j are visible, but they are absent in $C(j;s)$ (apart from minor fluctuations). The red power-law curve, represented by the formula $c(j) = \frac{A}{(j_0+j)^{1+\gamma}} + B$, with exponent $\gamma = 1.621$, amplitude $A = 2.222$, shift parameter/threshold $j_0 = 70.94$, and the background parameter $B = 4.048 \times 10^{-5}$, indeed fits well with the upper black empirical curve (as indicated by the sufficiently small square root of residual sum-of-squares of 2.768×10^{-5}). These quantities may also depend on the scale of s. Furthermore, the black dotted curve defines the plateau of the upper black curve, which is located at a height determined by B. The red dashed horizontal straight line indeed represents the asymptotic line for the case of entirely uncorrelated events (given by Eq. (5.75)). The black dashed upper line is drawn at height $C(j = 0; s) = \langle F^2(s) \rangle$. Despite the small amplitudes of $C(j;s)$ and $K(j;s)$, they allow for the crucial distinction between both types of autocorrelations

$$C_{uncorr}(s) = \left\{ \frac{1}{100} \sum_{i=201}^{300} \frac{1}{2N_s} \left\{ \sum_{\nu=1}^{N_s} \mid Y^{far}(i; \nu, s) \mid + \sum_{\nu=N_s+1}^{2N_s} \mid Y^{back}(i; \nu, s) \mid \right\} \right\}^2 . \tag{5.75}$$

The expression above was derived directly from (5.74), assuming that $i \gg 1$ and the process is asymptotically non-autocorrelated. Consequently, even with a small j, the factorization of expression (5.74) remains unimpeded, signifying no observed asymptotic dependence of the autocorrelation function on j.

However, since the black dotted horizontal line surpasses the red dashed horizontal line, we infer the presence of residual autocorrelation, indicating non-factorizing autocorrelation in the process. This suggests the existence of structure induced by nonlinear autocorrelation.

Let us also note that, in the light of the considerations made in Sect. 3.7.7, bilinear short-term autocorrelations can coexist with long-term nonlinear autocorrelations, as shown in Fig. 5.9. For this, it is sufficient for the autocorrelation of signs to be of the short-term type. However, this does not answer the fundamental question about the relationship between multifractality and nonlinear autocorrelations.

> The extension of MF-DFA to MF-DACA, which integrates autocorrelations into direct multifractal analysis, represents a notable advantage.

5.3.1.5 Stage Four: Q-Fluctuation Function and Q-Autocorrelation Function

Definition of q-fluctuation function. The q-fluctuation function is defined as follows:

$$F_q(s) \stackrel{\text{def.}}{=} \left\{ \frac{1}{2N_s} \left\{ \sum_{\nu=1}^{N_s} \left[F^{far}(\nu, s)\right]^{q/2} + \sum_{\nu=N_s+1}^{2N_s} \left[F^{back}(\nu, s)\right]^{q/2} \right\} \right\}^{1/q} , \tag{5.76}$$

where q is an actual number. For $q = 2$ we obtain $\left[F_q(s)\right]^q = F^2(s)$. In case $q \to 0$, we will return in the next paragraph.

Our main objective is to examine the relationship between the q-fluctuation function and the interval size s across various values of q.

In the limit $q \to 0$, the expressions on the left-hand side (LHS) and right-hand side (RHS) of Eq. (5.76) can be reformulated into alternative, more insightful forms,

$$\begin{aligned} &LHS: \\ &\left[F_q(s)\right]^q = \exp\left[q \ln F_q(s)\right] \approx 1 - q \ln F_q(s) \\ &RHS: \\ &\frac{1}{2N_s}\sum_{\nu=1}^{2N_s}\exp\left[\frac{q}{2}\ln F^2(\nu, s)\right] \approx 1 - \frac{q}{4N_s}\sum_{\nu=1}^{2N_s}\ln F^2(\nu, s). \end{aligned} \quad \begin{aligned} (5.77) \\ (5.78) \end{aligned}$$

Since $LHS = RHS$ so, ultimately,

$$F_{q\to 0}(s) = \exp\left[\frac{1}{4N_s}\sum_{\nu=1}^{2N_s}\ln F^2(\nu, s)\right], \tag{5.79}$$

which is the expression we are looking for.

Definition of the lag-dependent q-autocorrelation function. We define the non-linear time/lag- and scale-dependent q-autocorrelation function, analogous to the definition of the q-fluctuation function provided by Eq. (5.76), in the following form:

$$C_q(j; s) \stackrel{\text{def.}}{=} \left\{\frac{1}{2N_s}\left\{\sum_{\nu=1}^{N_s}\left[C^{far}(j; \nu, s)\right]^{q/2} + \sum_{\nu=N_s+1}^{2N_s}\left[C^{back}(j; \nu, s)\right]^{q/2}\right\}\right\}^{1/q}, \tag{5.80}$$

where for $q = 2$ we obtain $\left[C_q(j; s)\right]^q = C(j; s)$, as we used in Eq. (5.74) a simplified notation.

Note that, akin to the q-fluctuation function described above for $q \to 0$, we can reformulate Eq. (5.74) into an equivalent form

$$C_{q=0}(j; s) = \exp\left[\frac{1}{4N_s}\sum_{\nu=1}^{2N_s}\ln C(j; \nu, s)\right]. \tag{5.81}$$

We integrate Eq. (5.74) over a time lag j to explore the scaling behavior of the q-autocorrelation function on scale s, with an exponent dependent on q. Our primary focus lies in understanding its nonlinear dependence on q.

As evident, MF-DACA constitutes a multiscale approach. Analogous to MF-DFA, it incorporates autocorrelation, thereby introducing the multiscale aspect of fluctuations. Hence, these approaches complement each other, providing a holistic perspective on multifractality. It's worth mentioning that a statistical perspective of multifractality, including the cascade model and a strategy based on the Hölder exponent, are also available for a comprehensive understanding.

5.3.1.6 Stage Five: Scaling of the q-Fluctuation and q-Autocorrelation Functions

Scaling of the q-fluctuation function. We are interested in investigating the power-law dependence of the q-fluctuation function on s, which is characterized by an expression of the form,

$$F_q(s) \propto s^{h(q)}, \tag{5.82}$$

where $h(q)$ represents the so-called generalized Hurst exponent (we have included considerations on this exponent and also on the Rényi dimensions in Appendix 20); the proportionality coefficient, not explicitly stated here, may generally depend on q and could be a slowly varying function of scale s. (This coefficient should satisfy additional properties discussed later.) The function $F_q(s)$ was presented in Fig. 5.10 for three different values of q.

We examine the relationship described by Eq. (5.82) when the time series $\{x_k\}_{k=1}^{N}$ exhibits nonlinear long-term correlations characterized by a power-law decay. As introduced earlier, the extended MF-DFA, namely MF-DACA, directly investigates this phenomenon.

In Fig. 5.10 (left panel), we depict the plot of Eq. (5.82) using MF-DFA based on Alphabet Inc. 5-minute stock quotes. on the NASDAQ. Consequently, based on MF-DACA, we present the plot of Eq. (5.84) in the right panel (in the same figure). The

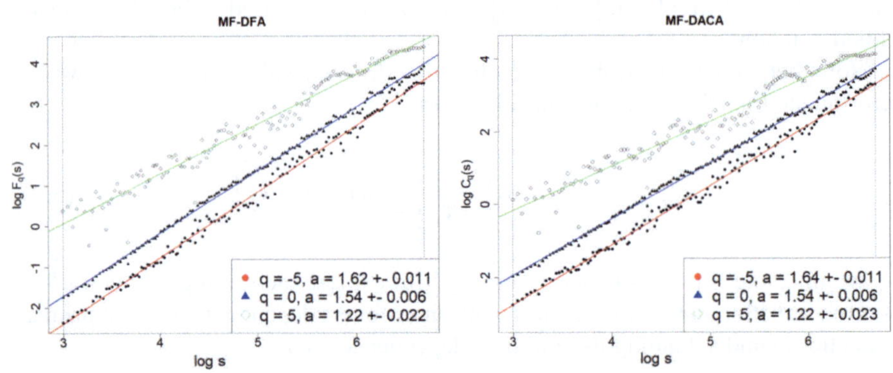

Fig. 5.10 Dependence of the q-fluctuation function, $F_q(s)$, given by Eq. (5.82) (left panel) and the q-autocorrelation function, $C_q(s)$, given by Eq. (5.84) (right panel), on the scale s (that is, the time scale horizon) are presented in log-log form for three selected values of $q = -5, 0, 5$, for the Google quotations shown in Fig. 5.7. The differences in the straight-line slopes (adjusted to the empirical data) are relatively small (within the adopted scale of the figure) but distinct (i.e., more significant than the standard deviations of the corresponding slopes, see $a = h(q)$ values in the legend). Of course, this approach is an instruction that allows us to directly determine the generalized Hurst exponent $h(q)$ (see Eq. (5.82) and Fig. 5.11). The systematic disparities observed between the graphical representations in both panels are minimally significant; nonetheless, they manifest with greater clarity for $q < 0$ as depicted in Figs. 5.11 and 5.12

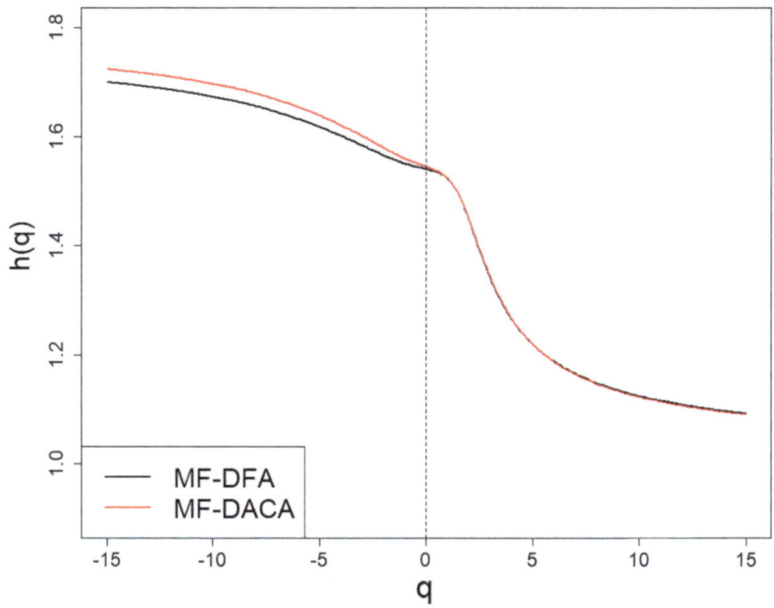

Fig. 5.11 Dependence of the generalized Hurst exponent $h(q)$ on q for the empirical data presented in Fig. 5.7. One can see (on the scale of the figure) small but systematic differences between the results obtained under MF-DFA and MF-DACA for $q < 0$. At the point, $q = 1$, both curves coincide, which is the result of the standardization of generalized scale exponents $\tau(q)$ and $\tau_{AC}(q)$ (see upper panels Fig. 5.12) and in consequence the lower panels (in the same figure)

resulting dependence of the generalized Hurst exponent, $h(q)$, on q is illustrated in the subsequent Fig. 5.11. In addition, we include the corresponding $h_{AC}(q)$ defined below for comparative analysis.

For completeness, in Fig. 5.12, we compare (also for the 5-minute stock quotations of Alphabet Inc. on the NASDAQ) the multifractal characteristics: $\tau(q)$ (upper panels), $D(q)$ (lower left panel), and $\alpha(q)$ (lower right panel) with the corresponding quantities: $\tau_{AC}(q)$, $D_{AC}(q)$, and $\alpha_{AC}(q)$ defined below. It is noteworthy that Rényi dimensions necessitate separate numerical calculations[16] near the point $q = 1$.

The formula defined for the generalized scaling exponent, $\tau(q)$, is as follows:

$$\tau(q) \overset{\text{def.}}{=} q\, h(q) - h(q = 1). \tag{5.83}$$

[16] The challenge in numerically computing $D(q = 1)$ directly relates to the definition of $D(q) = \tau(q)/(q - 1)$. We expand $\tau(q)$ into a Taylor series at $q = 1$ to address this challenge. Consequently, $\tau(q) = \tau(1 + (q - 1)) \approx \tau(q = 1) + (q - 1)\,\alpha(q = 1) = (q - 1)\,\alpha(q = 1)$ due to the derivative $\alpha(q) = \frac{d\tau(q)}{dq}$ and the standardization $\tau(q = 1) = 0$. Thus, we derive $D(q = 1) = \alpha(q = 1)$. A similar derivation applies to $D_{AC}(q)$.

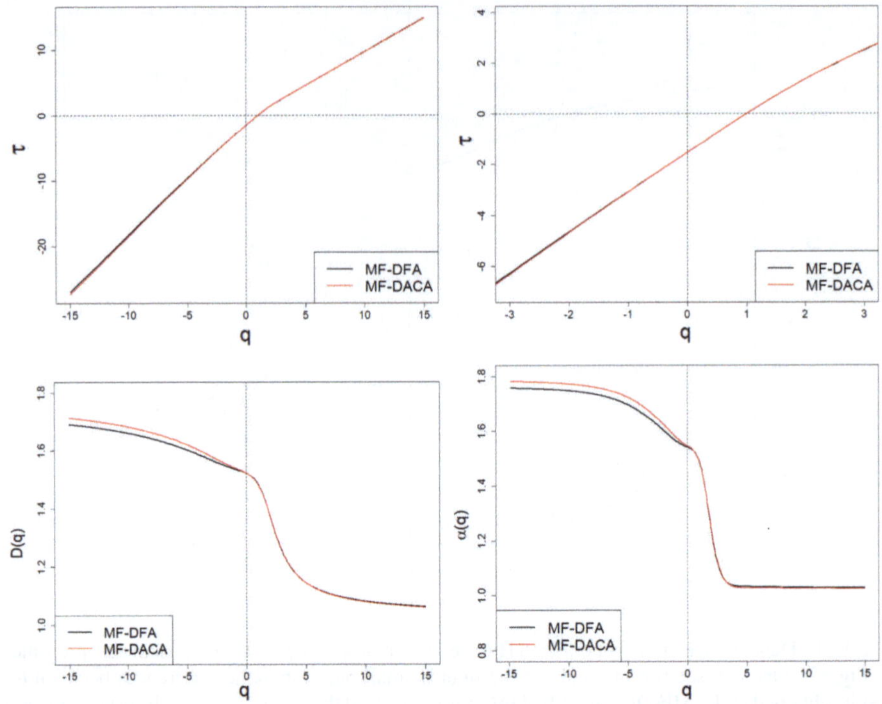

Fig. 5.12 Significant characteristics of multifractality for the empirical data presented in Fig. 5.7. The black curves of all graphs are based on the q-fluctuation function (MF-DFA), while the red ones are based on the q-autocorrelation function (MF-DACA). (i) Upper panels: dependence of the generalized exponent $\tau(q)$ on $-15 \leq q \leq 15$ (left panel) and $-3 \leq q \leq 3$ (right panel). The nonlinear behavior of both curves is visible. (ii) Lower panels (for $-15 \leq q \leq 15$): dependence of Rényi dimensions $D(q)$ (left panel) and $\alpha(q)$ (right panel) versus q. The differences between black and red curves are residual for all graphs

This exponent is calibrated as follows: $-\tau(q = 0) = h(q = 1) = D(q = 0)$ (see the multipanel Fig. 5.12) and becomes zero at $q = 1$. This standardization procedure allows one to compare multifractalities, such as those obtained by MF-DFA and MF-DACA. The autocorrelation counterpart of $\tau(q)$, also in a standardized form, is defined by Eq. (5.86) below.

Finally, we determined the singularity spectrum, $f(\alpha)$, (see Sect. 5.1.5), and compared it in Fig. 5.13 with the corresponding singularity spectrum $f_{AC}(\alpha)$ defined below.

Scaling of the q-autocorrelation function. We now define (by analogy with Eq. (5.76)) the full/integral form of the q-autocorrelation function,

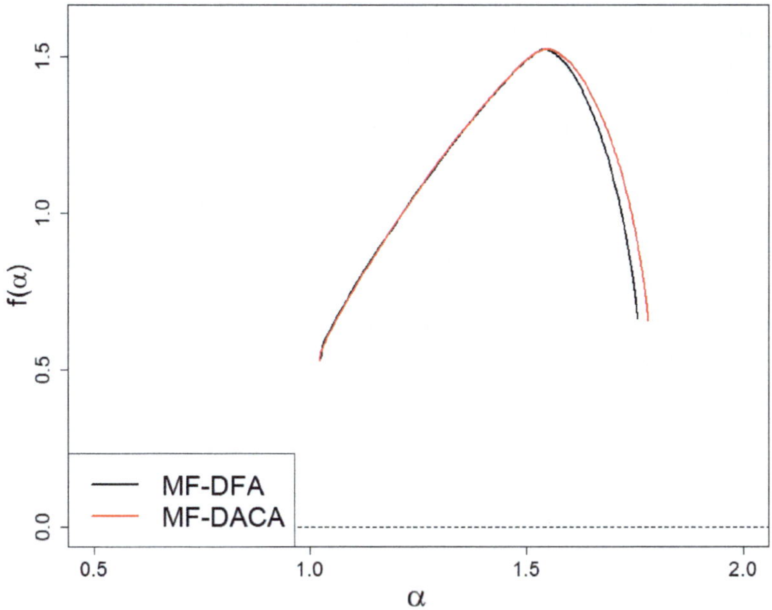

Fig. 5.13 Dependence of the singularity spectrum f on the singularity α for the five-minute quotes of Alphabet Inc. on the NASDAQ stock exchange in the range $1.0 \leq \alpha \leq 1.80$. The spread of these curves is determined (to a good approximation) by the vertical spread of the $\alpha(q)$ and $\alpha_{AC}(q)$ curves shown in the multipanel Fig. 5.12. The black curve is based on the q-fluctuation function, while the red one is on the q-autocorrelation function. For $\alpha \leq \alpha_{AC}(q=0)$, both the left branches of the spectrum of singularities $f(\alpha)$ and $f_{AC}(\alpha)$ overlap in the resolution power of the graph. In contrast, there is a little systematic difference between the two right branches for $\alpha > \alpha(q=0)$. This means that the impact of small (in terms of amplitudes) fluctuations and correlations is slightly different for these branches. All these behaviors result from the relationship $\alpha(q)$ and α_{AC} versus q shown in the multipanel Fig. 5.12. The result presented here means that, to the lowest-order approximation, both the MF-DFA and MF-DACA approaches are equivalent. However, as we do not have mathematical proof, this must be checked every time we examine empirical data

$$C_q(s) = \frac{1}{s}\sum_{j=0}^{s-1} C_q(j; s), \tag{5.84}$$

enabling the study of its power-law dependence on the s scale.

We examine the dependence of $C_q(j; s)$ on s for fixed values of j and then the integral $C_q(s)$ versus s for three characteristic values of q.

Therefore, we are looking for the relationship of the form,

$$C_q(s) \propto s^{h_{AC}(q)}, \tag{5.85}$$

where $h_{AC}(q)$ is (in general) a nonlinear function of q, which we can obtain from empirical data (in analogy to the Hurst exponent $h(q)$ in Eq. (5.82)). Not explicitly presented, the proportionality coefficient may depend on q and be a slowly varying function of s. The dependences $C_q(s)$ versus s for three example values of q are presented in Fig. 5.10 (right panel).

Now we can find in the standardized form, in analogy to $\tau(q)$,

$$\tau_{AC}(q) \stackrel{\text{def.}}{=} q\, h_{AC}(q) - h_{AC}(q=1), \tag{5.86}$$

assuming our usual calibration $-\tau_{AC}(q=0) = h_{AC}(q=1) = D_{AC}(q=0)$. On this basis, we build all the characteristics of multifractality.

Comparison of the corresponding straight lines in Figs. 5.14 and 5.10 (right panel) suggest (at most) a very weak dependence of the slope of the straight lines on j. We also reach a similar conclusion for larger values of j. Therefore, we assume that

$$C_q(j;s) = A_q(j) s^{h_{AC}(q)} \tag{5.87}$$

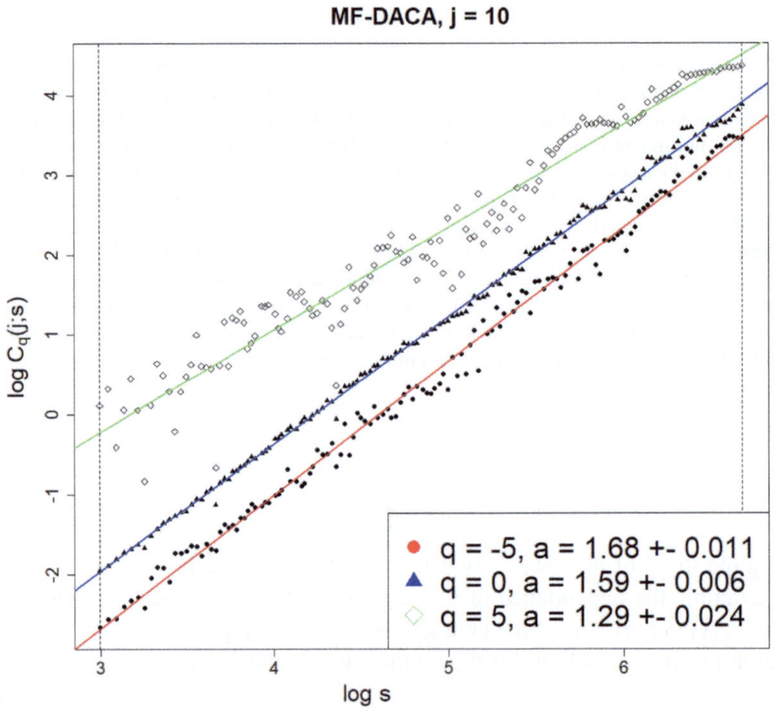

Fig. 5.14 Dependence of the partial q-autocorrelation function, $C_q(j=10;s)$, given by Eq. (5.80), versus s on the log-log scale for three values of q. As one can see, this typical dependence is a power-law increase, as we expected. We obtain similar results for larger values of j. This result is the starting point for the study of multifractality from the perspective of autocorrelations

hence

$$C_q(s) = \hat{A}_q\, s^{h_{AC}(q)}, \tag{5.88}$$

where

$$\hat{A}_q = \frac{1}{s}\sum_{j=0}^{s-1} A_q(j); \tag{5.89}$$

moreover, we no longer consider the (presumably weak) dependence of $\hat{A}_q$ on the scale s.

Additionally, it is worth noting that $C_{q=2}(j; s) = c(j) = \frac{A}{(j_0+j)^{1+\gamma}} + B$, where $c(j)$ was referenced in the caption of Fig. 5.9 with a simplified notation $c(j)$ instead of $c(j; s)$. The self-consistency condition mandates that $A = const_1\, s^{h_{AC}(q=2)}$ and $B = const_2\, s^{h_{AC}(q=2)}$, thus resulting in $A_{q=2}(j) = \frac{const_1}{(j_0+j)^{1+\gamma}} + const_2$.

The comparative analysis delineated herein juxtaposes the multifractal characteristics—$h(q)$ with $h_{AC}(q)$, $\tau(q)$ with $\tau_{AC}(q)$, $D(q)$ with $D_{AC}(q)$, $\alpha(q)$ with $\alpha_{AC}(q)$, and $f(\alpha)$ with $f_{AC}(\alpha_{AC})$—intimates that the multifractality discerned via the fluctuation methodology and the autocorrelation paradigm could be regarded congruent, although within a margin of reasonable precision. Nevertheless, in the absence of a definitive mathematical proof substantiating this equivalence, it is prudent for empirical investigations to employ both the Multifractal Detrended Fluctuation Analysis (MF-DFA) and the Multifractal Detrended Autocorrelation Analysis (MF-DACA) techniques. This is a significant result from the view of multifractality, as the importance/contribution of fluctuations and nonlinear autocorrelations is the same.

It can be seen that MF-DACA was developed based on two approaches to multifractality: the fluctuation method proposed by Kantelhardt et al. [7] and the autocorrelation method suggested by Bogachev, Eichner, and Bunde [6], along with the foundational work of Grassberger and Procaccia [4] as well as Schuster [1].

The paper by Rak and Zięba accurately describes Multifractal Flexibly Detrended Fluctuation Analysis (MFF-DFA) [8] and its distinguishing features. However, it is essential to clarify that MFF-DFA is not merely an alternative approach but a refinement of traditional MF-DFA. Although both methods share similar foundational principles, MFF-DFA enhances the analysis by introducing detrending polynomials tailored to specific time intervals, thereby discretizing empirical data more effectively. This adaptability allows for varying polynomial degrees in different intervals, enhancing the method's versatility in analyzing complex datasets. Moreover, the flexibility in the polynomial application can extend to other analysis methods, such as MF-DACA, offering broader applicability in exploring polynomial fits across different methodologies. In MF-DFA, the subsequent time intervals and the corresponding datasets partially overlap, making polynomial fits more stable and accurate.

Based on the preceding discussion, it can be inferred that approaches to multifractality may roughly be categorized into the following partially overlapping groups: fluctuation multifractality, autocorrelation (acknowledging the presence of cross-correlation multifractality, although not discussed in our textbook), statistical methods (basing only on the statistical sum introduced in Sect. 5.3.2 below), and those based on the Hölder function (or time-dependent Hurst exponent). In Sects. 8 and 11, we explore it using representative examples, and in this, the model of diffusion on a Gaussian substrate, multifractality generated by fluctuating inter-event time intervals characteristic of jump-diffusion processes.

5.3.2 *Relation Between the Fluctuation and Partition Functions*

To address the inquiry regarding the connection between the q-fluctuation function and the partition function, we will employ the axiomatic definition of the latter, the scaling form of the q-fluctuation function (5.82), and the interpretation of the generalized scaling exponent as provided by Eq. (5.83). Thus, we can propose a standardized partition function in the following form:

$$Z_q(s) \stackrel{\text{def.}}{=} Z_{q=0}(s) \cdot \left[F_q(s)\right]^q , \tag{5.90}$$

where $Z_{q=0}(s) = s^{-h(q=1)}$ and $\left[F_q(s)\right]^q = s^{qh(q)}$. The proportionality coefficient, mentioned when discussing Eq. (5.82), is assumed to be equal to 1 for simplicity, as we are mainly interested in the scaling exponent. In general, it can be a slowly varying function of s and depend on q so that it equals 1 for $q = 0$ and 1.

It can be seen that Eq. (5.90) leads to a modification/extension of axiom (2) from Sect. 5.1.1 of the partition function. That is, axiom (2) is satisfied here only in the particular case $h(q = 1) = D(q = 0) = 1$ and $s = 1/N$. However, if the time series domain is fractal (or, from the point of view of MF-DFA and/or MF-DACA, imitates a fractal), then $D(q = 0) \neq 0$. Therefore, we expand axiom (2) to the form: $Z_{q=0} = s^{-D(q=0)}$ and denote axiom (2').

We now consider the opportunity to generalize/modify axioms (3) as outlined in Sect. 5.1.1. This is feasible because $Z_q(s)$ can be expressed in the following manner:

$$Z_q(s) = \frac{Z_{q=0}(s)}{2N_s} Z_q^{nonst}(s) = \frac{Z_{q=0}(s)}{2N_s} \sum_{\nu=1}^{2N_s} [b_\nu(s)]^q , \tag{5.91}$$

where thermodynamic (unnormalized) probability,

$$b_\nu(s) = \begin{cases} \left[F^{far}(\nu, s)\right]^{1/2}, & 1 \leq \nu \leq N_s \\ \left[F^{back}(\nu, s)\right]^{1/2}, & N_s + 1 \leq \nu \leq 2N_s. \end{cases} \tag{5.92}$$

Thus, $Z_q^{nonst}(s) = \sum_{\nu=1}^{2N_s} [b_\nu(s)]^q$ serves as an analog to the statistical thermodynamic partition function, i.e., it is a non-standardized partition function. The definition (5.91) allows the partition function to be standardized so that $Z_{q=1}(s) = 1$ is always there (that is, it satisfies axiom (1)), which makes comparison of different multifractality easier. Therefore, the coarse-grained normalized probabilities $p_\nu(s)$, escort distribution of b_ν, can now be defined as follows:

$$p_\nu(s) = \frac{Z_{q=0}(s)}{2N_s} \cdot b_\nu(s), \tag{5.93}$$

where $1 \leq \nu \leq 2N_s$ indexes the segments (the index ν can now be observed to serve various roles). Hence,

$$Z_{q\to-\infty}(s) \approx \frac{Z_{q=0}(s)}{2N_s} \cdot \left[b^{min}(s)\right]^q$$
$$Z_{q\to\infty}(s) \approx \frac{Z_{q=0}(s)}{2N_s} \cdot \left[b^{max}(s)\right]^q, \tag{5.94}$$

where $b^{min,max}(s)$ denote, respectively, the minimum and maximum value of $b_\nu(s)$ for a given number of segments N_s and scale s. Hence, $p^{min,max}(s) = \frac{Z_{q=0}(s)}{2N_s} \cdot b^{min,max}(s)$, is the minimum or maximum value, respectively of coarse-grained probabilities. The equations in (5.94) extend/modify the axiom (3), so we denote it as an axiom (3').

The consequence of the equations in (5.94) is the appropriate and convenient forms of the extreme values $\alpha(q) = \frac{d\tau(q)}{dq} = \frac{d}{dq}\frac{\ln Z_q(s)}{\ln s}$. Namely,

$$\alpha_{\text{min,max}} = \frac{\ln b^{\text{max,min}}}{\ln s}, \tag{5.95}$$

because $\alpha_{\text{min}} < \alpha_{\text{max}}$, which results from the graph of the concave function $\tau(q)$ versus q in Fig. 5.12. These extreme values constitute the limits/edges of the concave spectrum of dimensions $f(\alpha)$ versus α.

Continuing along this line of reasoning, one can introduce the energy analog of a system's microstate as $E_\nu = -\ln b_\nu$, where $1 \leq \nu \leq 2N_s$ indexes the analogs of the microstates of the system, where E_ν can be positive, negative, or vanishing.

We can also define, for both types of partition functions mentioned above, the escort distribution of b of order q,

$$P_q(\nu, s) = \frac{[b_\nu(s)]^q}{\sum_{\nu=1}^{2N_s} [b_\nu(s)]^q}. \tag{5.96}$$

A change in the order q will alter the weights by which the "events" $1 \leq \nu \leq 2N_s$ contribute to the escort distribution. The possibility of changing q serves as a scanning tool for the structure of both partition functions. This is evident from the limiting equations in (5.94). In both of these limiting cases, $P_{q\to\mp\infty}(\nu, s) \to 1$.

It is essential to clarify that we utilize another standardized partition function representation in addition to the two above; we will employ it in the next section.

By substituting $[F_q(s)]^q$ with $[C_q(s)]^q$, $h(q = 1)$ with $h_{AC}(q = 1)$, and replacing the exponent of $\tau(q)$ with $\tau_{AC}(q)$ in expression (5.90), we derive analogous results.

In conclusion, this subsection emphasizes that the MF-DFA method facilitates practical (quantitative) multifractal analysis of any random walk, particularly those exhibiting anomalous diffusion.

5.3.2.1 Comments on Definitions of Anomalous Diffusion and Multifractality

The more general definition of one-dimensional, stationary anomalous diffusion, in contrast to that explored in the literature, is as follows:

$$\left\langle [X(t_2) - X(t_1)]^2 \right\rangle \propto | t_2 - t_1 |^{\eta(|t_2 - t_1|)}, \tag{5.97}$$

wherein the anomalous diffusion exponent $\eta(|t_2 - t_1|)$ for the second moment delineates a function of $|t_2 - t_1|$, analogous to an Hölder exponent, which exhibits a gradual variation with $|t_2 - t_1|$. Moreover, the interrelation between this exponent and the order q of the absolute q-moment, expressed as $\eta(q; |t_2 - t_1|)$, typically manifests a non-linear dependence on q. Within the scope of our textbook, anomalous diffusions arise predominantly in scenarios where the generalized diffusion exponent, $\eta(q; |t_2 - t_1|)$, is invariant in terms of the temporal disparity $|t_2 - t_1|$. This underscores the inherent presence of time-independent multifractality. Nevertheless, the methodologies expounded in our textbook facilitate the exploration of time-dependent multifractality, particularly in contexts involving a slowly evolving temporal dependence or when the system exhibits stationarity over sufficiently long time intervals.

The definition of multifractality is manifestly articulated within scholarly discourses. In particular, economic analyses conventionally acknowledge multifractality solely under the condition that the temporally variant Hurst exponent (corresponding to the Hölder exponent) predominates the dynamics of the time series [9]. Conversely, in the physical literature, multifractality is characterized by an exponent that exhibits nonlinear dependence on the order of fluctuations, as outlined by the nonlinear scaling fluctuation law or nonlinear multi-scaling [10].

The bifurcation of multifractality into two definitions is articulated and generally noninterchangeable. This text adopts the physical interpretation of multifractality, as expounded in the latter definition. However, in Sect. 3.7.1, we explicated the conditions under which the antiquated exponent engenders statistical (physical) multifractality. This delineation relies on intricate statistical methodologies, in which the parameters that define the underlying statistical distributions are considered stochastic variables.

Control and Review Questions and Exercises

5.1 Exercise
List the sources of multifractality and the characteristics/signatures of multifractality.

5.2 Question
Can a single singularity present in an invariant measure lead to bi-fractality, and if so, why?

5.3 Problem and Questions
Will bi-fractality turn into tri-fractality if the domain of the invariant measure is extended to the segment $[-1, 1]$ and the invariant measure to the form $\rho(x) = \frac{1}{2}(1 - \gamma) \mid x \mid^{1-\gamma}$, where $0 < \gamma < 1$? Will this happen if the left-sided singularity differs from the right-sided one and both positive exponents of the powers are still less than 1? If so, what will the spectrum of fractal dimensions look like then?

5.4 Exercise and Question
Discuss the role of statistical sum in the construct of multifractality, for example, for the measure mentioned in item (2). What role does the singular component play in this sum?

5.5 Question
Does the invariant measure given by Eq. (5.1) allow the introduction of multifractality, for which the domain of the variable q is given by $q \in [0, 1] \bigcup \infty$?

5.6 Exercise and Question
Explain the role of Saddle Point Approximation in multifractality extraction. When can SPA not be used?

5.7 Exercise
Provide the properties of the spectrum of dimensions.

5.8 Questions
What form of partition function leads to multifractality and why?

5.9 Questions
What does the linear and non-linear relationship of the scaling exponent $\tau(q)$ versus q mean? What does the asymptotic linear dependence on q of this exponent mean?

5.10 Exercise and Questions
Provide the physical meaning of fractal dimensions: $D(q=0)$, $D(q=1)$, and $D(q=2)$. Under what conditions is $D(q=0)$ a natural number, and under what conditions is it a fractional number?

5.11 Exercise
Find the relationship between $\tau'(q')$ and $\tau'(q)$. Having a concave curve $\tau(q)$ versus q, please draw these three curves.

5.12 Problem
Prove that in the case where the output curve is of the concave type, its counterparts are either of the convex, concave, or straight line type, depending on whether $\mid \tau(q=0) \mid$ is less than 1, greater than 1 or equal to 1.

5.13 Problem and Questions
What shape $\tau(q)$ versus q corresponds to the arms of $f(\alpha)$ versus α oriented upwards and which oriented downwards and why? Which orientation is appropriate for building a thermodynamic formalism and why?

5.14 Questions
What is the basic function on which the MF-DFA formalism is built? How does this formalism differ from the MF-DACA formalism? What is the role of the q-fluctuation function in the MF-DFA formalism, and what is the role of the q-autocorrelation function in the MF-DACA formalism?

5.15 Questions
Can the MF-DFA and MF-DACA methods be used for multifractal analysis of non-stationary time series? If so, why? If not, why not?

5.16 Exercise
Show that the function given by (5.90) satisfies all the axioms (1), (2'), and (3') of the partition function.

5.17 Exercise
Prove that for standardized MF-DFA we have $\tau(q=0) = -h(q=1)$, $\tau(q=1) = 0$, $D(q=0) = h(q=1)$, $D(q=1) = \alpha(q=1)$, $f(\alpha(q=0)) = -\tau(q=0) = h(q=1) = D(q=0)$, and $f(\alpha(q=1)) = \alpha(q=1)$.

5.18 Question
Is it possible to find an example of a fluctuation multifractality for which there is no autocorrelation multifractality or vice versa?

5.19 Exercise and Question
Provide the definitions of standardized and non-standardized partition functions and present their scaling. Discuss the role of the q parameter. Is the analogy with thermodynamic β complete?

5.20 Exercise and Question

Define multifractal anomalous diffusion. Are MF-DFA and MF-DACA suitable for studying this type of diffusion?

5.21 Problem

Prove the hypothesis that MF-DFA is equivalent to MF-DACA or provide an example that disproves this hypothesis.

References

1. Schuster, H.G.: Deterministic Chaos—An Introduction. Second Revised Edition. VCH Verlagsgesellschaft, Weinheim (1988). https://doi.org/10.1002/BBPC.198800268
2. Dauxois, T., Ruffo, S., Arimondo, E., Wilkens, M. (eds.): Dynamics and Thermodynamics of Systems with Long-Range Interactions. Lecture Notes in Physics, vol. 602. Springer, Berlin (2002). https://link.springer.com/book/10.1007/3-540-45835-2
3. Tsallis, C.: Non-additive entropy and non-extensive statistical mechanics—an overview after 20 years. Braz. J. Phys. **39**(2a), 337–356 (2009). https://doi.org/10.1590/S0103-97332009000400002
4. Grassberger, P., Procaccia, I.: On the characterization of strange attractors. Phys. Rev. Lett. **50**, 346–349 (1983). https://doi.org/10.1103/PhysRevLett.50.346
5. Paladin, G., Valpiani, A.: Anomalous scaling laws in multifractal objects. Phys. Rep. **156**(4), 147–225 (1987). https://doi.org/10.1016/0370-1573(87)90110-4
6. Bogachev, M.I., Eichner, J.F., Bunde, A.: On the occurrence of extreme events in long-term correlated multifractal data sets. Pure Appl. Geophys. **165**, 1195–1207 (2008). https://doi.org/10.1007/s00024-008-0353-5
7. Kantelhardt, J.W., Zschiegner, S.A., Koscielny-Bunde, E., A. Bunde, A., S. Havlin, S.: Multifractal detrended fluctuation analysis of nonstationary time series. Physica A **316**, 87–114 (2002). https://doi.org/10.1016/S0378-4371(02)01383-3
8. Rak, R., Zięba, P.: Multifractal flexibly detrended fluctuation analysis. Acta Phys. Pol. B **46**, 1925–1938 (2015). https://doi.org/10.5506/APhysPolB.46.1925
9. Ayache, A., Véhel, J.L.: Generalized multifractional brownian motion: definition and preliminary results. In: Dekking, M., Véhel, J.L., Lutton, E., Tricot. C. (eds.) Fractals—Theory and Applications in Engineering, pp. 16–32. Springer, Berlin (1999). https://inria.hal.science/inria-00578657
10. Peltier, R.-F., Véel, J.L.: Multifractional brownian motion: definition and preliminary results. HAL Open Sci. Id: inria-00074045. https://inria.hal.science/inria-00074045

Chapter 6
Dispersive Transport and Diffusion

Abstract The Hierarchical Continuous-Time Random Flight (HCTRF) formalism enhances our understanding of anomalous transport and diffusion phenomena, particularly in amorphous and vitreous solids and conducting light-emitting organic polymers. This formalism is grounded in the canonical random trap or valley model, in which the disorder of the energy landscape is characterized by an exponential distribution. The corresponding mean residence times within the traps exhibit power-law behavior. This disorder pattern is typical of various amorphous materials, including those with commercial applications, which renders the HCTRF formalism highly applicable. Furthermore, by leveraging Extreme Value Theory (EVT), we demonstrate that rare or extreme stochastic events significantly influence the transport and diffusion processes of these materials. This chapter employs a hierarchical waiting-time distribution and investigates biased HCTRF on a regular lattice with random depth traps. We integrate this model with statistical EVT and the Weierstrass random walk framework. The HCTRF can be seen as a variant of the Continuous-Time Random Walk (CTRW) formalism [1–4], encompassing both random walks and flights.

6.1 Introduction and Motivation

The relaxation phenomena observed in condensed and soft matter relate to the transport and/or diffusion of atoms, particles, carriers, defects, excitons, and complexes [5] (and references therein). Transport and diffusion are paradigms for the irreversible behavior exhibited by many ordered and disordered systems. A defining feature of disordered systems is the intricate temporal pattern in which the Debye relaxation no longer adheres. The statement we quote from Scher and Montroll [6] aptly captures the direct correlation between the physics of anomalous transient-time dispersion in an amorphous substance and its practical applications. The development of photocopying machines and scanners has spurred research on amorphous materials, some exhibiting anomalous transport and diffusion properties.

M. Chorowski et al., *Anomalous Stochastics*, Understanding Complex Systems,
https://doi.org/10.1007/978-3-031-80392-5_6

The theory of carrier transport in certain amorphous insulators, like the commercially used vitreous As_2Se_3, and some amorphous charge-transfer complexes of organic polymers, such as the commercially used trinitrofluorenone mixed with polyvinyl carbazole (TNF-PVK), serves as exemplary instances for:

(i) Continuous-time random flights and walks and
(ii) Broad- or long-tailed waiting-time distribution between steps.

Based on point (ii), it can be inferred that the stochastic process exhibits an anomalous diffusion (as elaborated in the following). Consequently, the process and its velocity & increments demonstrate autocorrelations.

In a more precise context, the comprehensive representation of dispersive transport and diffusion [7], as observed in canonical experiments on transient current in an amorphous medium (induced by flash-light [6, 8–12] or voltage pulse [13] and related references), is indeed encapsulated by the formalism of Hierarchical Continuous-Time Random Flight [3, 6, 14–16]. In this framework, a distinction is made between particle flights (instantaneous) and walks (requiring time to move between traps)[1].

The primary objective of this chapter is to articulate this using the Extreme Value Theory (EVT) [2, 17, 18]. Such an approach reveals that rare or extreme (stochastic) events are crucial in regulating the transport and diffusion observed in the abovementioned materials.

This chapter is divided into three informal parts. The first part (Sect. 6.2) briefly presents the experiments of Sharfe, Gill, and Pfister, characterizing/defining the problem—they open the field of anomalous diffusion within canonical physics.

The second part (Sect. 6.3) encompasses the valley model of non-Brownian and non-Gaussian (anomalous) transport and diffusion. In other words, it extends beyond the Central Limit Theorem (CLT), delving into the Generalized Central Limit Theorem (GCLT) of Lévy-Khintchine and arriving at the generalized master equation (with memory) through the Continuous-Time Random Walk (CTRW) formalism.

Part three (Sect. 6.4) continues the preceding sections, focusing on accidental trapping and its influence on anomalous transport and diffusion. This exploration leads us seamlessly to the scaling equation and its solution in the Weierstrass process, which is highly beneficial in later sections of our textbook.

This chapter paves the way for multifractality induced directly by inter-event/pausing times fluctuations.

---

[1] We distinguish between particle flights and walks as the former are instantaneous while the latter need some time to move between the traps.

6.2 Sharfe, Gill, and Pfister Experiments

A notable experiment that observed anomalous dispersive transport was initially conducted in 1970 by Sharfe [8]. In 1974, G. Pfister furthered this research by examining the relationship and analyzing the pressure-dependent nature of anomalous transport in the sample [10]. Both authors investigated the frequency-dependent photoconductivity in amorphous $\alpha - As_2Se_3$ by measuring the decay over time of the photocurrent induced by a short-duration light pulse.

The measurement system is relatively simple. Its pivotal component is a sample composed of photosensitive semiconductors with hole conductivity between two electrodes. One of these electrodes (in gold) is semi-transparent. Pulsed illumination facilitates the generation of conductive holes in the sample. Subsequently, because of the (modest) potential difference applied to the electrodes, these holes traverse the opposing electrode, generating a hole current that diminishes over time. The intensity of this current, denoted as $I(t)$, is measured over time. The results, displayed in relative units on a log-log scale, are depicted in Fig. 6.1, revealing two distinct regions of the power law dependence of the current on time t. Interestingly, the slopes of both straight lines add up to -2.

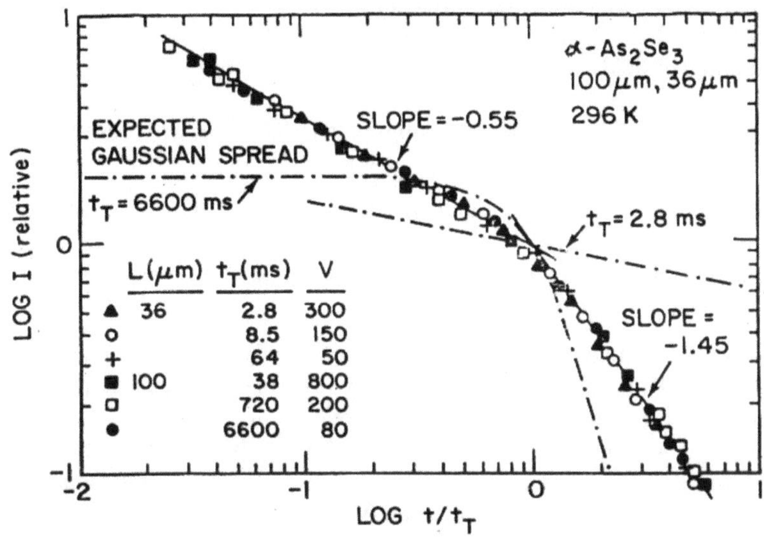

Fig. 6.1 The non-exponential photocurrent relaxation observed in amorphous $\alpha - As_2Se_3$ is depicted in a scaled or collapsed form (in the log-log scale). Two different regimes of power law relaxation are discernible below and above $t = t_T$, where t_T denotes the transition from one transport & diffusion regime to another, shown in the legend specifically to each sample. The independent scaling variable is t/t_T. As indicated in the legend, measurements were performed for six different sample/film thicknesses, L, and different applied potential differences, V. Reprinted with permission from [1].

The result obtained has a more general significance. In 1972, Gill [9] also observed a similar effect in the organic complex of trinitrofluorenone and poly-n-vinylcarbazole.

A primary objective of this chapter is to elucidate the observed phenomenon, which plays a crucial role in understanding the phenomenon known as anomalous diffusion, especially multifractal anomalous diffusion.

6.3 Wandering in Continuous Time

The models that describe transport and diffusion on disordered substrates are based on the Continuous-Time Random Walk formalism. The significant simplification in these models is that the disordered energetic landscape of the substrate can be described by an exponential distribution and incorporated into a regular lattice. In this work, we consider random instantaneous hops/jumps (or flights) of a single particle between regularly displaced valleys that have, however, different depths; the mountain peaks have all at the same energy level when a bias (an external (drift) force, potential or field) is absent (which justifies the name of the model as the valley model).

This section introduces a model of a single carrier's continuous-time random jump wandering (or random hops/flights). We examine two complementary scenarios.

(1) In the absence of an external field (its potential is shown schematically in Fig. 6.2).
(2) In the presence of an external force causing drift (the potential is shown schematically in Fig. 6.3).

The introduction of mathematical formalism (commonly known by its generic name, the Continuous-Time Random Walk formalism [16] enabled the description of wandering at any given time, which marked a breakthrough in the theory of random processes. This advancement extended the field of stochastic processes beyond the Central Limit Theorem, reaching non-Brownian and non-Gaussian stochastic processes, transcending the realm of Brownian/Wiener motion. It notably encompassed memory processes, facilitating incorporating Lévy processes into physics.

The potentials depicted in Figs. 6.2 and 6.3 form the foundation of the widely used valley model of (random) wandering, a model we continue to analyze in the absence and presence of an external drift-inducing constant force F.

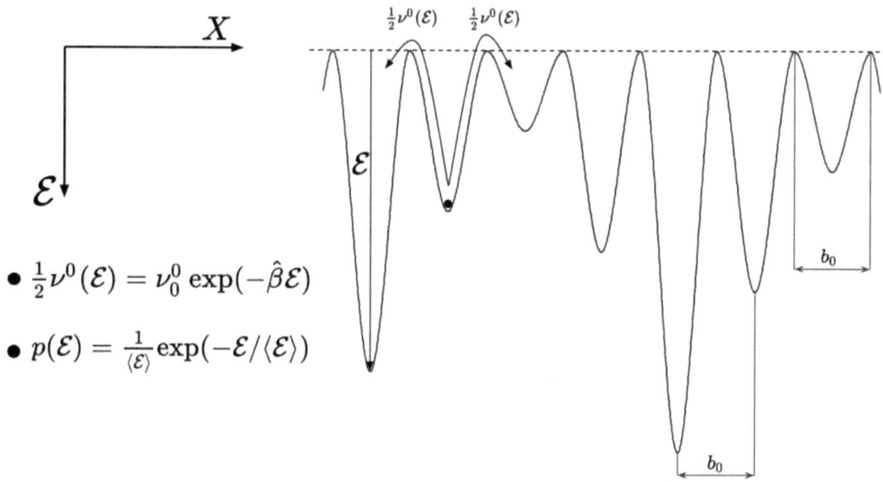

Fig. 6.2 The thermally activated over-barrier wandering of the carrier (black small circle) in the valley potential without an external force (hence, the reference level of the potential is parallel to the X axis). The only element of amorphousness here comes from the random depth of the potential/energy minima $\mathcal{E}$

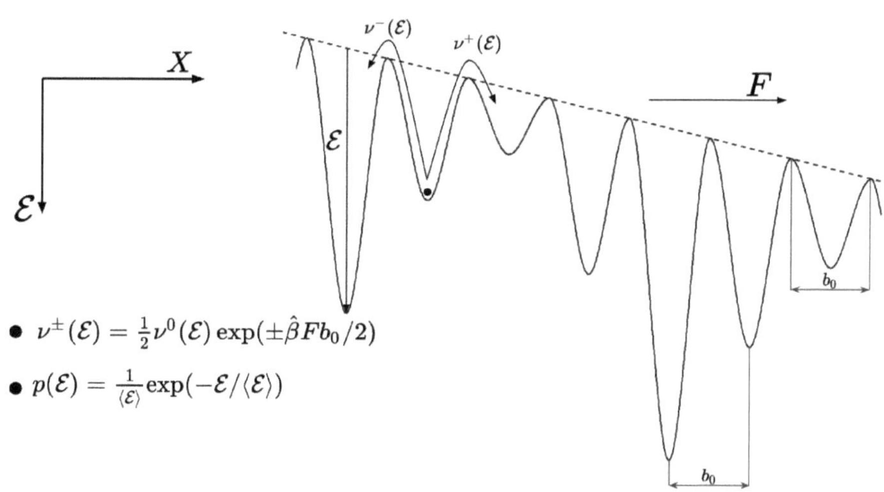

Fig. 6.3 The thermally activated over-barrier wandering of the carrier (black small circle) in the valley potential in the presence of an external driving force F. Hence, the reference level of the potential is inclined at a non-zero angle to the (horizontal) X axis. In this case $\nu^+(\mathcal{E}) > \nu^-(\mathcal{E})$, with the constraint given by Eq. (6.19), as $\hat{\beta}Fb_0 \ll 1$. The graph has significantly enlarged the angle of inclination of the potential to make it more visible

6.3.1 Basic Quantities: Wandering Without Drift

We start with the quantitative formulation of the model by introducing:

(1) The probability density $\Phi_{\mathcal{E}}(t)$ which represents the probability of the survival of the molecule within a specific potential valley with depth $\mathcal{E}$ for a minimum period of t. This survival duration is measured from the initial moment the molecule entered the valley, extending at least until time t (in other words, the molecule may persist for a longer duration in the potential valley, but certainly not for a shorter one),

and related to it

(2) atomic waiting time (sharp) PDF in the potential valley, $\phi_{\mathcal{E}}(t)$.

The function $\phi_{\mathcal{E}}(t)$ is precisely defined as the probability density that indicates the survival of the atom in any (arbitrary) selected potential valley with a depth of $\mathcal{E}$ exactly up to the time t.

In other words, it signifies the moment the atom departs from this valley. The connection between both functions can be derived from the above definitions:

$$\Phi_{\mathcal{E}}(t) = \int_t^{\infty} dt' \phi_{\mathcal{E}}(t') = 1 - \int_0^t dt' \phi_{\mathcal{E}}(t'), \tag{6.1}$$

where writing the second equality, we used the normalization condition,

$$\int_0^{\infty} dt \phi_{\mathcal{E}}(t) = 1. \tag{6.2}$$

This condition implies that within a specified potential valley, a particle is guaranteed to survive for any duration. The $\phi_{\mathcal{E}}$ function is more operationally convenient, so we consider it the initial one in our model. However, frequently, especially in intermediate stages, we also utilize both functions. In this chapter, we adopt the waiting time distribution function in Poisson form,

$$\phi_{\mathcal{E}}(t) = \nu^0(\mathcal{E}) \exp(-\nu^0(\mathcal{E})t), \tag{6.3}$$

where $\nu^0(\mathcal{E})$ represents the frequency of the atom's jumps between adjacent potential valleys (in the one-dimensional case, these could be the two nearest neighbors, as illustrated in Fig. 6.2) in the absence of an external force causing drift. We maintain the assumption that these jumps are thermally activated, over-barrier jumps, and their frequency is determined by the following expression:

$$\nu^0(\mathcal{E}) = 2\nu_0^0 \exp\left(-\hat{\beta}\mathcal{E}\right), \tag{6.4}$$

where

$$\hat{\beta} = \begin{cases} (k_B T)^{-1} & \text{for the Hopf-Arrhenius (HA) law,} \\ (k_B \Theta)^{-1} & \text{for the Vogel-Tamm-Vulcher (VTF) law,} \end{cases} \tag{6.5}$$

where k_B is the Boltzmann constant, T is the absolute temperature, and the temperature difference $\Theta = T - T_g > 0$, here T_g is the transition to the glass phase. As one can see, the above formula applies to two classes of materials—the HA law, those that are not glasses or are in the distant phase from the glass transition, while the VTF law of materials near the glass transition point.

From the formulas (6.1) and (6.3) we get that the sojourn probability,

$$\Phi_{\mathcal{E}}(t) = \exp(-\nu^0(\mathcal{E})t). \tag{6.6}$$

We can now move on to building a separable, partial, space-time waiting time distribution function, denoted as $\psi_{\mathcal{E}}(x, t)$ (please note that this notation is unrelated to the wave function in quantum mechanics). To achieve this, we need to introduce an additional element,

(3) spatial distribution of atom displacements $p(x)$

defined as the probability density of the particle moving along the vector x (given the particle's one-dimensional motion, we've omitted the notation '**x**' for simplicity). Naturally, it satisfies the normalization condition, expressed as,

$$\int_{-\infty}^{\infty} dx p(x) = 1. \tag{6.7}$$

From the discussion above, it follows that the distribution $p(x)$ can be expressed in a straightforward form,

$$p(x) = \frac{1}{2}[\delta(x - b_0) + \delta(x + b_0)], \tag{6.8}$$

where b_0 is the constant distance (lattice constant) between adjacent valleys (or maxima) of the potential. Therefore, this type of distribution allows only symmetric spatial jumps between adjacent valleys at a distance of b_0.

Now, we can write the space-time relationship in a separable form,

$$\psi_{\mathcal{E}}(x, t) = \phi_{\mathcal{E}}(t) p(x). \tag{6.9}$$

As one can observe, the aforementioned partial waiting time distribution function can be interpreted as the probability density of the following sequence of events: first, the atom survives at a given location (i.e., a potential valley with a depth of $\mathcal{E}$) until the moment t, and then, precisely at instant t, it moves (or more precisely, jumps) by the vector x. The separability of the partial distribution function is imposed here by the separability of both stochastic variables: the displacement x and the time t (or, more precisely, the time interval). It's worth noting that time appears here as a random variable, which, as we will see, doesn't change its role in any way. Assuming the separability of the function $\psi_{\mathcal{E}}(x,t)$ implies that two fundamentally different events, such as waiting and moving the particle, are statistically independent of each other. This assumption seems natural for elementary processes in which the atom undergoes many oscillations at a given potential minimum, thus forgetting its origin and previous events. Later in the textbook, nonseparable walks are also discussed.

The normalization conditions (6.2), (6.7) and the definition (6.9) result in the necessary normalization condition,

$$\int_0^{\infty} dt \int_{-\infty}^{\infty} dx \psi_{\mathcal{E}}(x,t) = 1. \tag{6.10}$$

In a more general case, if the distribution function $\psi_{\mathcal{E}}(x,t)$ were not separable, we would need to impose the condition (6.10) as the necessary normalization. Furthermore, based on the definitions in (6.1) and (6.9), along with the normalization condition (6.7), we obtain that:

$$\Psi_{\mathcal{E}}(t) = \int_t^{\infty} dt' \int_{-\infty}^{\infty} dx \psi_{\mathcal{E}}(x,t') = 1 - \int_0^{t} dt' \int_{-\infty}^{\infty} dx \psi_{\mathcal{E}}(x,t'), \tag{6.11}$$

where the definitions of the functions $\Psi_{\mathcal{E}}(t)$ and $\Phi_{\mathcal{E}}(t)$ are analogous. In the separable case, which applies to wandering in the absence and presence of an external field, as will be shown below, we have $\Psi_{\mathcal{E}}(t) = \Phi_{\mathcal{E}}(t)$. It's worth emphasizing that the equalities presented in (6.11) are general, independent of the separability property of the distribution function $\psi_{\mathcal{E}}(x,t)$, and result solely from its definition and the normalization condition (6.10).

6.3.2 *Space-Time Distribution Function of Waiting Times in the Presence of Drift*

The presence of a systematic drift (induced by an external force acting on the wandering atom, as shown in Fig. 6.3) alters, as we observe, the spatial component of the spatiotemporal waiting time distribution function $\psi_{\mathcal{E}}(x,t)$ and requires the expansion of the relation (6.1). However, we can still employ the (probability density) $\psi_{\mathcal{E}}(x,t)$ as the fundamental distribution function.

In the initial step, we introduce the frequency $\nu^{\pm}(\mathcal{E})$ of the carrier jumping between adjacent potential valleys (refer to Fig. 6.3) in the direction aligned with the drift (positive sign) and in the opposite direction (negative sign), respectively. Following the interpretation of the distribution function, it now adopts the following form,

$$\psi_{\mathcal{E}}(x,t) = \psi_{\mathcal{E}}^{+}(x,t) + \psi_{\mathcal{E}}^{-}(x,t), \tag{6.12}$$

where we introduced the marking

$$\psi_{\mathcal{E}}^{\pm}(x,t) = \nu^{\pm}(\mathcal{E})\delta(x \mp b_0)\exp(-\nu(\mathcal{E})t), \tag{6.13}$$

resulting from the existence of drift, with the total jump frequency,

$$\nu(\mathcal{E}) = \nu^{+}(\mathcal{E}) + \nu^{-}(\mathcal{E}) \tag{6.14}$$

does not depend on the orientation of the applied external force. It is because by expanding the formula (6.4), we get,

$$\begin{aligned}\nu^{\pm}(\mathcal{E}) = {} & \nu_0^0 \exp\left(-\hat{\beta}(\mathcal{E} \mp Fb_0/2)\right) = \frac{1}{2}\nu^0(\mathcal{E})\exp(\pm\hat{\beta}Fb_0/2) \\ & \Rightarrow \nu(\mathcal{E}) = \nu^0(\mathcal{E})\cosh(\hat{\beta}Fb_0/2).\end{aligned} \tag{6.15}$$

Moreover, we can express the directional frequencies as,

$$\nu^{\pm}(\mathcal{E}) = p^{\pm}\nu(\mathcal{E}), \tag{6.16}$$

where the weights

$$p^{\pm} = \frac{\nu^{\pm}(\mathcal{E})}{\nu(\mathcal{E})} = \frac{1}{2}\frac{\exp(\pm\hat{\beta}Fb_0/2)}{\cosh(\hat{\beta}Fb_0/2)}, \tag{6.17}$$

are the probabilities of selecting one of the two single-jump orientations–as one can see, the necessary normalization $p^+ + p^- = 1$ takes place.

In further considerations, we use the weak field approximation (consistent with experience), i.e., $\hat{\beta} F b_0 \ll 1$. Therefore, expanding $\cosh(\hat{\beta} F b_0)$ into a Taylor series and limiting ourselves to only the first two terms, we get

$$\cosh(\hat{\beta} F b_0) \approx 1 + \frac{1}{8}(\hat{\beta} F b_0)^2, \tag{6.18}$$

It is still possible to neglect the quadratic term in the above expression, which leads (based on (6.15)) to a useful relation,

$$\nu(\mathcal{E}) \approx \nu^0(\mathcal{E}); \tag{6.19}$$

this means, based on the expression (6.13), that

$$\psi_{\mathcal{E}}^{\pm}(x, t) = p^{\pm}\delta(x \mp b_0)\nu^0(\mathcal{E}) \exp(-\nu^0(\mathcal{E})t) = p^{\pm}\delta(x \mp b_0)\phi_{\mathcal{E}}(t). \tag{6.20}$$

Thus, the partial waiting time distribution function can be represented (based on (6.12)) still in separable form,

$$\psi_{\mathcal{E}}(x, t) = p(x)\phi_{\mathcal{E}}(t); \tag{6.21}$$

analogous to that (6.9) for the case of no external field, where

$$p(x) = p^+\delta(x - b_0) + p^-\delta(x + b_0) \tag{6.22}$$

is an extension of the expression (6.8) to the asymmetric case, taking into account the existence of an external field F. Hence, the structure factor of the biased random walk is defined as the Fourier transform of $p(x)$,

$$\tilde{p}(k) = \cos(b_0 k) - i(p_+ - p_-)\sin(b_0 k) \approx \cos(b_0 k) - \frac{i}{2}\hat{\beta} F b_0 \sin(b_0 k), \tag{6.23}$$

where k is a Fourier variable; the second component cannot be neglected here compared to the first.

In the case of a weak external field, the expression (6.17) simplifies to a linear form,

$$p^{\pm} \approx \frac{1}{2}\left(1 \pm \frac{\hat{\beta} F b_0}{2}\right), \tag{6.24}$$

particularly useful in linear response theory (calculating susceptibility, conductivity, and diffusion coefficient).

In the second step, armed with the formula (6.12) for the distribution function $\psi_{\mathcal{E}}(x,t)$ and the auxiliary expressions (6.13)–(6.24), we can employ the expanded definition (6.11) of the probability density $\Psi_{\mathcal{E}}(t)$ for the survival of the molecule (at least) for time t in a given potential valley with depth $\mathcal{E}$. This yields

$$\Psi_{\mathcal{E}}(t) = \exp(-\nu^0(\mathcal{E})t) = \Phi_{\mathcal{E}}(t) \tag{6.25}$$

i.e., the form is identical to that in the absence of an external field. Both basic formulas, (6.21) and (6.25) result from the partial separability of the waiting-time distribution function and the independence of the total jump frequency $\nu(\mathcal{E})$ from the external field.

With the probability densities (i.e., PDFs) introduced above, we will compute the propagator that describes the random walk process, both in the absence and presence of an external field causing drift.

6.3.3 *Single Particle Propagator*

The carrier waiting-time distribution and its probability of sojourn averaged over the disordered landscape of the potential energy are, together with the structure factor mentioned above, the relevant quantities to construct the full propagator considered in this subsection.

The primary objective of this subsection is to establish the propagator $\mathcal{P}(X, t \mid X_0, t_0)$—or, in simpler terms, the conditional probability density of locating a molecule/atom/particle/carrier at position X at time t, given that initially, this particle was at position X_0 at time t_0.

The molecule could have reached a specific potential valley at position X at time t due to the following processes:

(1) could last in a given potential valley in the position $X(= X_0)$ from the very beginning until the moment t, provided that this position happened to be the initial one—we describe such duration using a propagator $P_{\mathcal{E}}^{(0)}(X, t \mid X_0, t_0)$,
(2) could end up in position X in the result a single flight—this process is described by a propagator $P_{\mathcal{E}_0,\mathcal{E}}^{(1)}(X, t \mid X_0, t_0)$, or
(3) as a result of two subsequent flights separated by waiting in some potential valley—a propagator describes this process $P_{\mathcal{E}_0,\mathcal{E}_1,\mathcal{E}}^{(2)}(X, t \mid X_0, t_0)$, etc., in general,

(4) a molecule could have appeared in a given valley potential in position X at time t as a result of $n(\geq 1)$ flights, each of which was preceded by a (shorter or longer) wait in some potential valley—we describe this process using a partial propagator $P^{(n)}_{\mathcal{E}_0,\mathcal{E}_1,\dots,\mathcal{E}_{n-1},\mathcal{E}}(X, t \mid X_0, t_0)$.

In summary, we describe the above processes by a summary propagator,

$$P_{\mathcal{E}_0,\mathcal{E}_1,\mathcal{E}_2,\dots,\mathcal{E}}(X, t \mid X_0, t_0) = P^{(0)}_{\mathcal{E}}(X, t \mid X_0, t_0) + \sum_{n=1}^{\infty} P^{(n)}_{\mathcal{E}_0,\mathcal{E}_1,\dots,\mathcal{E}_{n-1},\mathcal{E}}(X, t \mid X_0, t_0), \quad (6.26)$$

describing the probability density of finding a molecule in position X at time t as a result of any stochastic process, i.e., continuing in the initial position (if $X = X_0$—the component with index $n = 0$ in the expression (6.26)) or as the result of a process consisting of any number of alternating waitings/pausings and flights (components with $n \geq 1$).

The question can now be raised about the relationship earlier introduced propagator $\mathcal{P}(X, t \mid X_0, t_0)$ with the above defined $P_{\mathcal{E}_0,\mathcal{E}_1,\mathcal{E}_2,\dots,\mathcal{E}}(X, t \mid X_0, t_0)$. To find this relationship, we write partial propagators in explicit form $P^{(n)}_{\mathcal{E}_0,\mathcal{E}_1,\dots,\mathcal{E}_{n-1},\mathcal{E}}(X, t \mid X_0, t_0),\ n = 0, 1, 2, \dots$, entering more convenient notation. That is,

$$P^{(0)}_{\mathcal{E}}(X, t) (\equiv P^{(0)}_{\mathcal{E}}(X, t \mid X_0, t_0)) = \delta(X - X_0)\Psi_{\mathcal{E}}(t - t_0), \quad (6.27)$$

next

$$P^{(1)}_{\mathcal{E}_0,\mathcal{E}}(X, t) (\equiv P^{(1)}_{\mathcal{E}_0,\mathcal{E}}(X, t \mid X_0, t_0)) = \int_0^t dt_1 \psi_{\mathcal{E}_0}(X - X_0, t_1 - t_0)\Psi_{\mathcal{E}}(t - t_1), \quad (6.28)$$

and

$$P^{(2)}_{\mathcal{E}_0,\mathcal{E}_1,\mathcal{E}}(X, t) (\equiv P^{(2)}_{\mathcal{E}_0,\mathcal{E}_1,\mathcal{E}}(X, t \mid X_0, t_0)) = \int_{-\infty}^{\infty} dx_1 \int_0^t dt_2 \int_0^{t_2} dt_1 \, \psi_{\mathcal{E}_0}(x_1 - X_0, t_1 - t_0)\psi_{\mathcal{E}_1}(X - x_1, t_2 - t_1)\Psi_{\mathcal{E}}(t - t_2), \quad (6.29)$$

etc., generally, we can write

$$P^{(n)}_{\mathcal{E}_0,\mathcal{E}_1,\dots,\mathcal{E}_{n-1},\mathcal{E}}(X, t) (\equiv P^{(n)}_{\mathcal{E}_0,\mathcal{E}_1,\dots,\mathcal{E}_{n-1},\mathcal{E}}(X, t \mid X_0, t_0)) = \int_{-\infty}^{\infty} dx_{n-1} \dots \int_{-\infty}^{\infty} dx_2 \int_{-\infty}^{\infty} dx_1 \int_0^t dt_n \int_0^{t_n} dt_{n-1} \dots \int_0^{t_2} dt_1 \, \psi_{\mathcal{E}_0}(x_1 - X_0, t_1 - t_0)\psi_{\mathcal{E}_1}(x_2 - x_1, t_2 - t_1) \dots \psi_{\mathcal{E}_{n-1}}(X - x_{n-1}, t_n - t_{n-1}) \, \Psi_{\mathcal{E}}(t - t_n),\ n = 1, 2, 3, \dots; \quad (6.30)$$

The expressions (6.28), (6.29), and (6.30) are essentially built on the assumption that the distribution function $\psi_{\mathcal{E}}(x_2 - x_1, t_2 - t_1)$ characterizes the state of a partial (locally) stationary state. Therefore, it depends on the difference in spatial and time variables, similar to conditions in the complete stationary state, which we will discuss later. Additionally, these expressions were formulated with the presumption that the first wait and the subsequent flight are governed by the same distribution function, which extends to successive pairs of events of this nature. We further explore this nuanced aspect of the wandering process in later discussions.

As is evident, propagators of the P type incorporate information on the depths of the potential valleys the molecule explores. On the other hand, propagators of the $\mathcal{P}$ type are obtained by averaging these propagators over the landscape of potential energy. We elaborate on the averaging procedure.

First, we assume that the depths of the potential valleys are statistically independent of each other which means that the distribution $p(\mathcal{E}_0, \mathcal{E}_1, \ldots, \mathcal{E}_n, \ldots)$ with which we average the partial propagators $P^{(n)}_{\mathcal{E}_0,\mathcal{E}_1,\ldots,\mathcal{E}_n}(X, t),\ n = 0, 1, 2, \ldots$, (and hence the summary propagator $P_{\mathcal{E}_0,\mathcal{E}_1,\ldots,\mathcal{E}_n,\ldots}(X, t)$) is factorized, i.e.,

$$p(\mathcal{E}_0, \mathcal{E}_1, \ldots, \mathcal{E}_n, \ldots) = p(\mathcal{E}_0)p(\mathcal{E}_1)\ldots p(\mathcal{E}_n)\ldots . \tag{6.31}$$

Due to the above, averaging the equalities (6.27) and (6.30) (thereby ((6.28) and (6.29)), we obtain, respectively,

$$\mathcal{P}^{(0)}(X, t)(\equiv \mathcal{P}^{(0)}(X, t \mid X_0, t_0)) = \delta(X - X_0)\Psi(t - t_0), \tag{6.32}$$

$$\mathcal{P}^{(n)}(X, t)(\equiv \mathcal{P}^{(n)}(X, t \mid X_0, t_0)) = \int_{-\infty}^{\infty} dx_{n-1} \ldots \int_{-\infty}^{\infty} dx_2 \int_{-\infty}^{\infty} dx_1$$
$$\int_0^t dt_n \int_0^{t_n} dt_{n-1} \ldots \int_0^{t_2} dt_1\ \psi(x_1 - X_0, t_1 - t_0)\psi(x_2 - x_1, t_2 - t_1)\ldots$$
$$\psi(X - x_{n-1}, t_n - t_{n-1})\Psi(t - t_n),$$
$$n = 1, 2, 3, \ldots, \tag{6.33}$$

where we introduced the following designations of average values,

$$\psi(x, t) = \int_0^{\infty} d\mathcal{E} p(\mathcal{E})\psi_{\mathcal{E}}(x, t),$$
$$\Psi(t) = \int_0^{\infty} d\mathcal{E} p(\mathcal{E})\Psi_{\mathcal{E}}(t),$$
$$\mathcal{P}^{(n)}(X, t) = \int_0^{\infty}\int_0^{\infty}\ldots\int_0^{\infty} d\mathcal{E}_0 d\mathcal{E}_1 \ldots d\mathcal{E}_n$$
$$p(\mathcal{E}_0)p(\mathcal{E}_1)\ldots p(\mathcal{E}_n)P^{(n)}_{\mathcal{E}_0,\mathcal{E}_1,\ldots,\mathcal{E}_n}(X, t),\ n = 1, 2, \ldots, \tag{6.34}$$

which we will further interpret and indicate how to implement them.

Finally, from Eqs. (6.26), and (6.31)–(6.34) we obtain the expression,

$$\mathcal{P}(X,t) = \sum_{n=0}^{\infty} \mathcal{P}^{(n)}(X,t), \tag{6.35}$$

which allows us to answer why, at the macroscopic level, we perceive certain types of wandering as having a singular (fractal or multifractal) character.

6.3.4 *Closed Form of Propagator*

A crucial technical question emerges: how to express (if possible) the propagator (6.35) in a closed form? The answer to this question is affirmative, which involves transitioning to Fourier and Laplace transforms.

Subsequently, Eq. (6.35) assumes the following form,

$$\tilde{\mathcal{P}}(k,s) = \sum_{n=0}^{\infty} \tilde{\mathcal{P}}^{(n)}(k,s), \tag{6.36}$$

where we used the definition of the Fourier-Laplace transform

$$\tilde{\mathcal{F}}(k,s) = \int_{-\infty}^{\infty} dX \exp(-ikX) \int_{0}^{\infty} dt \exp(-st)\mathcal{F}(X,t), \tag{6.37}$$

here $\mathcal{F}$ is any function that satisfies the theorem of Fourier and Laplace transforms inversion [19]. Using the definition of the propagator $\mathcal{P}^{(n)}$, $n = 0, 1, 2, \ldots$, and above given the Fourier-Laplace transform, we get the following,

$$\tilde{\mathcal{P}}^{(n)}(k,s) = \tilde{\Psi}(s)[\tilde{\psi}(k,s)]^n, \; n = 0, 1, 2, \ldots. \tag{6.38}$$

Hence and from (6.36), we obtain the sought, closed form of the summary propagator in the inverse space (i.e., in Fourier-Laplace variables),

$$\tilde{\mathcal{P}}(k,s) = \frac{\tilde{\Psi}(s)}{1-\tilde{\psi}(k,s)} = \frac{1}{s}\frac{1-\tilde{\psi}(k=0,s)}{1-\tilde{\psi}(k,s)}, \tag{6.39}$$

where, when deriving the second equality, we additionally used the Laplace transform averaged over $\mathcal{E}$ in Eq. (6.11).

Note that to derive the above formula, there was no need to assume the separability of the distribution function $\psi(x,t)$—in this case the formula (6.39) takes the form,

$$\tilde{\mathcal{P}}(k,s) = \frac{\tilde{\Phi}(s)}{1-\tilde{\phi}(s)\tilde{p}(k)} = \frac{1}{s}\frac{1-\tilde{\phi}(s)}{1-\tilde{\phi}(s)\tilde{p}(k)}, \tag{6.40}$$

where, similarly to (6.39), we used the (averaged over $\mathcal{E}$) Laplace transform of the formula (6.1). We have adopted here the separable form $\tilde{\psi}(k,s) = \tilde{p}(k)\tilde{\phi}(s)$, resulting from the separability $\psi(x,t) = p(x)\phi(t)$.

Now, we can accurately articulate the main objective of this chapter: the examination of the propagator defined by the expression (6.40) through the analysis of the waiting time distribution function $\tilde{\phi}$ and the structural factor of flights $\tilde{p}(k)$.

6.3.5 *Generalized Homogeneous Master Equation*

Equation (6.40) allows us to introduce the so-called integral memory kernel (or simply memory). To demonstrate this, let us write the right side of this equation as $1/[s-\tilde{\mathcal{K}}(k,s)]$, where, after simple algebraic transformations, we obtain that

$$\tilde{\mathcal{K}}(k,s) = [\tilde{p}(k)-1]\tilde{\varphi}(s) \tag{6.41}$$

is also given in separable form, whereby,

$$\tilde{\varphi}(s) = s\frac{\tilde{\psi}(s)}{1-\tilde{\psi}(s)} \tag{6.42}$$

is, as we will show, the memory we are looking for. We can now rewrite the Eq. (6.40) as follows,

$$s\tilde{\mathcal{P}}(k,s) - \tilde{\mathcal{P}}(k=0,t) = \tilde{\mathcal{K}}(k,s)\tilde{\mathcal{P}}(k,s) = [\tilde{p}(k)-1]\tilde{\varphi}(s)\tilde{\mathcal{P}}(k,s), \tag{6.43}$$

which allows us to move to the differential-integral form; when deriving (6.43) from (6.40) we used the initial condition

$$\mathcal{P}(X,t=0) = \delta(X) \equiv \tilde{\mathcal{P}}(k,t=0) \tag{6.44}$$

which the propagator satisfies. Note that the left side of the Eq. (6.43) is the Laplace transform of the time derivative of the propagator $\tilde{\mathcal{P}}(k, t)$ and the right side is the Laplace transform of the convolution of the time-dependent quantity $\tilde{\mathcal{K}}$ and $\tilde{\mathcal{P}}$. Therefore,

$$\frac{\partial}{\partial t}\tilde{\mathcal{P}}(k,t) = \int_0^t dt'\tilde{\mathcal{K}}(k,t-t')\tilde{\mathcal{P}}(k,t') \\ = \int_0^t dt'[\tilde{p}(k)-1]\varphi(t-t')\tilde{\mathcal{P}}(k,t'). \quad (6.45)$$

As observed, $\tilde{\mathcal{K}}$ serves as an integral memory kernel because (generally) it enables the current behavior of the propagator to depend not only on its present state (for time $t' = t$) but also on its past states (for time $t' < t$). The φ function qualifies as proper memory since it depends solely on time.

Note that the memory kernel $\tilde{\mathcal{K}}$ has a form factored into a spatial part, $\tilde{p}(k) - 1$, and a temporal part, $\varphi(t - t')$. This follows directly from the separable representation of the waiting time distribution in the form (6.9). The memory $\varphi(t - t')$ that appears here is fully decoupled from the spatial activity of a particle. It is a pure, independent memory, independent of the spatial structure of a random walk, while dependent only on the structure of the waiting times and their mutual dependencies and autocorrelations. As can be seen, it is possible to consistently reconcile space-time independence with memory.

The initial equality in (6.45) is termed the generalized homogeneous master equation[2] (GHME). In contrast, the second is termed the separable GHME. In this second equation, we execute the inverse Fourier transform transitioning to the spatial variable. Therefore,[3]

$$\frac{\partial}{\partial t}\mathcal{P}(l,t) = \int_0^t dt'\varphi(t-t')\sum_{l'(\neq l)}\left[p(l-l')\mathcal{P}(l',t) - p(l'-l)\mathcal{P}(l,t)\right], \quad (6.46)$$

where along the way, we assumed that we were dealing with a space/network/lattice that was homogeneous/uniform, and we used normalization/standardization,

[2] Equation (6.45) is a particular case of the generalized inhomogeneous master equation considered in Appendix A.6.

[3] The transition from continuous space representation to discrete space presented here is a mental shortcut. Using discrete representation is more convenient technically and does not reduce the generality of the conclusions.

$$\sum_{l'(\neq l)} p(l'-l) = 1. \tag{6.47}$$

Here, $p(l'-l)$ represents the probability per unit time of a molecule jumping from the lattice node l to l'. Restricting this probability to that defined by the formula (6.22) for a one-dimensional network, we derive for $\mid l'-l \mid = b_0$,

$$p(l'-l) = \begin{cases} p^+ \text{ for } l' > l, \\ p^- \text{ for } l' < l, \end{cases} \tag{6.48}$$

while for $\mid l'-l \mid \neq b_0$ the probability $p(l'-l) = 0$. This means we are dealing with molecule jumps only between the nearest neighboring sites in the network. Generalizing equation (6.46) to a larger number of dimensions (both continuous and discrete space) does not pose any major difficulties.

6.3.5.1 Examples of Specific Memory

This paragraph presents three typical examples of distinct characteristic waiting-time distributions ϕ and the corresponding memories, φ.

A. Exponentially decaying distribution of waiting times

Let us decompose $\phi(t) = \frac{1}{\tau}\exp\left(-\frac{t}{\tau}\right) \equiv \tilde{\phi}(s) = \frac{1/\tau}{s+1/\tau}$. Hence and using the Eq. (6.42) we get $\tilde{\varphi}(s) = \frac{1}{\tau} \equiv \varphi(t) = \frac{1}{\tau}\delta(t)$. This means that the system has no memory.

B. Exponentially decaying specific memory

Let the memory $\varphi(t) = a\exp(-\lambda t) \equiv \tilde{\varphi}(s) = \frac{a}{s+\lambda}$. Hence and from the Eq. (6.42) we get that $\tilde{\phi}(s) = \frac{\tilde{\varphi}(s)}{s+\tilde{\varphi}(s)} = \frac{\frac{a}{s+\lambda}}{s+\frac{a}{s+\lambda}} = \frac{a}{s^2+\lambda s+a} = \frac{a}{(s-s_-)(s-s_+)}$, where $s_\mp = -\frac{\lambda}{2} \mp \frac{1}{2}\sqrt{\lambda^2-4a}$.

In the next step, we decompose $\tilde{\phi}(s)$ into simple fractions (two here), obtaining $\tilde{\phi}(s) = a\left(\frac{A}{s-s_-} + \frac{B}{s-s_+}\right) = aA\left(\frac{s-s_+}{(s-s_-)(s-s_+)}\right)$, where we obtained $B = -A$ by comparison with the initial form $\tilde{\varphi}(s)$. Another comparison of this type leads to $A = \frac{1}{s_- - s_+}$. Hence we get that $\tilde{\phi}(s) = \frac{a}{s_- - s_+}\left(\frac{1}{s-s_-} - \frac{1}{s-s_+}\right)$. Finally (after performing the inverse Laplace transformation of the function $\tilde{\phi}(s)$ to the time domain) we obtain $\phi(t) = \frac{a}{\sqrt{(\lambda/2)^2-a}}\exp(-\lambda/2\, t)\sinh\left(\sqrt{(\lambda/2)^2-a}\right)$, with $\lambda/2 > \sqrt{a}$, i.e., there is a relationship (of the threshold type) between the amplitude of the actual memory and its relaxation rate.

The question about the evolution equation built based on this memory can be obtained from the GME given by Eq. (6.46). To do this, we differentiate this equation side by side, obtaining the following intermediate differential form,

$$\frac{\partial^2 \mathcal{P}(l,t)}{\partial t^2} = a\left[\sum_{l'(\neq l)} p(l-l')\mathcal{P}(l',t) - \mathcal{P}(l,t)\right] - \lambda\frac{\partial \mathcal{P}(l,t)}{\partial t}, \tag{6.49}$$

which we obtained by reusing the Eq. (6.46). As one can see, the amplitude a has the dimension of $time^{-2}$, and the memory relaxation rate λ has the dimension of the inverse of time. Next, note that the expression in square brackets can be expressed as follows:

$$\begin{aligned}\sum_{l'(\neq l)} p(l-l')\mathcal{P}(l',t) - \mathcal{P}(l,t) &= \frac{\mathcal{P}(l+b_0,t) - 2\mathcal{P}(l,t) + \mathcal{P}(l-b_0,t)}{b_0^2} \\ &\approx \nabla^2\mathcal{P}(X,t)\mid_{X=l},\end{aligned} \tag{6.50}$$

by using the definition (6.48) and limiting ourselves (for simplicity) to isotropic space, i.e., when $p^- = p^+ = 1/2$. By combining the Eqs. (6.49) and (6.50), we finally obtain the drift-free wave-diffusion equation in the continued form (e.g., when $b_0 \ll X$) called the telegraphists' equation,

$$\frac{\partial^2 \mathcal{P}(X,t)}{\partial t^2} + \lambda\frac{\partial \mathcal{P}(X,t)}{\partial t} = \frac{ab_0^2}{2}\nabla^2\mathcal{P}(X,t). \tag{6.51}$$

Let us now consider two extreme cases: (i) when the second (diffusive) term on the left side of the Eq. (6.51) is much larger than the first one and (ii) conversely, when the first (wave) term is much greater than the second.
Ad. (i) In this case, the Eq. (6.51) simplifies into a diffusion equation,

$$\frac{\partial \mathcal{P}(X,t)}{\partial t} = D\nabla^2\mathcal{P}(X,t), \tag{6.52}$$

here the diffusion coefficient $D = \frac{b_0^2}{2\tau}$, where we have introduced a more convenient notation $\tau = \frac{\lambda}{a}$.
Ad. (ii) In this case, the Eq. (6.51) simplifies into a wave equation,

$$\frac{\partial^2 \mathcal{P}(X,t)}{\partial t^2} - v^2\nabla^2\mathcal{P}(X,t) = 0, \tag{6.53}$$

where the velocity of the probability wave $v = \sqrt{\frac{D}{\tau_m}}$, with $\tau_m = \frac{1}{\lambda}$.

In summary, we can assert that the diffusion phenomenon manifests itself for extended (macroscopic) times, i.e., when $\tau_m \gg 1$. However, for brief times, i.e., when $\tau_m \ll 1$, we are dealing with a probability wave, which signifies that the memory in the system has not yet decayed.

C. Relaxation of the power law of the waiting time distribution

Let the waiting time distribution function decay according to power-law for long times, i.e., let $\phi(t) = \frac{A}{t^{1+\alpha}}$, dla $t \gg 1$, where $0 < \alpha < 1$. Then, the Laplace transform of this distribution, for $| s | \ll 1$, is: $\tilde{\phi}(s) = \mathcal{L}[\phi(t)](s) = \mathcal{L}\left[-\frac{A}{\alpha}\frac{d}{dt}\frac{1}{t^{\alpha}}\right](s) = 1 - A\frac{\Gamma(1-\alpha)}{\alpha}s^{\alpha}$, as distribution $\phi(t)$ is normalized, which translates into convergence $\tilde{\phi}(s \to 0) \to 1$. Using the relationship (6.42) between $\tilde{\phi}$ and $\tilde{\varphi}$, we obtain, for $| s | \ll 1$, that $\tilde{\varphi}(s) = \left(A\frac{\Gamma(1-\alpha)}{\alpha}\right)^{-1} s^{1-\alpha} = \left(A\frac{\Gamma(1-\alpha)\Gamma(\alpha)}{\alpha}\right)^{-1} \mathcal{L}\left[\frac{d}{dt}t^{\alpha-1}\right](s)$. Hence, $\varphi(t) = \left(A\frac{\Gamma(1-\alpha)\Gamma(\alpha)}{\alpha(\alpha-1)}\right)^{-1} \frac{1}{t^{2-\alpha}}$.

In conclusion, we can assert that if the distribution of waiting times follows a power-law decay over time, the actual memory also adheres to this temporal function, and vice versa.

6.3.6 *First Moment*

To elucidate the canonical experiments by Sharfe, Gill, and Pfister involving pulse-induced photocurrent in an amorphous material, as mentioned at the beginning of this chapter, we must compute the first moment $\langle X(t)\rangle$. To achieve this, let us take note that:

$$\langle \tilde{X}(s)\rangle = -i\nabla_k\tilde{\mathcal{P}}(k,s)\mid_{k=0} = -\frac{i}{s}\frac{1}{1-\tilde{\psi}(k=0,s)}\nabla_k\tilde{\psi}(k,s)\mid_{k=0}, \tag{6.54}$$

where we used the formula (6.39) along the way. For the separable distribution function $\tilde{\psi}(k,s)$ the above formula simplifies to the form,

$$\langle \tilde{X}(s)\rangle = -i\nabla_k\tilde{\mathcal{P}}(k,s)\mid_{k=0} = -\frac{i}{s}\frac{\tilde{\phi}(s)}{1-\tilde{\phi}(s)}\nabla_k\tilde{p}(k)\mid_{k=0}, \tag{6.55}$$

where we have now applied the formula (6.40). In the following, we will use the explicit form $\tilde{p}(k)$ and $\tilde{\phi}(s)$ to represent in the explicit form the expression (6.55) and thus explain upper part of the graph in Fig. 6.1.

Example. Suppose the probability $p(x)$ is given by (6.22); therefore, the random walk structural factor takes the form,

$$\tilde{p}(k) = (p^{+} + p^{-})\cos(k) + i(p^{+} - p^{-})\sin(k). \tag{6.56}$$

Hence and from the formula (6.55) we obtain a simple expression,

$$\langle \tilde{X}(s) \rangle = (p^+ - p^-)\frac{1}{s}\frac{\tilde{\phi}(s)}{1-\tilde{\phi}(s)}, \tag{6.57}$$

to which we will return later in Sect. 6.4.4, after establishing the explicit dependence of $\tilde{\phi}$ on the variable s.

6.4 Accidental Trapping

The discussions in this chapter unfold in two stages. *In the first stage*, we formulate the *continuous distribution function of waiting times* as a weighted average, considering the energy "landscape" of the substrate. *In the second stage*, we scrutinize associated functions closely tied to it, such as its first moment, which represents *the average waiting time*. This lets us easily differentiate between the Poisson and Lévy processes.

6.4.1 Waiting Time Distribution Function as a Complex Statistics

We examine the incremental movement of a single molecule within the potential depicted in Fig. 6.2. This particular form of random motion is called the random-trap or valley model.

We make an initial assumption that the probability density for the occurrence of a valley with a specified depth $\mathcal{E}$ in the local potential minimum follows the exponential decay law (i.e., it adheres to the Poisson distribution),

$$p(\mathcal{E}) = A \exp\left(-\frac{\mathcal{E}}{\langle \mathcal{E} \rangle}\right), \tag{6.58}$$

where A is the normalization constant, which is easy to calculate from the normalization condition,

$$\int_0^\infty p(\mathcal{E})d\mathcal{E} = 1. \tag{6.59}$$

By substituting the expression (6.58) into this condition and performing straightforward transformations, we find that $A = 1/\langle \mathcal{E} \rangle$. Naturally, the normalization condition (6.59) stems from the fact that $p(\mathcal{E})$ represents the probability density for a randomly selected potential valley to possess a specific depth $\mathcal{E}$; hence, the certainty that it will have some depth requires this normalization. At this juncture, it's reasonable to inquire about the physical significance of the constant $\langle \mathcal{E} \rangle$. To elucidate this, let's observe that:

$$\langle \mathcal{E} \rangle = \int_0^\infty \mathcal{E} p(\mathcal{E}) d\mathcal{E}, \tag{6.60}$$

which implies that $\langle \mathcal{E} \rangle$ represents the average depth of the potential valley. Throughout the preceding discussions, we implicitly presumed that valley depths are unbounded. However, it is crucial to acknowledge that, in reality, this is not the case. We need to consider that the depth of the valley falls within the range $0 < \mathcal{E} \leq \mathcal{E}_{\max}$, where $\mathcal{E}_{\max} (\gg \langle \mathcal{E} \rangle)$ denotes the maximum depth any chosen potential valley can attain. Extending the above considerations to accommodate this more realistic perspective is straightforward. Nevertheless, the fundamental conclusions remain consistent across both approaches.

The expression (6.58) depicting the distribution $p(\mathcal{E})$ is one of the two most commonly employed for the statistical portrayal of the energy "landscape" in disordered systems, including amorphous or glassy ones; the other is simply the Gaussian distribution. The disorder evident in the potential-valley depth distribution may arise from the random arrangement of various atoms (constituting the crystal lattice) at the nodes of a given lattice. This disorder is related to the composition, not the geometry of the lattice (i.e., the lattice constant remains constant from node to node). Both distributions characterize the static properties of the energy landscape and are associated with the materials themselves, not the wandering particle. The distribution (6.58) is more straightforward to apply as it is a single-parameter distribution, unlike the Gaussian distribution, which, in addition to the mean value, incorporates the dispersion of a random variable. This chapter focuses on disordered materials characterized by the exponential distribution (6.58).

We persist in assuming (the second assumption) that the wandering process is of a over-barrier, thermally activated nature. This implies that the probability of a molecule making a jump per unit of time from one potential valley to another—essentially, the frequency of jumps between adjacent valleys—is given by,

$$\nu^0(\mathcal{E}) = \gamma_0 \exp\left(-\hat{\beta}\mathcal{E}\right) = \gamma_0 \gamma^{\frac{\mathcal{E}}{\Delta}} \tag{6.61}$$

where $\gamma_0 (= 2\nu_0^0)$ is the frequency of vibrations (fundamental) in a given potential valley, and

$$\gamma = \exp\left(-\hat{\beta}\Delta\right), \tag{6.62}$$

where Δ here is the (physical) unit of energy (e.g., the minimum distance between discrete energy levels).

The formula (6.61) shows that the average waiting (staying) time of a molecule in the selected potential valley is

$$\tau^0(\mathcal{E}) = \frac{1}{\nu^0(\mathcal{E})}. \tag{6.63}$$

The above formula (supported by the formula (6.61)) shows, as it should be, that the deeper the potential valley, the longer the molecule's residence time in it.

The third assumption is the Poisson shape of the function waiting time distribution, $\phi_{\mathcal{E}}(t)$, which is defined as the sharp probability density that a stray molecule will survive in a given potential valley with a depth of $\mathcal{E}$ exactly for the time t (i.e., after this time, it will leave it), which means that

$$\phi_{\mathcal{E}}(t) = \nu^0(\mathcal{E}) \exp(-\nu^0(\mathcal{E})t). \tag{6.64}$$

As can be seen from (6.61), the function $\phi_{\mathcal{E}}(t)$, treated as a function of the variable $\mathcal{E}$, is an exponentially skinned exponent.

Our goal is to calculate the following closed-form weighted average,

$$\phi(t) = \int_0^\infty p(\mathcal{E})\phi_{\mathcal{E}}(t)d\mathcal{E}, \tag{6.65}$$

which is, of course, the average function of the time distribution expectations that meet the normalization condition,

$$\int_0^\infty dt\phi(t) = 1, \tag{6.66}$$

and playing a fundamental role in the wandering model in continuous-time (see Chap. 6.3) in amorphous or disordered systems, as well as, for example, in the aging process of glasses [20].

The above distribution function represents the average density of the probability that a stray particle will survive in any valley exactly for time t. Average after valley depths (i.e., after the variable $\mathcal{E}$) you can be achieved by at least two different approaches.

The first approach is to consider the behavior of many independent molecules in a sample (corresponding to a dilute network gas) and then average over an ensemble of these molecules. This approach is closer to experimental implementation than the second approach below.

The second approach involves (mentally) generating a vast number of stochastic replicas, systems that are macroscopically identical to the initial one but differ microscopically. Each replica comprises a single wandering molecule and an energy landscape that forms its environment, drawn from a specified distribution $p(\mathcal{E})$. Averaging over $\mathcal{E}$ in the expression (6.65) can now be conducted easily across the constructed statistical ensemble. So let us substitute the expression (6.58) and (6.64) into (6.65) using (6.61). We obtain a complex distribution of the form,

$$\phi(t) = \frac{1}{\langle \mathcal{E} \rangle} \int_0^\infty d\mathcal{E} \exp\left(-\frac{\mathcal{E}}{\langle \mathcal{E} \rangle}\right) \gamma_0 \exp\left(-\hat{\beta}\mathcal{E}\right) \exp\left(-\gamma_0 \exp\left(-\hat{\beta}\mathcal{E}\right) t\right). \quad (6.67)$$

The above integral can be evaluated in three significantly different ways.

The first method (direct calculation) involves performing an auxiliary variable conversion,

$$\begin{aligned} y &= \gamma_0 \exp\left(-\hat{\beta}\mathcal{E}\right) t, \\ dy &= -\hat{\beta} y d\mathcal{E}, \end{aligned} \quad (6.68)$$

which, when combined with the Eq. (6.67) leads to the following sequence of transformations,

$$\begin{aligned} \phi(t) &= -\frac{1}{\hat{\beta}\langle \mathcal{E} \rangle} \frac{1}{t} \int_{\gamma_0 t}^{0} \exp\left(-\frac{\mathcal{E}}{\langle \mathcal{E} \rangle}\right) \exp(-y) dy \\ &= \frac{\gamma_0 \alpha}{(\gamma_0 t)^{1+\alpha}} \int_0^{\gamma_0 t} \left[\gamma_0 \exp\left(-\hat{\beta}\mathcal{E}\right) t\right]^\alpha \exp(-y) dy \\ &= \frac{\gamma_0 \alpha}{(\gamma_0 t)^{1+\alpha}} \int_0^{\gamma_0 t} y^\alpha \exp(-y) dy = \frac{\gamma_0 \alpha}{(\gamma_0 t)^{1+\alpha}} \mathbf{\Gamma}_{low}(1+\alpha, \gamma_0 t), \end{aligned} \quad (6.69)$$

where the exponent $\alpha = 1/\hat{\beta}\langle \mathcal{E} \rangle > 0$, while $\mathbf{\Gamma}_{low}(1+\alpha, \gamma_0 t)$ is the lower incomplete gamma function (here dependent on the argument $\gamma_0 t$), which has the following property that is decisive for this derivation,

$$\mathbf{\Gamma}_{low}(1+\alpha, \gamma_0 t \to \infty) = \mathbf{\Gamma}(1+\alpha). \quad (6.70)$$

The plot of the $\phi(t)$ functions given by formulas (6.69), (6.71), and (6.72) are presented in Fig. 6.4. In other words, in the case where $\gamma_0 t \gg \max[\alpha, 1]$ we obtain the asymptotic form of the distribution $\phi(t)$,

$$\phi(t) \approx \frac{\gamma_0 \alpha}{(\gamma_0 t)^{1+\alpha}} \mathbf{\Gamma}(1+\alpha), \quad (6.71)$$

which is crucial for our further consideration. We got it after taking into account all three elements (6.58), (6.61) and (6.64) gives the power-law decay of the distribution function (6.71) in time. As we can see, the power-law decay of the distribution function occurs for any $\alpha \geq 0$, but, as we will see below in Sects. 6.4.3 and 6.4.4, only the case $\alpha < 1$ is fascinating.

The second method, presented in Appendix A.7, is to express the exponential function of the variable t using the Mellin transform [21] and then using the method

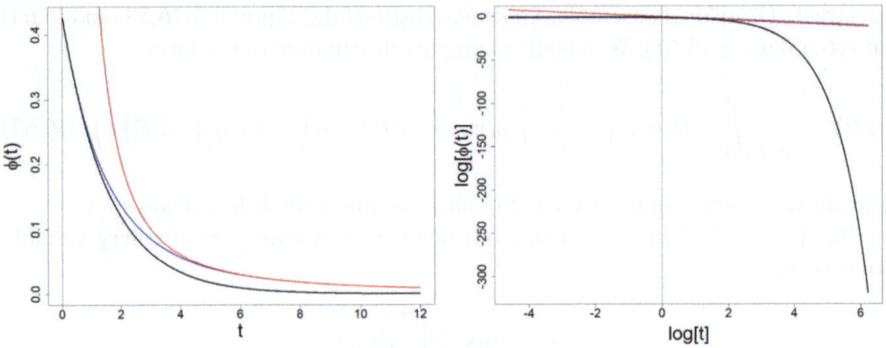

Fig. 6.4 Power-law dependence of the probability distribution function of waiting times ϕ on time t in linear (the left panel) and log-log (the right panel) scales given by the formulae (6.69) (the blue curve) and (6.71) (the red curve) for exponent $\alpha = 0.75$ and $\gamma_0 = 1$. For comparison, the exponential dependence of ϕ on time t given by the expression (6.72) (black line) for $\alpha = 0.75$ is also shown. As expected, the blue curve lies between its extremes: the black curve and the red curve

of calculating contour integrals in the complex plane through residuals [22]. This approach has also been used in a *third method* treating the inverse Laplace transform of the waiting time distribution function; we discussed it in Sect. 6.4.3.

Let us now consider the behavior of the distribution function $\phi(t)$ for short times, i.e., for the case when $\gamma_0 t \ll 1$. By expanding the exponential function in the integrand function of the expression (6.69), then integrating term by term and limiting ourselves to quadratic terms in $\gamma_0 t$, we obtain $\phi(t)$ in the exponential form (see Appendix A.8),

$$\phi(t) \approx \frac{\gamma_0 \alpha}{1+\alpha} \exp\left(-\frac{1+\alpha}{2+\alpha}\gamma_0 t\right); \tag{6.72}$$

it follows immediately that $\phi(t=0) = \frac{\gamma_0 \alpha}{1+\alpha}$.

6.4.2 *Related Quantities*

The Laplace transform is beneficial in our further considerations,

$$\tilde{\phi}(s)(\stackrel{\text{def.}}{=} \mathcal{L}_s(\phi(t))) = \int_0^\infty dt \exp(-ts)\phi(t), \tag{6.73}$$

which we consider for $s \to 0$, which, according to Tauberin's theorem (see Appendix A.10), corresponds to the situation of asymptotically long times. The calculations performed in Appendix A.11 gave, in the case where $\alpha < 1$, a non-holomorphic dependence on s in the neighborhood of $s = 0$ both for

$$\tilde{\phi}(s) \approx 1 - \frac{1}{\gamma_f'}\left(\frac{s}{\gamma_0}\right)^{\alpha}, \tag{6.74}$$

where dimensionless coefficient γ_f'

$$\gamma_f' = \frac{\sin(\pi\alpha)}{\pi\alpha}, \tag{6.75}$$

was determined in Appendix A.11 by the formula (A.101), as well as

$$\tilde{\Phi}(s) \approx \frac{1}{\gamma_0\gamma_f'}\left(\frac{\gamma_0}{s}\right)^{1-\alpha}. \tag{6.76}$$

As we showed earlier, both functions are essential in the continuous-time walk model. Let's note that from (6.76), we immediately obtain an asymptotic time dependence of the form,

$$\Phi(t) \approx \frac{1}{\Gamma(1-\alpha)\gamma_f'}\frac{1}{(\gamma_0 t)^{\alpha}}, \tag{6.77}$$

which is the shortest way to obtain the asymptotic function $\Phi(t)$. We will derive below a generalization of the expression (6.74), covering both normal and anomalous forms.

6.4.3 Time Equation of Discrete Scaling

We now answer the question of what scaling equation governs $\phi(t)$ given by the general equation (6.67). We warn the reader that the answer will be approximate.

The scaling equation for a given function is derived from:

(1) operation of linear rescaling of the independent variable,
(2) the linear response of the function itself to this rescaling.

First, the equation (6.67) must be reduced to the following, more convenient form,

$$\phi(t) = \left(1 - \frac{1}{M}\right) \int_0^{\infty} \frac{1}{M^{\xi}} \gamma_0 \gamma^{\xi} \exp\left(-\gamma_0 \gamma^{\xi} t\right) d\xi \tag{6.78}$$

by subjecting variables and constants to an appropriate non-linear transformation. Therefore, we introduced the variable $\xi = \mathcal{E}/\Delta$ and the constant $M = \exp(\Delta/\langle\mathcal{E}\rangle)$. Furthermore, we recall that the constant γ is given by (6.62). We continue to assume that $\Delta \ll \langle\mathcal{E}\rangle$ and $\Delta \ll 1/\hat{\beta}$, which allows us to show that integral heterogeneity reduces to algebraic and does not depend on γ, i.e., on the scaling factor of the independent variable s. Moving from the Eq. (6.67) to (6.78) is crucial because, as we show below in Sects. 6.4.8 and 6.4.10, it allows us to extract the hierarchical Weierstrass structure of waiting time distribution $\phi(t)$ and examine the consequences of this fact.

In the Laplace representation, the expression (6.78) takes the approximate form,

$$\tilde{\phi(s)} \approx \left(1 - \frac{1}{M}\right) \int_0^{\infty} \frac{1}{M^{\xi}} \frac{1}{1 + \frac{s}{\gamma_0 \gamma^{\xi}}} d\xi, \tag{6.79}$$

where we gradually used the approximation $1 - \frac{1}{M} \approx \frac{\Delta}{\langle\mathcal{E}\rangle} (= \ln M)$.

Note that the Laplace transform $\tilde{\phi}(s)$ satisfies non-uniform scaling equation,

$$\tilde{\phi}(\gamma^{-1}s) = M\tilde{\phi}(s) - M \ln M \int_0^1 \frac{1}{M^{\xi}} \frac{1}{1 + \frac{s}{\gamma_0 \gamma^{\xi}}} d\xi. \tag{6.80}$$

We only consider the asymptotic case in time, which corresponds (by Tauberin's theorem) to $s \to 0$. Hence, the heterogeneity takes an approximate, simpler form,

$$\begin{aligned} M \ln M \int_0^1 d\xi \frac{1}{M^{\xi}} \frac{1}{1 + \frac{s}{\gamma_0 \gamma^{\xi}}} &\approx M \ln M \int_0^1 \frac{1}{M^{\xi}} \left(1 - \frac{s}{\gamma_0 \gamma^{\xi}}\right) d\xi \\ &= M - 1 - \frac{s}{\gamma_0} \frac{\ln M}{\ln(M\gamma)} M \left(1 - \frac{1}{M\gamma}\right) \\ &\approx (M - 1)\left(1 - \frac{s}{\gamma_0}\right), \end{aligned} \tag{6.81}$$

because in this case $s/\gamma_0\gamma \ll 1$. Thanks to (6.81), the non-uniform scaling Eq. (6.80) can be rewritten as,

$$\tilde{\phi}(\gamma^{-1}s) = M\tilde{\phi}(s) - (M - 1)\left(1 - \frac{s}{\gamma_0}\right), \tag{6.82}$$

which is easier to solve.

6.4.3.1 Solution of the Eq. (6.82)

We look for the solution to the Eq. (6.82) in the form of the sum of two components,

$$\tilde{\phi}(s) = \tilde{\phi}_{reg}(s) + \tilde{\phi}_{sing}(s), \tag{6.83}$$

where $\tilde{\phi}_{reg}(s)$ is the general, regular solution of the non-homogeneous equation (6.80) and $\tilde{\phi}_{sing}(s)$ is the solution special, singular homogeneous equation,

$$\tilde{\phi}_{sing}(\gamma^{-1}s) = M\tilde{\phi}_{sing}(s). \tag{6.84}$$

The form of the *general solution* is imposed by the heterogeneity of the Eq. (6.80). Because we treat this heterogeneity (see (6.81)) therefore, up to square terms in the variable s we have,

$$\tilde{\phi}_{reg}(s) \approx 1 - \frac{1}{\gamma'}\frac{s}{\gamma_0} \tag{6.85}$$

where coefficient

$$\gamma' = \frac{1 - \frac{1}{M\gamma}}{1 - \frac{1}{M}}, \tag{6.86}$$

please do not confuse the markings γ_0, γ, γ' and γ'_f in the following sections.

One can easily check (by substituting into the Eq. (6.83)) that *solution singular* is of the form

$$\tilde{\phi}_{sing}(s) \approx -Q\left(-\frac{\ln s}{\ln \gamma}\right)\left(\frac{s}{\gamma_0}\right)^{\alpha}, \tag{6.87}$$

where $Q(\ldots)$ is the log-periodic function whose period is equal to 1; this function is defined by Eq. (3.10) (and helpully by Eq. (3.11)) in Sect. 3.2. However, the zero-order of the Fourier expansion of the log-periodic function is equal to the dimensionless coefficient γ'_f, which is given by (6.75); a systematic method to find the singular solution, and therefore this coefficient[4] we have given in the Appendix A.11 along with the form of the controlling exponent $\alpha = -\ln M/\ln \gamma (= 1/(\hat{\beta}\langle\mathcal{E}\rangle))$.

Finally, the solution of Eq. (6.80) for $s \to 0$ takes the approximate form as follows,

$$\tilde{\phi}(s) \approx 1 - \frac{1}{\gamma'_f}\left(\frac{s}{\gamma_0}\right)^{\alpha} - \frac{1}{\gamma'}\frac{s}{\gamma_0}, \tag{6.88}$$

[4] More precisely, it is not a constant coefficient but a periodic function of the variable $\ln s$ with a period equal to $-\ln \gamma$—given here the coefficient is only a zero approximation of the expression depending on the variable $\ln s$, as discussed in the Appendix A.11.

which is the desired generalization of the expression (6.74). If all terms of the Taylor expansion of the integrand function in the inhomogeneity of the scaling equation (6.80) were taken into account (for $| s |< \gamma_0\gamma$), then the regular solution $\tilde{\phi}_{reg}(s)$ would be of a power series of the variable s. This means that the exact solution of $\tilde{\phi}(s)$ could be written (symbolically) as,

$$\tilde{\phi}(s) = 1 - \frac{1}{\gamma_f'}\left(\frac{s}{\gamma_0}\right)^{\alpha} - \frac{1}{\gamma'}\frac{s}{\gamma_0} + \Theta(s^2), \tag{6.89}$$

where $\Theta(s^2)$ is the remainder (power series) of order not less than s^2. The solution (6.89) requires discussion. As can be seen, the normal and anomalous worlds co-exist—the dominance of one of them for an asymptotically long time is determined by the controlling exponent α. This is discussed below.

6.4.4 Lévy Distribution Versus Poisson Distribution

Let us consider two main cases, which we present as follows:

$$\tilde{\phi}(s) \approx \begin{cases} 1 - \frac{1}{\gamma_f'}(\frac{s}{\gamma_0})^{\alpha} \approx \frac{1}{1+\frac{1}{\gamma_f'}\left(\frac{s}{\gamma_0}\right)^{\alpha}} \approx \\ \exp\left(-\frac{1}{\gamma_f'}\left(\frac{s}{\gamma_0}\right)^{\alpha}\right), & \text{for singular case, i.e., } \alpha < 1; \\ 1 - \frac{1}{\gamma'}\frac{s}{\gamma_0} \approx \frac{1}{1+\frac{1}{\gamma'}\frac{s}{\gamma_0}} \approx \exp\left(-\frac{1}{\gamma'}\frac{s}{\gamma_0}\right), & \text{for regular case, i.e., } \alpha > 1. \end{cases} \tag{6.90}$$

As one can see, the obtained singular solution is identical to the one obtained earlier—see Sect. 6.4.2, expression (6.74). Therefore, the asymptotic time dependence of the distribution function is given by the formula (6.71) of Sect. 6.4.1.

Still, the regular asymptotic-time solution is obtained by directly inverting the Laplace transform in the expression (6.90) for the case $\alpha > 1$ [19]; this leads to the Poisson distribution,

$$\phi(t) \approx \gamma_0\gamma' \exp\left(-\gamma_0\gamma' t\right), \tag{6.91}$$

and hence the average life time at the minimum potential $\langle t\rangle = \frac{1}{\gamma_0\gamma'}$.

> We've got ourselves two flavors of distributions here. When $\alpha < 1$, it is a long-time tail of the distribution function, capturing the essential phenomena. When $\alpha > 1$, it is all about the body of the distribution function, with $\alpha = 1$ as the boundary between these two distinct worlds.

6.4.4.1 Explicit Mean Displacement

Now, we can obtain the general formula for the average time-dependent displacement of the particle along the direction of the external field. This is the essential quantity that characterizes the drift of each particle.

From (6.40) we obtain in the Laplace domain

$$\langle X\tilde{(s)}\rangle = i\frac{d}{dk}\tilde{P}(k,s)\mid_{k=0} = \langle x\rangle \cdot \frac{1}{s} \cdot \frac{\tilde{\phi}(s)}{1-\tilde{\phi}(s)}, \tag{6.92}$$

where the single-hop mean displacement $\langle x\rangle = b_0 \cdot (p_+ - p_-)$. From (6.89) and (6.92) we obtain for $\mid s \mid \ll \gamma_0$,

$$\langle X\tilde{(s)}\rangle \; = \langle x\rangle \cdot \frac{1}{s} \cdot \frac{1}{\frac{1}{\gamma_f'}\left(\frac{s}{\gamma_0}\right)^{\alpha} + \frac{1}{\gamma'}\frac{s}{\gamma_0}}. \tag{6.93}$$

The obtained result (6.93) allows the express mean displacement in explicit form for two opposite cases; first, its Laplace transform,

$$\langle \tilde{X}(s)\rangle \approx \begin{cases} \langle x\rangle \dfrac{\gamma_f'}{\gamma_0}\dfrac{1}{\left(\frac{s}{\gamma_0}\right)^{1+\alpha}}, \text{ for the singular case, i.e., } \alpha < 1; \\ \langle x\rangle \dfrac{\gamma'}{\gamma_0}\dfrac{1}{\left(\dfrac{s}{\gamma_0}\right)^{2}}, \text{ for the regular case, i.e., } \alpha > 1. \end{cases} \tag{6.94}$$

and hence, asymptotic dependence on time,

$$\langle X(t)\rangle \approx \begin{cases} \langle x\rangle \frac{\gamma_f'}{\Gamma(1+\alpha)}(\gamma_0 t)^{\alpha}, \text{ for the singular case } \alpha < 1; \\ \frac{\langle x\rangle}{\langle t\rangle}t, \qquad \text{for the singular case } \alpha > 1. \end{cases} \tag{6.95}$$

where $\langle x\rangle = b_0(p^+ - p^-)$ and $\boldsymbol{\Gamma}(\cdots)$ denotes well-known gamma-Euler function.

Finally, from (6.95), one can determine the lifting speed, which is proportional to the photocurrent density,

$$\frac{d}{dt}\langle X(t)\rangle = \langle V(t)\rangle \approx \begin{cases} \langle x\rangle \frac{\gamma_f'\gamma_0}{(\gamma_0 t)^{(1-\alpha)}}, \text{ for } \alpha < 1; \\ \frac{\langle x\rangle}{\langle t\rangle}, \qquad \text{for } \alpha > 1. \end{cases} \tag{6.96}$$

Hence, we've cracked the code on the upper section of the power-law decay of the photocurrent over time, as illustrated in Fig. 6.1. The challenge lies in untangling the lower relationship, where the cathode's absorbing impact takes center stage and still awaits a comprehensive explanation.

6.4.4.2 Explicit Mean-Square Displacement

The diffusion process's main stochastic characteristic is the mean-square displacement involving infinitely many steps by a walker or a time-dependent displacement variance.

At first, we derive this quantity in the Laplace domain,

$$
\begin{aligned}
\langle \tilde{X}^2\rangle(s) &= -\frac{d^2\tilde{P}(k,s)}{dk^2}\Big|_{k=0} = \frac{1}{s}\frac{\tilde{\phi}(s)}{1-\tilde{\phi}(s)}\left(\langle x^2\rangle + 2\,\langle x\rangle^2\cdot\frac{\tilde{\phi}(s)}{1-\tilde{\phi}(s)}\right) \\
&\approx \langle x^2\rangle\,\frac{1}{s}\,\frac{1}{\frac{1}{\gamma_f'}\left(\frac{s}{\gamma_0}\right)^{\alpha}+\frac{1}{\gamma'}\frac{s}{\gamma_0}} + 2\,\langle x\rangle^2\,\frac{1}{s}\left(\frac{1}{\frac{1}{\gamma_f'}\left(\frac{s}{\gamma_0}\right)^{\alpha}+\frac{1}{\gamma'}\frac{s}{\gamma_0}}\right)^2 \\
&\approx \begin{cases} \langle x^2\rangle\dfrac{\gamma_f'}{\gamma_0}\dfrac{1}{\left(\frac{s}{\gamma_0}\right)^{\alpha+1}} + 2\left(\langle x\rangle\gamma_f'\right)^2\dfrac{1}{\gamma_0}\dfrac{1}{\left(\frac{s}{\gamma_0}\right)^{2\alpha+1}}, & \text{for } \alpha<1, \\[2ex] \langle x^2\rangle\dfrac{\gamma'}{\gamma_0}\dfrac{1}{\left(\frac{s}{\gamma_0}\right)^{2}} + 2\left(\langle x\rangle\,\gamma'\right)^2\dfrac{1}{\gamma_0}\dfrac{1}{\left(\frac{s}{\gamma_0}\right)^{3}}, & \text{for } \alpha>1, \end{cases}
\end{aligned}
\tag{6.97}
$$

where $\langle x^2\rangle = b_0^2$.

Next, from (6.97) and (6.95), we obtain for the asymptotic time, i.e., for $t \gg \tau_0$, where $\tau_0 = 1/\gamma_0$,

$$
\langle X^2(t)\rangle - \langle X(t)\rangle^2 \approx \begin{cases} \langle x^2\rangle\gamma_f'\,\frac{1}{\alpha}\,\frac{1}{\Gamma(\alpha)}\left(\frac{t}{\tau_0}\right)^{\alpha} + \left(\langle x\rangle\gamma_f'\right)^2 & \\ \times\frac{1}{\alpha}\left[\frac{2}{\Gamma(2\alpha)} - \frac{1}{\alpha\Gamma(\alpha)^2}\right]\left(\frac{t}{\tau_0}\right)^{2\alpha}, & \text{for } \alpha<1, \\ \langle x^2\rangle\frac{t}{\langle t\rangle}, & \text{for } \alpha>1. \end{cases}
\tag{6.98}
$$

The time-dependence of the mean-square displacement below and above the threshold $\alpha = 1$ differ essentially. Moreover, in the anomalous case $\alpha < 1$, there is a significant difference between diffusion in the presence of drift and its absence (for asymptotically long times). In the first case, the variance is proportional to $t^{2\alpha}$, which is the square of the drift, and in the second, to t^{α}. Thus, in the first case, the noise (standard deviation) to signal (drift) ratio is constant, i.e., we are dealing with drift-governed fluctuations—this is equivalent to the vital fluctuation-dissipation theorem.

6.4.4.3 Asymptotic Form of the Propagator

For $\frac{|s|}{\gamma_0} \ll 1$ and $| k\, b_0 | \ll 1$ the propagator (given by (6.40)) can assume the following explicit form,

$$\tilde{P}(k,s) = \frac{1}{s}\frac{1}{1+[1-\tilde{p}(k)]\,\frac{\tilde{\phi}(s)}{1-\tilde{\phi}(s)}}$$
$$\approx \frac{1}{s}\frac{1}{1+[1-\tilde{p}(k)]\,\frac{1}{\frac{1}{\gamma_f'}\left(\frac{s}{\gamma_0}\right)^{\alpha}+\frac{1}{\gamma'}\frac{s}{\gamma_0}}}$$
$$\approx \begin{cases} \dfrac{1}{s}\dfrac{1}{1+[1-\tilde{p}(k)]\,\dfrac{1}{\dfrac{1}{\gamma_f'}\left(\dfrac{s}{\gamma_0}\right)^{\alpha}}}, & \text{for } \alpha < 1 \\ \dfrac{1}{s+[1-\tilde{p}(k)]\,\dfrac{1}{\langle t\rangle}}, & \text{for } \alpha > 1. \end{cases} \tag{6.99}$$

Also, we used the explicit asymptotic form of the waiting-time distribution (6.88) to get the second and third (approximate) equality. In the Fourier and time domain, the above relation transforms still into the relatively simple form,

$$\tilde{P}(k,t) \approx \begin{cases} E_{\alpha}\left(-[1-\tilde{p}(k)]\gamma_f'\left(\frac{t}{\tau_0}\right)^{\alpha}\right), & \text{for } \alpha < 1 \\ \exp\left(-[1-\tilde{p}(k)]\,\frac{t}{\langle t\rangle}\right), & \text{for } \alpha > 1. \end{cases} \tag{6.100}$$

where $E_{\alpha}(\cdots)$ is the well known Mittag-Leffler function [7], called sometimes the generalized exponent,

$$E_{\alpha}(x) = \sum_{n=0}^{\infty}\frac{x^n}{\Gamma(1+n\alpha)}. \tag{6.101}$$

The Fourier transformation of the second relation in (6.100) into the real space gives the well-known shifted Gaussian. The analogous transformation for $\alpha < 1$ is unknown in a closed form. However, it can be expressed in the integral form in terms of the non-shifted Gaussian and the weight given by the corresponding Fox H function as an integrand (for details, see [7] and the refs. therein).

6.4.5 Fractal Random Wandering Without Drift

One of the intriguing instances of a stochastic process, elucidated by the propagator $\tilde{P}(k,t)$ (refer to Eq. (6.100)), is a random walk in fractal time.

Let us dive into this propagator for $\alpha < 1$, represented by the upper expression in (6.100). This formulation aligns with the fractal diffusion equation (3.169), as derived in Sect. 3.9, under the condition that the random walk structural factor $\tilde{p}(k)$ lacks a singular component. In other words, it consists solely of the regular part (also discussed in Sect. 7.4 for further clarification). A reasonable approximation for $\tilde{p}(k)$ can be expressed as follows: $\tilde{p}(k) \approx 1 - \frac{1}{2}\langle x^2\rangle k^2$. This approximation holds for $\langle x^2\rangle k^2 \ll 1$, and for simplicity, we have assumed the absence of drift in this context, $\langle x\rangle = 0$. Then for $\alpha < 1$, the propagator[5] or a single relaxation k-mode,

$$\tilde{P}(k,t) \approx E_\alpha\left(-\frac{\langle x^2\rangle}{2}k^2\gamma_f'\left(\frac{t}{\tau_0}\right)^\alpha\right) = E_\alpha\left(-D_\alpha k^2 t^\alpha\right), \qquad (6.102)$$

decays according to the Mittag-Leffler function (or the generalized exponential function). The fractal diffusion coefficient is expressed as $D_\alpha = \frac{1}{2}\frac{\langle x^2\rangle}{(\tau_0)^\alpha}\gamma_f'$, as a reflection of the anomalous diffusion process that we are dealing with here. The expression (6.102) is also a solution of the fractional diffusion equation (3.169) in Sect. 3.9. This presents a quintessential case of a non-Brownian and non-Gaussian (nBnG) stochastic process because time is a fractional quantity here, and the propagator is (generally) non-Gaussian.

Furthermore, the distribution (6.102) exhibits shape instability due to the non-standard exponent $\alpha \neq 1$. In this context, the pivotal role of the generalized exponent function, precisely the Mittag-Leffler function, becomes apparent. Expression (6.102) bridges both approaches to anomalous diffusion: physical from the HCTRF for the valley model side and heuristic from the fractional diffusion equation side.

It can be demonstrated (refer to Appendix A.1) that for significantly long times (i.e., when $\alpha D_\alpha k^2 \gg 1$), the k-mode propagator (6.102) adopts the shape of a decaying power function:

[5] The propagator appearing here is equivalent to the Fourier transform $\tilde{f}(k,t)$ of the function $f(x,t)$ appearing in the fractal subdiffusion equation (3.169).

$$\tilde{P}(k,t) \approx \frac{1}{D_\alpha \mathbf{\Gamma}(1-\alpha)} \frac{1}{k^2} \frac{1}{t^\alpha}. \tag{6.103}$$

Therefore, the exponential relaxation (or Debye) observed in diffusive modes during Fickian (normal) diffusion has been replaced by a *slow relaxation* (power law, non-Debye) in subdiffusion fractals.

Bear in mind that when $D_\alpha k^2 t^\alpha \ll 1$, the Mittag-Leffler function approximates well to an ordinary exponential, yielding a Gaussian relaxation k-mode,

$$\tilde{P}(k,t) \approx \exp\left(-\frac{D_\alpha}{\mathbf{\Gamma}(1+\alpha)} k^2 t^\alpha\right). \tag{6.104}$$

In this approximation, the expression (6.104) characterizes a non-Brownian and Gaussian (nBG) stochastic process that exhibits shape instability. We are dealing here with anomalous diffusion within the normal fractal Brownian motion (nfBm) of John Nash & Benoit Mandelbrot. Notably, the Mittag-Leffler function in Eq. (6.102) interpolates between exponential and power-law forms.

From the information provided above (i.e., Eq. (6.104)), we derive the singular variance of the aggregate random variable,

$$\sigma_X^2(t) = \langle X^2(t) \rangle = \frac{2D_\alpha}{\mathbf{\Gamma}(1+\alpha)} t^\alpha, \tag{6.105}$$

identical to (6.98) (for $\langle x \rangle = 0$ and $\alpha < 1$), as it should be. This nonlinear dependence of the total variance on time combined with the form of the propagator (6.102) indicates that we are dealing with a random walk in fractal time.

To conclude this subsection (and also Sect. 3.9), it is worth providing a useful form of the propagator $P(X,t)$ convenient for numerical calculations [7] (see Appendix A.1 for details),

$$P(X,t) = \frac{1}{\sqrt{4\pi D_\alpha t^\alpha}} \sum_{n=0}^{\infty} \frac{(-1)^n}{n!\,\mathbf{\Gamma}(1-(n+1)\alpha/2)} \left(\frac{|X|}{4D_\alpha t^\alpha}\right)^n. \tag{6.106}$$

Ultimately, Fig. 6.5 juxtaposes the subdiffusion propagator $P(X,t)$ (the left panel)—illustrated with an exponent $\alpha = 0.45$—with a Gaussian distribution $G(X,t)$ (the right panel). The fundamental distinctions between them are discernible.

Figure 6.6 also illustrates the formula (6.106) compared to the Gaussian curve for a fixed time $t = 1$. The main feature of the formula (6.106) are curves whose vertex has the cusp shape, as can be seen in this plot and in Fig. 6.5. It must be said that non-Gaussian curves in both figures are drawn in such a way that (except for the

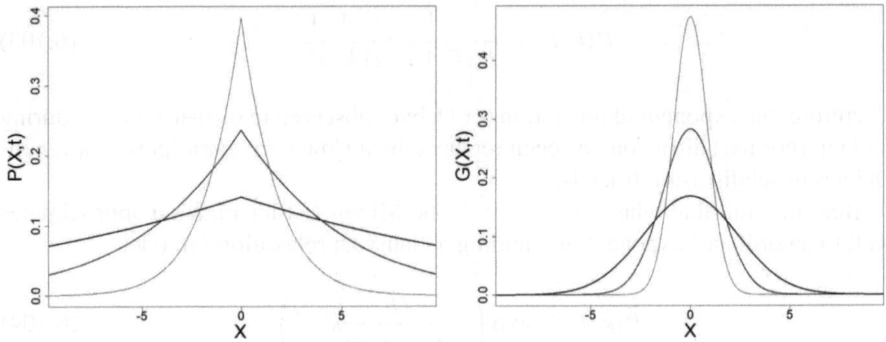

Fig. 6.5 The propagator $P(X, t)$ (the left panel), as defined by the formula (6.106), serves as the solution to the fractal diffusion equation (3.169) (see Appendix A.1 for details) with the exponent $\alpha = 0.45$ and fractional diffusion coefficient $D_\alpha = 1$. We examine three moments: $t = 0.1, 1.0, 10.0$ (from the top). The shape of the blade is notably distinct, especially near the initial moment $t = 0$. Beyond the shape, the distributions are further set apart by their power-law decay, setting them apart from the Gaussian distribution (the right panel). The graphs in this figure effectively highlight the distinctions between the sharp or cusp-like propagator and softly outlined or bell-shaped Gaussian distribution, which also serves as a solution to the fractal-time diffusion equation

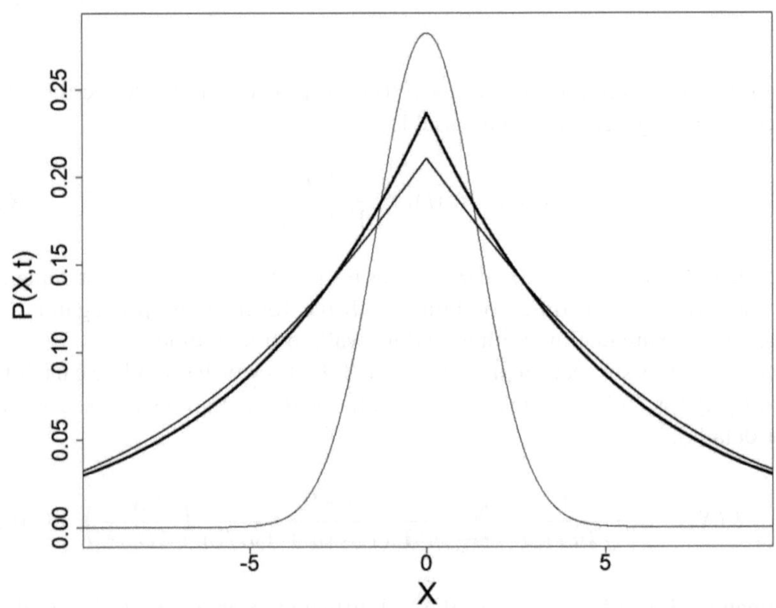

Fig. 6.6 The propagator $P(X, t)$ for three characteristic cases for $D_\alpha = 1$ and t = 1. Going from the bottom: The exponent $\alpha = 0.45$ applies to the curve whose peak is the lowest. The exponent $\alpha = 0.65$ refers to a curve whose peak lies in the middle. However, the curve with the highest peak is a zero-centered Gaussian with variance $\sigma^2 = 2D$

curve whose vertex is the lowest) they go beyond the range of applicability of the formula (6.106), i.e., meeting the condition $\frac{|X|}{4D_\alpha t^\alpha} < 1$. Despite this, the course of the curves is as expected (i.e., it does not show any breaks, faults, or divergences).

It is crucial to highlight that the propagator forms presented in this chapter were derived within the Hierarchical Continuous-Time Random Flight formalism framework. This framework establishes a physical foundation for applying anomalous diffusion in random-walk physics [23], particularly in the context of chaotic dynamics [24].

On the other hand, the cornerstone is the fractal diffusion equation (3.169) [7, 25] discussed in the Sect. 3.9.

6.4.6 *Lévy Fractal Diffusion Equation*

Yet another extreme, equally fascinating example is the Lévy flight process. It occurs in a regular time scenario when $\alpha > 1$, with the random walk structural factor $\tilde{p}(k)$ incorporating a singular component ruled by exponent $\beta < 2$ (it refers to Sect. 7.4 and there Eq. (7.29) with the required definitions). In such cases, for an exponent $\beta < 2$ and $\alpha > 1$, the lower branch in expression (6.100) transforms into:

$$\tilde{P}(k,t) \approx \exp\left(-D'_f \mid k \mid^\beta \cdot \frac{t}{\langle t \rangle}\right), \tag{6.107}$$

defining the characteristic function of a centered, symmetric (i.e., in the absence of drift) Lévy distribution in continuous time. It can be verified that the distribution defined in this manner satisfies the following Lévy fractal diffusion equation,

$$\frac{\partial P(X,t)}{\partial t} = D'_f \; {}_{-\infty}D^\beta_X P(X,t), \tag{6.108}$$

where ${}_{-\infty}D^\beta_X$ is the Weyl fractional derivative/operator presented in Appendix A.1. Thanks to its property (A.12), it can be easily shown that Eq. (6.107) is a solution of the Lévy fractal diffusion equation (6.108).

It is worth knowing that the solution of Eq. (6.108) in space and time can be conveniently expressed in a closed form using the Fox H-function [7]:

$$P(X,t) = \frac{1}{\beta \mid x \mid} H^{11}_{22}\left[\frac{\mid X \mid}{(D_\beta t)^{1/\beta}} \middle| \begin{array}{l} (1, \frac{1}{\beta}), (1, 1/2) \\ (1, 1), (1, 1/2) \end{array}\right]. \tag{6.109}$$

It should be emphasized that both extreme examples (presented in the previous Sect. 6.4.5 and in this subsection) are the basis for two significantly different categories of stochastic processes: a category of fractalizing time and a category of fractalizing space. Both do not satisfy the Central Limit Theorem, which has been intensively developed in recent decades. These processes have already found many essential applications in broadly understood physics (including geophysics, biophysics, and astrophysics) and beyond (e.g., biology, economics, and sociology).

The generalization of the above-mentioned extreme categories is an equation related to the fractal treatment of space and time. We considered such a situation in the useful example of the Weierstrass random wandering [26]. The stationary version of this wandering allowed the construction of a rich stationary phase diagram of anomalous diffusion as a competition between pausing/waiting times and space propagations. Considering phase transitions between different diffusion phases (singular and regular) in the spatial-temporal shape exponents' space seems to be an interesting research direction for the future.

6.4.7 Average Waiting Time Versus Ergodicity Breaking

We're pursuing an answer to a fundamental question: What's the average waiting time (lifespan)

$$\langle t \rangle = \int_0^\infty t\phi(t)dt, \tag{6.110}$$

of a wandering molecule in any valley (minimum) of the potential? Let's delve into the critical scenario of $\alpha < 1$. It is evident, upon substituting (6.71) into (6.110) (and leveraging the fact that the function $\phi(t)$ is bounded), that $\langle t \rangle = \infty$. Likewise, a quick check reveals that any moment $\langle t^m \rangle = \infty,\ m = 2, 3, \ldots$. This implies that the average waiting time (life) of a molecule in any valley (minimum) of the potential, along with its dispersion (statistical spread), is infinite.

This raises crucial questions: How does the above theoretical result manifest in real-world experiences? To tackle this query, let's focus on a chosen time interval Δt during which we observe the particle's meanderings. Suppose that the particle $n(\gg 1)$ relocates within this timeframe, enabling us to measure its lifetime t_j, n times; $j = 1, 2, \ldots$, in successively visited potential valleys. Based on this, we can calculate the average waiting time of the molecule (an estimate of the expected value) as $\langle t(\Delta t) \rangle \approx 1/n \sum_{j=1}^{n} t_j$. Naturally, this value hinges on the duration of the Δt interval during which observations were conducted. This relationship can manifest in two ways. The average waiting time is finite if the exponent $\alpha > 1$. According to

Bernoulli's law of large numbers, as the length of the time interval increases $\Delta t \to \infty$, the estimate above $\langle t(\Delta t \to \infty)\rangle$ tends to settle at a finite plateau, stabilizing at the level of its expected value.

If $\alpha < 1$, then the average waiting time is infinite, and an extension of the observation time doesn't stabilize obtained results. In contrast, a longer observation interval Δt increases the likelihood of a rare event occurring in an exceptionally long waiting time at an exceedingly deep local potential minimum. This occurrence can drastically increase the mean (estimate) $\langle t(\Delta t)\rangle$ as Δt increases.

In simpler terms, an infinite mean waiting time translates into an ever-increasing estimate of the value of $\langle t(\Delta t)\rangle$ as Δt grows. Moreover, this scenario prompts crucial questions about ergodicity and the system's ability to reach and maintain a state of statistical equilibrium—topics we'll explore further.

6.4.8 *Weierstrass Representation of the Waiting-Time Distribution*

Let's introduce the initial definition, which provides the desired distribution function in an infinite form through the superposition of Poisson functions, to achieve the Weierstrass representation of the waiting-time distribution function.

$$\phi(t) = \sum_{j=0}^{\infty} v_j \gamma_j \exp(-\gamma_j t), \tag{6.111}$$

where

$$\tau_j = \frac{1}{\gamma_j}. \tag{6.112}$$

This refers to the average residence time of a molecule in the j-th potential valley (note that a specific depth corresponds to a unique index j regardless of the valley's location). The weights v_j satisfy the normalization condition,

$$\sum_{j=0}^{\infty} v_j = 1. \tag{6.113}$$

As one can see, introducing discrete representation involves numbering the potential valleys from the shallowest to the deepest.

The formula (6.113) results directly from the normalization condition for a distribution function of the form,

$$\int_0^\infty \phi(t)dt = 1. \tag{6.114}$$

Still, under the influence of the representation (6.67), we allow the simplest variant in which the weight ratio in subsequent rows $j = 0, 1, 2, \ldots$, is a decreasing function and independent of the row, i.e.,

$$\frac{v_{j+1}}{v_j} = \frac{1}{M} < 1; \tag{6.115}$$

this means that the M parameter acts as a coefficient stochastic similarity (here, the stochastic weakening coefficient) of the stochastic temporal structure. It follows directly from (6.113) and (6.115) that

$$v_j = \left(1 - \frac{1}{M}\right)\frac{1}{M^j}. \tag{6.116}$$

Moreover, in the spirit of geometric dependency (6.115), we assume that the ratio

$$\frac{\gamma_{j+1}}{\gamma_j} = \gamma < 1, \tag{6.117}$$

is also independent of the order (generation) j; the dimensionless coefficient γ acts as a time similarity coefficient, while γ_0 is the frequency characterizing the process at the level of the initial zero generation. From (6.117) it immediately follows that

$$\gamma_j = \gamma_0\gamma^j, \; j = 0, 1, 2, \ldots, \tag{6.118}$$

and from Eq. (6.112)

$$\tau_j = \tau_0\tau^j = \frac{1}{\gamma_0\gamma^j}. \tag{6.119}$$

By substituting the expression (6.116) and (6.118) into the definition (6.111), we obtain the following form of the waiting-time distribution function in the Weierstrass form, ready for further analysis,

$$\phi(t) = \gamma_0\left(1 - \frac{1}{M}\right)\sum_{j=0}^{\infty}\frac{1}{M^j}\gamma^j \exp(-\gamma_0\gamma^j t). \tag{6.120}$$

Note that the condition (6.115) and (6.117) guarantee that the distribution $\phi(t)$ has a finite value if

$$\frac{\gamma}{M} < 1. \tag{6.121}$$

To enhance the understanding of the physical implications of the superposition (6.120), it's worth noting the correspondence between the continuous representation of the waiting time distribution function (6.67) and its discrete counterpart (6.120), as listed in the Table 6.1.

The self-similar hierarchical structure of waiting times derived from the probability density function (PDF) defined by Eq. (6.120) is depicted in Fig. 6.7 after appropriate ordering (of the sequence of straight line segments). Fluctuations in the order and the number of segments representing these waiting times have been eliminated. An example is shown for $M = 3$ and $\tau = 2$. The depicted structure exhibits a pyramidal cascade pattern of waiting times, characterized by segments of the appropriate lengths. Hence, the figure exclusively illustrates the regularized aspect of the hierarchical probabilistic or noisy structure. A systematic analysis of such structures, encompassing probability distribution functions and Weierstrass wanderings, is expounded upon in the subsequent subsections and chapter.

The stochastic process under consideration, spanning all orders (generations), prompts inquiry into an appropriate physical time unit—a topic we will shortly investigate. It is crucial to emphasize that the regularity of the stochastic process at each level j does not necessarily imply the regularity of the overall process. In other words, it does not guarantee a single, finite average waiting time for the molecule.

Table 6.1 Relations between representations

Continuous representation	Discrete representation
$\int_0^\infty d\mathcal{E}/\Delta$	$\sum_{j=0}^\infty$
$\xi \overset{\text{def.}}{=} \mathcal{E}/\Delta$	j
$\exp(\Delta/\langle\mathcal{E}\rangle)$	M
$\Delta/\langle\mathcal{E}\rangle$	$1 - \frac{1}{M}$
$\exp(-\hat{\beta}\Delta)$	γ
$\gamma_0(\exp(-\hat{\beta}\Delta))^\xi$	$\gamma_0\gamma^j$

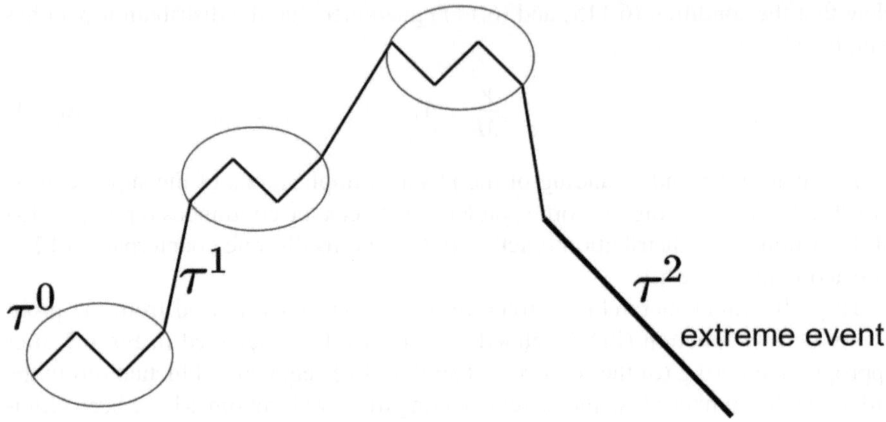

Fig. 6.7 An ordered (regularized) part of the stochastic hierarchy of two-dimensional zig-zag segments τ_j (the art view analogous to that in Fig. 7.1) given by Eq. (6.119), where $\tau_0 = 1$. They are average waiting times depicted schematically. They are drawn from the discretized & hierarchical distribution function $\phi(t)$ given by Eq. (6.120). This is illustrated by considering $M = 3$ and $\tau = 2$. Furthermore, the emphasis is placed on the properties of self-similarity and scaling while introducing the graphical concept of a rare extreme event. The self-similarity of this trajectory is visible if we treat the ovals as single points (i.e., areas viewed from a distance). At the same time, we increase the number of sections making up this trajectory M times

6.4.9 *The Time Equation of Discrete Scaling an Alternative Approach*

Let us introduce the Laplace transform of the waiting-time distribution function,

$$\tilde{\phi}(s) = \int_0^\infty dt \exp(-ts)\phi(t) = \left(1 - \frac{1}{M}\right)\gamma_0 \sum_{j=0}^{\infty} \left(\frac{\gamma}{M}\right)^j \frac{1}{s + \gamma_0\gamma^j}, \quad (6.122)$$

this makes it easy to see that the following discrete scaling equation is fulfilled,

$$\tilde{\phi}(\gamma^{-1}s) = M\tilde{\phi}(s) - (M-1)\frac{\gamma_0}{s+\gamma_0} \approx M\tilde{\phi}(s) - (M-1)\left(1 - \frac{s}{\gamma_0}\right); \quad (6.123)$$

the last approximation becomes valid when assuming $\frac{|s|}{\gamma_0} \ll 1$. The equation above is the same as the one in (6.82) (found in the Sect. 6.4.3) but derived differently—specifically, within the framework of CTRW. This shows that the Weierstrass function in (6.120) and the one derived as a solution of (6.82) are identical. This marks a crucial symbiosis of two independently developed formalisms in this textbook: the CTRW formalism and the Weierstrass function—the hierarchical flight formalism.

The solution to the Eq. (6.123) can be found (as usual) in the form of a sum,

$$\tilde{\phi}(s) = \tilde{\phi}_{reg}(s) + \tilde{\phi}_{sing}(s), \tag{6.124}$$

where $\tilde{\phi}_{reg}(s)$ is the general, regular solution of the non-homogeneous equation (6.123) and $\tilde{\phi}_{sing}(s)$ is the solution special, singular of a homogeneous equation

$$\tilde{\phi}_{sing}(\gamma^{-1}s) = M\tilde{\phi}_{sing}(s). \tag{6.125}$$

The form of the *general solution* is imposed by the heterogeneity of the Eq. (6.123). By expanding it into a Taylor series, we obtain an alternating power series in the variable s, which allows us to determine the coefficients of the power series (also in the variable s), which is a regular solution. Down to linear terms, we can write, with a good approximation for $| s | \ll \gamma_0$,

$$\tilde{\phi}_{reg}(s) \approx 1 - \frac{s}{\gamma'}, \tag{6.126}$$

where the generalized hop/jump probability per unit time,

$$\gamma' = \gamma_0 \frac{1 - \frac{1}{M\gamma}}{1 - \frac{1}{M}}. \tag{6.127}$$

Note that $\gamma' > 0$ if and only if $M\gamma > 1$, which corresponds to the situation for which there is a finite expected value,

$$\langle t \rangle = \int_0^\infty t\phi(t)dt = \left(1 - \frac{1}{M}\right)\frac{1}{\gamma_0}\sum_{j=0}^{\infty}\frac{1}{(M\gamma)^j} = \frac{1}{\gamma'}. \tag{6.128}$$

This can be easily checked using the homogeneous equation (6.125) that the desired *shape of the singular solution* is as follows,

$$\tilde{\phi}_{sing}(s) \approx -\frac{s^\eta}{\gamma_f'} \tag{6.129}$$

where exponent

$$\eta = -\frac{\ln M}{\ln \gamma} = \alpha; \tag{6.130}$$

determines the proportionality coefficient, i.e., the fractal rate of the transition per unit time. Whereas, γ_f' in the Eq. (6.129) requires a more subtle approach using the Mellin transform and the residual method for calculating integrals (in the complex plane).

Finally, using (6.124), (6.126) and (6.129) we get for $s \to 0$,

$$\tilde{\phi}(s) \approx 1 - \frac{s^\alpha}{\gamma_f'} - \frac{s}{\gamma'} \approx \frac{1}{1 + \frac{s^\alpha}{\gamma_f'} + \frac{s}{\gamma'}}. \tag{6.131}$$

The simplification of the above formula depends on the value of the exponent α, namely,

$$\tilde{\phi}(s) \approx \begin{cases} 1 - \frac{s}{\gamma'} \approx \frac{1}{1 + \frac{s}{\gamma'}}, & \text{for the regular case, i.e., } \alpha > 1 \\ 1 - \frac{s^\alpha}{\gamma_f'} \approx \frac{1}{1 + \frac{s^\alpha}{\gamma_f'}}, & \text{for the anomalous case, i.e., } \alpha < 1; \end{cases}$$

the case of marginally $\alpha = 1$ requires a different, more advanced approach (using the Mellin transform and integration by residuals)—this is discussed later in the textbook.

For the regular case ($\alpha > 1$) $\tilde{\phi}(s) \approx \tilde{\phi}_{reg}(s)$, i.e., for small s, the regular solution dominates, as opposed to the anomalous case (which requires discussion). Regular solution means that after performing the inverse Laplace transform, the distribution function

$$\phi(t) \approx \gamma' \exp(-\gamma' t) \tag{6.132}$$

is given, for asymptotically long times t, by the Poisson function.

In the anomalous case ($\alpha < 1$), after performing the inverse Laplace transformation, the distribution function for asymptotically long times decays power-wise, i.e.,

$$\phi(t) \approx \frac{\alpha}{\Gamma(1-\alpha)} \frac{1}{\gamma_f'} \frac{1}{t^{1+\alpha}}. \tag{6.133}$$

It's straightforward to verify (using (6.133) and the bounded nature of $\phi(t)$) that in this scenario, the expected value $\langle t \rangle = \infty$. As one can observe, the decompositions provided by (6.132) and (6.133) are fundamentally distinct—our focus is on discussing the latter.

6.4.10 *Pictorial Analysis of Rank Ordering*

The main goal of this subsection is to show the decisive role of rare events in Hierarchical Continuous-Time Random Walk (HCTRW) for asymptotic time steps (see Sect. 6, especially Sects. 6.4.3 and 6.4.8 for detail). We will achieve this goal by introducing the dependence of the characteristics of this wandering on the number of time steps $L \gg 1$. In other words, we treat considerations in Sects. 6.4.10 and 9.3 as a necessary complement and extension of Sects. 6.4.3 and 6.4.8.

To make our analysis more convenient, we treat variable $\mathcal{E}/\Delta$ as a discrete one $j,\ j = 0, 1, 2, \ldots,$ (just like it happened in Sects. 6.4.3 and 6.4.8; see also Table 6.1), which is possible as Δ can always be assumed to be sufficiently small (i.e., by assuming $\Delta \ll \langle\mathcal{E}\rangle$). We assume that $x \equiv \tau$ is our random variable distributed according to the power law defined by the second expression in (9.25). We are reminding that we defined $M = \exp\left(\frac{\Delta}{\langle\mathcal{E}\rangle}\right)$; hence, with a good approximation, $\frac{\Delta}{\langle\mathcal{E}\rangle} \approx 1 - \frac{1}{M}$, which makes the transformation to the discrete distribution,

$$\rho(\mathcal{E}) \Rightarrow \hat{\rho}(j) = \left(1 - \frac{1}{M}\right) \cdot \frac{1}{M^j},\ j = 0, 1, 2, \ldots, \tag{6.134}$$

and the definition of a rare event PDF,

$$\hat{\rho}(j_{\max}) = \left(1 - \frac{1}{M}\right) \cdot \frac{1}{M^{j_{\max}}}, \tag{6.135}$$

clear.

6.4.10.1 Hierarchical Waiting-Time Distribution in a Discrete Representation

Note that our hierarchical waiting-time distribution, $\psi(t)$ (which is the basic function of the HCTRW formalism; see Sect. 6 for detail), assumes, within the discrete representation introduced above, the following useful form

$$\psi(t) = \sum_{j=0}^{\infty} \hat{\rho}(j) \cdot \psi_j(t), \tag{6.136}$$

where the conditional Poisson waiting-time distribution,

$$\psi_j(t) = \frac{1}{\tau_0 \tau^j} \cdot \exp\left(-\frac{t}{\tau_0 \cdot \tau^j}\right), \tag{6.137}$$

and $\hat{\rho}(j)$ is a fundamental weight in these considerations. (This discretized $\psi(t)$ still conserves the normalization and scaling). For example, the sojourn time can be easily calculated by using the weight,

$$\langle t \rangle = \sum_{j=0}^{\infty} \hat{\rho}(j) \cdot \langle t \rangle_j, \ \langle t \rangle_j = \tau_j = \int_0^{\infty} t \cdot \psi_j(t) dt = \tau_0 \cdot \tau^j. \tag{6.138}$$

Note that the partial (or a single scale) mean residence time $\langle t \rangle_j, \ j = 0, 1, 2, \ldots,$ is always finite but the total residence time is finite only when $\alpha > 1$ and equal to

$$\langle t \rangle = \tau_0 \cdot \frac{1 - \frac{1}{M}}{1 - \frac{\tau}{M}}; \tag{6.139}$$

otherwise, it diverges. Besides, in combination with the condition (6.121), the finite values of $\phi(t)$ and $\langle t \rangle$ occur simultaneously when $M\gamma > \max[1, \gamma^2]$ (typically $\gamma < 1$). (Note that in our considerations both the notations τ_0, τ and $\gamma_0 (= 1/\tau_0)$, $\gamma (= 1/\tau)$ are used.)

Hence, to obtain $\langle t \rangle$ finite, the weight $\hat{\rho}(j)$ must converge sufficiently quickly with the increase of variable j (or the scale level).

It is decisive for our present considerations that the ratio of successive weights

$$\frac{\hat{\rho}(j+1)}{\hat{\rho}(j)} = \frac{1}{M}, \tag{6.140}$$

be already j-independent. This means that in every single step, the residence of a carrier in a trap with a single-scale sojourn time $\tau_0 \cdot \tau^j$ (as well in state as hierarchy level) j is N times more likely than those of the next larger order $j + 1$. Hence, one expects (on average) that the walker will visit N^j traps having the shortest sojourn time τ_0 before he encounters a sufficiently deep trap with a single-scale mean residence time $\tau_j = \tau_0 \cdot \tau^j, \ j = 1, 2, \ldots$

6.4.10.2 Practical Goal: a Ranking

Let's go back to Fig. 6.7; the schematic illustration of this essential observation is given in the form of hierarchically ordered one-dimensional time intervals or mean residence (sojourn) times in the corresponding traps. Here,

(i) we neglect (due to Bernoulli's Law of Large Numbers) the fluctuation of the number of hierarchy levels as well as their succession (as we calculate the summarized quantities), and
(ii) plot only the length of the average time-intervals $\langle t \rangle_j, \ j = 0, 1, 2, \ldots$

As can be seen, we transformed the stochastic hierarchy into its deterministic representation. This makes it easier to realize our practical aims, namely to discuss

(1) the rank ordering of residence times,
(2) the finite-size effect as the scaling of characteristic quantities with the size of the hierarchy.

From Fig. 6.7, one gets the useful relation between the size of hierarchy L and the number of its levels $j(\gg 1$ and $\tau,\ M > 1)$,

$$L(j) = M^j + M^{j-1} + \cdots + M^1 + M^0 \approx \frac{1}{1 - 1/M} \cdot M^j. \tag{6.141}$$

The quantity $L(j)$ is the total number of steps after which the walker encountered the trap with single-level/scale mean sojourn time $\tau_0 \cdot \tau^j$.

Now, we can set the rank $n = L(j)$ and look for the corresponding sojourn time as a function of n ranked according to its decreasing amplitude. Hence, we can write the one-to-one correspondence in the form: $n = L(j) \Leftrightarrow \tau(n) = \tau_0 \cdot \tau^{j_{\max}-j}$, where $j_{\max}$ is related to the total number of observations L. By using relation (6.141) we can write

$$L = L(j_{\max}) \approx \frac{1}{1 - 1/M} \cdot M^{j_{\max}}. \tag{6.142}$$

Dividing the (6.141) and (6.142) equations side by side, we calculate exponent $j_{\max} - j$, and by introducing it into the formula for $\tau(n)$ given the above, we finally find the searched rank dependence

$$\tau(n) = \tau_0 \cdot \tau^{j_{\max}-j} = \tau_0 \cdot \left(\frac{L}{n}\right)^{1/\alpha}, \tag{6.143}$$

which is (for fixed large L) the power-law of rank n with exponent $-1/\alpha$. We will return to an in-depth analysis of this equation in Sect. 9.3 in the context of Extreme Value Theory.

6.4.10.3 Empirical Verification of the Tail

The rank relation (6.143) is beneficial in identifying the nature of the tails of probability distributions. The single-step procedure is as follows. We sort, in decreasing order, the series of observed random variables (for example, τ's) and then draw a plot of $\tau(n)$ as a function of n. If power law distributes the variables, this graph should be a straight line in a log-log plot, with a slope given by exponent $-1/\alpha$ (as shown, e.g., by expression (6.143)).

6.4.10.4 Decisive Role of Rare Events

Our subsequent goal is realized in connection with rare events. We prove that the (mean) total time the carrier remains in the traps encountered during the L steps obeys the same scaling law as a rare event.

First, from (6.134) and (6.135) we easily obtain the discrete complementary cumulative distribution function,

$$\hat{\rho}_{\geq}(\Lambda) = \sum_{j=\Lambda}^{\infty} \hat{\rho}(j) = \frac{1}{M^{\Lambda}} \Rightarrow \hat{\rho}_{\geq}(\Lambda_{\max}) = \frac{1}{M^{\Lambda_{\max}}} = \frac{1}{L}$$
$$\equiv \Lambda_{\max} = \frac{\ln L}{\ln M}, \tag{6.144}$$

where Λ is the lower threshold for j variable. By the way, let us note that the largest single event of $\lambda_{\max}$, observed whenever in a single long series of $L \gg 1$ i.i.d. random variables, we call rare. This is an event that (in general) is even rarer than an extreme event ("black swan"). The last two equivalent equalities in (6.144) define the rare event in agreement with the definition (9.14). We will come back to this in Sect. 9.2 where we discuss an alternative approach to Extreme Value Theory

Hence, using the last equality in (6.144) we have

$$\tau^{\Lambda_{\max}} = L^{1/\alpha}. \tag{6.145}$$

By combining relations (6.142) and (6.135), we find that $j_{\max}$ is the rare event in the stronger sense given by (9.15). Thus,

$$\tau^{j_{\max}} = \left(1 - \frac{1}{M}\right)^{1/\alpha} \cdot L^{1/\alpha}, \tag{6.146}$$

which means that the difference $\Lambda_{\max} - j_{\max} = -\ln(1 - 1/M)/\ln M > 0$ for $1 < M < \infty$, is an unsignificant constant. Therefore, $\tau^{j_{\max}}$ can also be viewed as a rare event and $\Lambda_{\max}$ as an extreme event, as these are quantities belonging to the same equivalence class.

The following sum gives the mean total time mentioned above

$$\frac{t_{tot}}{\tau_0} \approx M^0 \tau^{j_{\max}} + M^1 \tau^{j_{\max}-1} + \cdots + M^{j_{max-1}} \tau^1 + M^{j_{\max}} \tau^0$$
$$= M^{j_{\max}} \frac{(\frac{\tau}{M})^{j_{\max}+1} - 1}{\frac{\tau}{M} - 1} \approx \begin{cases} \frac{1}{1-\frac{M}{\tau}} \cdot \tau^{j_{\max}}, & \text{for } \alpha < 1 \\ \frac{1}{1-\frac{\tau}{M}} \cdot M^{j_{\max}}, & \text{for } \alpha > 1. \end{cases} \tag{6.147}$$

By introducing Eqs. (6.146) and (6.142) into (6.147) we obtain the important relations

$$\frac{t_{tot}}{\tau_0} \approx \begin{cases} \frac{(1-\frac{1}{M})^{1/\alpha}}{1-\frac{M}{\tau}} \cdot L^{1/\alpha}, & \text{for } \alpha < 1 \\ \frac{1-\frac{1}{M}}{1-\frac{\tau}{M}} \cdot L, & \text{for } \alpha > 1. \end{cases} \tag{6.148}$$

Note that both relations (6.147) and (6.148) distinguish two essentially different ranges of exponent α (the marginal case $\alpha = 1$ is not considered here). For the first range ($\alpha < 1$) we find t proportional to the rare event, i.e., it scales with the number of steps L in the same manner as the rare event given by Eqs. (6.145) and (6.146). It is the main result of this section. The proportionality coefficient is called the fractional (dimensionless) residence time. For the opposite, regular case, the similar coefficient is simply the residence time given above (cf. Equation (6.139)).

Now, it is easy to calculate the dependence of the meantime, $\langle t\rangle$, used by the walker for a single step on L. For the asymptotic long L, one can write the following average single time-step calculated along the L-step trajectory

$$\begin{aligned}\frac{t_{tot}}{\tau_0}\frac{1}{L} = \frac{\langle t\rangle}{\tau_0} &\approx \frac{M^{j_{\max}}}{L}\tau^0 + \frac{M^{j_{\max}-1}}{L}\tau^1 + \cdots + \frac{M^0}{L}\tau^{j_{\max}} \\ &\approx \begin{cases} \frac{\left(1-\frac{1}{M}\right)^{1/\alpha}}{1-\frac{M}{\tau}} L^{\frac{1}{\alpha}-1}, & \text{for } \alpha < 1 \\ \frac{1-\frac{1}{M}}{1-\frac{\tau}{M}}, & \text{for } \alpha > 1, \end{cases}\end{aligned} \tag{6.149}$$

Of course, this result can be obtained straightforwardly from expression (6.148) by dividing it simply by L.

6.4.10.5 The Scaling Paradox

It is useful to list two questions related to rare events. The first question is: How does the (direct) distance between successive rare events increase with L? Again from (6.141) we obtain

$$\Delta L(j) = L(j+1) - L(j) = M^{j+1} = (M-1)\cdot L, \tag{6.150}$$

i.e., this distance increases linearly with L regardless of scaling.

The second question concerns the relationship between the value of potential rare events and their (direct) difference. Directly from Eq. (6.146) we find that this ratio is simply equal to τ independently of L while their (direct) difference,

$$\tau_0 \cdot [\tau^{l_{\max}+1} - \tau^{l_{\max}}] = \tau_0 \cdot (\tau - 1) \cdot \left(1 - \frac{1}{M}\right)^{1/\alpha} L^{1/\alpha}, \ \tau > 1, \tag{6.151}$$

scales with L as a single rare event. The comparison of results (6.150) and (6.151) indicates a paradoxical situation: nonlinear scaling is revealed for selected variables.

Control & Review Questions and Exercises

6.1 Exercise
State the difference between walks and flight models.

6.2 Question
What is the difference between transport and diffusion?

6.3 Exercise
List the materials in which anomalous transport and diffusion are observed.

6.4 Exercise
What type of amorphism does HCTRF use?

6.5 Exercise & Question
Name the base PDF on which the HCTRF formalism is based. What physical sense does a sharp PDF make?

6.6 Exercise & Question
Discuss the multi-step/multi-scale propagator structure. What role do waiting-time distributions play in it?

6.7 Exercise
Derive the relationship between the memory kernel and the waiting-time distribution.

6.8 Exercise
Build a separable space-time waiting-time distribution. Write it for the exponential substrate. Derive its asymptotic closed form. Based on this, proceed analogously for sojourn probability.

6.9 Problem
Prove that for a small value of drift force, jump rates in the presence of this force and its absence are equal to an accuracy of a small higher order. Compare the spatio-temporal waiting-time distributions for both cases.

6.10 Problem
Derive the formula for an n-step propagator and, hence, for a full propagator in both real and dual (Fourier-Laplace) space.

6.11 Problem
Derive the formula for the memory kernel appearing in the generalized homogeneous master equation.

6.12 Question
What form of memory kernel generates the telegraphists' equation? When do we deal with Fick's diffusion equation and the wave equation?

6.13 Question
Which form of waiting time distribution leads to the memory kernel's long-term dependence on time, and which leads to memory deficiency?

6.14 Problem

Specify the conditions under which the winning in the Weierstrass extension of the Petersburg game is finite and under which it is infinite and why. Which limit theorems will break down in the latter case—mention at least two.

6.15 Exercise & Question

Derive the condition for a finite PDF of waiting times, $\phi(t)$. What condition should be met so that the average residence time $\langle t \rangle$ is also finite?

6.16 Exercise

Output the scaling condition with the number of time steps L average single time step.

References

1. Montroll, E.W., Shlesinger, M.F.: On the wonderful world of random walks. In: Lebowitz, J.L., Montroll, E.W. (eds.) Nonequilibrium Phenomena II. From Stochastics to Hydrodynamics. Studies in Statistical Mechanics, vol. XI, pp. 1–121. North-Holland, Amsterdam (1984)
2. Kozłowska, M., Kutner, R.: Anomalous transport and diffusion versus extreme value theory. Phys. A **357**, 282–304 (2005). https://doi.org/10.1016/j.physa.2005.06.03
3. Weiss, G.H.: A primar of random walkology. In: Bunde, A., Shlomo Havlin, S. (eds.) Fractals in Science. Second Printing, Springer, Berlin (1995). https://doi.org/10.1007/978-3-662-11777-4
4. Uchaikin, V., Sibatov, R.: Fractional Kinetics in Solids: Anomalous Charge Transport in Semiconductors, Dielectrics, and Nanosystems. World Scientific, New Jersey (2013). https://doi.org/10.1142/9789814355438-0001
5. Zumofen, G., Klafter, J., Blumen, A.: Models for Anomalous Diffusion. In: Richert, R., Blumen, A. (eds.) Disorder Effects on Relaxation Processes: Glasses, Polymers, Proteins, pp. 251–278. Springer, Berlin (1994). https://doi.org/10.1007/978-3-642-78576-4-8
6. Scher, H., Montroll, E.W.: Anomalous transient-time dispersion in amorphous solids. Phys. Rev. B **15**, 2455–2477 (1975). https://doi.org/10.1103/PhysRevB.12.2455
7. Metzler, R., Klafter, J.: The random walk's guide to anomalous diffusion: a fractional dynamics approach. Phys. Rep. **339**(1), 1–77 (2000). https://doi.org/10.1016/S0370-1573(00)00070-3
8. Scharfe, M.E.: Transient photoconductivity in vitrous As_2Se_3. Phys. Rev. B **2**, 5025–5034 (1974). https://doi.org/10.1103/PhysRevB.2.5025
9. Gill, W.D.: Drift mobilities in amorphous charge-transfer complexes of trinitrofluorenone and poly-n-vinylcarbazole. J. Appl. Phys. **43**, 5033–5040 (1972). https://doi.org/10.1063/1.1661065
10. Pfister, G.: Pressure-dependent electronic transport in amorphous As_2Se_3. Phys. Rev. Lett. **33**, 1474–1477 (1974). https://doi.org/10.1103/PhysRevLett.33.1474
11. Pfister, G.: Dispersive low-temperature transport in α-selenium. Phys. Rev. Lett. **36**, 271–273 (1976). https://doi.org/10.1103/PhysRevLett.36.271
12. Pfister, G., Scher, H.: Time-dependent electronical transport in amorphous solids: As_2Se_3. Phys. Rev. B **15**, 2062–2083 (1977). https://doi.org/10.1103/PhysRevB.15.2062
13. Campbell, A.J., Bradley, D.D.C., Lidzey, D.G.: Space-charge limited conductivity with traps in poly(phenylene vinylene) light emitting diodes. J. Appl. Phys. **82**, 6326–6342 (1997). https://doi.org/10.1063/1.366523
14. Weiss, G.H., Rubin, R.J.: Random walks: theory and selected applications. In: Prigogine, I., Rice, S.A. (eds.) Advances in Chemical Physics, vol. 52, pp. 364–505, Chap. 5. Willey, New York (2007). https://doi.org/10.1002/9780470142769.ch5

15. Bouchaud, J.-P., Georges, A.: Anomalous diffusion in disordered media: statistical mechanics, models, and physical applications. Phys. Rep. **195**(4&5), 127–293 (1990). https://doi.org/10.1016/0370-1573(90)90099-N
16. Haus, J.W., Kehr, K.W.: Diffusion in regular and disordered lattices. Phys. Rep. **150**(5& 6), 263–406 (1987). https://doi.org/10.1016/0370-1573(87)90005-6
17. Bouchaud, J.-P., Potters, M.: Theory of Financial Risks. From Statistical Physics to Risk Management. Second reprint, Cambridge University Press, Cambridge (2001)
18. Embrechts, P., Klüppelberg, C., Mikosh, T.: Modelling Extremal Events for Insurence and Finance. Springer, Berlin (1997)
19. Gradshteyn, I.S., Ryzhik, I.M.: Table of Integrals, Series, and Products. In: Jeffrey, A., Daniel Zwillinger, D. (eds.), 7th edn, Elsevier & Academic Press, Amsterdam (2007)
20. Mounthus, C., Bouchaud, J.-P.: Models of traps and glass phenomenology. J. Phys. A: Math. Gen. **29**, 3847–3869 (1966). https://doi.org/10.1088/0305-4470/29/14/012
21. Bateman, H., Erdélyi, A.: Tables of Integral Transforms, vol. 1. McGraw-Hill, New York (1954)
22. Stein, E.M., Shakarchi, R.: Complex Analysis: Princeton Lectures in Analysis, vol. II. Princeton University, Princeton & Oxford (2003)
23. Klafter, J., Zumofen, G.: Probability distributions for continuous-time random walks with long tails. J. Phys. Chem. **98**(30), 7366–7370 (1994). https://doi.org/10.1021/j100081a022
24. Zumofen, G., Klafter, J.: Scale-invariant motion in intermittent chaotic systems. Phys. Rev. E **47**(2), 851–863 (1993). https://doi.org/10.1103/physreve.47.851
25. Hilfer, R. (ed.): Applications of Fractional Calculus in Physics. World Scientific, Singapore (2000). https://doi.org/10.1142/3779
26. Gubiec, T., Klamut, J., Kutner, R.: Multi-phase long-term autocorrelated diffusion: stationary continuous-time weierstrass walk versus flight. In: Grech, D., Miśkiewicz, J. (eds.) Simplicity of Complexity in Economic and Social Systems, Chap. 4, pp. 55–88, Springer Proceedings in Complexity. Springer, Berlin (2021). https://doi.org/10.1007/978-3-030-56160-4-4

Chapter 7
Fractal Wanderings

Abstract In this chapter, we explore fractal wandering through the lens of a particularly notable example—Weierstrass wandering. These walks elucidate the fundamental mechanism behind the emergence of algebraically diminishing long-range "tails" in both probability distributions and correlation functions. As demonstrated, this phenomenon is rooted in *rare, extreme events* generated, e.g., by the *stochastically self-similar (hierarchical) structure of the trajectory* of the process (i.e., the wandering of a test particle). Viewed broadly, this trajectory exhibits a stochastic fractal structure. The presence of rare, extreme events introduces a novel avenue of exploration in statistical physics and chaotic dynamics. This path diverges from traditional approaches and, despite notable progress, remains an area of active and intense research. In this chapter, we employ the Weierstrass spatial function. In Chap. 6, we considered the Weierstrass time function. From this perspective, Chaps. 6 and 7 are complementary, covering the space-time of random events.

7.1 Weierstrass Flights

Weierstrass[1] flights present a specific case of Weierstrass wanderings, i.e., a stochastic process that may occur when

(1) displacements of a single molecule can be treated as immediate, so-called Weierstrass flights

as well as those in which

(2) The speed of movement of the particle is finite, the so-called Weierstrass walks.

Describing the first case is considerably simpler due to the fewer degrees of freedom in characterization. The focal point of our discussion lies in the layout. The stochastic process known as the fly-by type is alternatively referred to as *hopping* or

[1] We use the term "Weierstrass wandering" here as an extension of "Weierstrass walks," a concept introduced by Montroll and Schlesinger [1]. In our textbook, we prefer the term "random wandering" over "random walk" because the former applies to both walks and flights of test particles, as required.

M. Chorowski et al., *Anomalous Stochastics*, Understanding Complex Systems,
https://doi.org/10.1007/978-3-031-80392-5_7

jumping process. This term finds extensive application in condensed matter physics, particularly solid-state physics, where it depicts phenomena like diffusion and ionic conduction.

7.2 Definitions and Interpretations

Let's establish the spatial component, denoted as $p(x)$, representing the probability density (process) of a test particle in motion along the vector x during a single *flight*. We will focus on one-dimensional flights to simplify the mathematical discussion. Flights in spaces with more dimensions are addressed in Appendix A.12. In particular, we delve into spherically symmetric Weierstrass flights, also known as Rayleigh-Pearson walks, as a direct use of the one-dimensional scenario.

Let us present the following initial definition, elucidating the discrete-time kinetics of a molecule's single (arbitrarily chosen) displacement along the x vector.

$$p(x) = \frac{1}{2}\sum_{j=0}^{\infty} w_j[\delta(x - b_j) + \delta(x + b_j)], \tag{7.1}$$

which is a direct extension of the formula (6.8) from Sect. 6.3.1 limited to $j = 0$, i.e., taking into account only hops to the nearest neighbors; here the weights $w_j,\ j = 0, 1, 2, \ldots$, satisfy the normalization condition,

$$\sum_{j=0}^{\infty} w_j = 1, \tag{7.2}$$

denote the probability of the molecule moving a distance b_j in generation j. Of course, the condition (7.2) immediately follows the normalization condition for the probability density function $p(x)$,

$$\int_{-\infty}^{\infty} p(x)dx = 1. \tag{7.3}$$

Let us observe that the factor of $1/2$ preceding the sum represents the probability of the molecule opting for one of the two permissible orientations of the displacement vector x. As each orientation is equally probable, $p(x)$ characterizes random movement in isotropic space, indicating the absence of an external field.

Let's highlight that the definition (7.1) applies to hops in discrete time. Consequently, the total displacement of the molecule, denoted as $X(n)$, after n time steps, can be expressed straightforwardly,

$$X(n) = \sum_{k=1}^{n} x_k, \tag{7.4}$$

the index k numbers the individual displacements of the test particle in subsequent time steps. We will return to the analysis of the total displacement when discussing the variable-step Weierstrass flight.

We still allow only the simplest hierarchical/geometrical variant of wandering, in which the ratio of weights in subsequent generations $j = 0, 1, 2, \ldots$, is a decreasing function and independent of the generation, i.e.,

$$\frac{w_{j+1}}{w_j} = \frac{1}{N} < 1, \tag{7.5}$$

which means that the N parameter acts as a coefficient of the stochastic similarity (here stochastic weakening coefficient) of the stochastic structure. It follows immediately from (7.2) and (7.5) that

$$w_j = \left(1 - \frac{1}{N}\right)\frac{1}{N^j} \tag{7.6}$$

decreases exponentially with order j. Still, in the spirit of dependency (7.5), we assume that

$$\frac{b_{j+1}}{b_j} = b > 1, \tag{7.7}$$

b acts as the geometric similarity coefficient (sometimes called the geometric gain coefficient of the stochastic structure). Similarly to the weights, (7.7) immediately results in a power-law relationship,

$$b_j = b_0 b^j, \tag{7.8}$$

Here, b_0 represents a constant, unit zero-order flight length. At specific points, we will adopt the convention $b_0 = 1$ to streamline mathematical discussions. Assuming that longer flights are less probable due to increased resistance to movement, i.e., assuming that $b > 1$. On the contrary, the scenario where $b < 1$ might apply to trajectories meandering in phase spaces with hierarchical traps, commonly encountered in nonlinear dynamic problems with chaotic behavior. The case of $b = 1$ is considered marginal and will not be addressed here.

The expression (7.6) and (7.8) shows that test particle flights can be grouped into generations both according to the frequency of their occurrence and the length of the flights, which allows the output formula to be rewritten (7.1) in the form,

$$p(x) = \frac{1}{2}\left(1 - \frac{1}{N}\right)\sum_{j=0}^{\infty}\frac{1}{N^j}[\delta(x - b_0 b^j) + \delta(x + b_0 b^j)]$$
$$= \frac{1}{2}\left(1 - \frac{1}{N}\right)\sum_{j=0}^{\infty}\frac{1}{N^j}\delta(\mid x \mid - b_0 b^j). \tag{7.9}$$

As evident from the formula (7.9), it permits test particle flights of arbitrary lengths. The term "flights" is employed to describe the meandering of a single molecule in this context, as both the overarching definition (7.1) and the formula (7.9) characterize the molecule's immediate movement between successive turning points (with the stops demarcating the beginning and end of a single flight). It's crucial to highlight that all the conclusions articulated in this chapter also extend to the previously mentioned spherical Weierstrass flights, as detailed in Appendix A.12. For instance, the discourse here regarding formulas (7.1) and (7.9) is equally applicable to formula (A.106) and its associated definitions found in the Appendix A.12.

7.2.1 Asymptotic Form of a Single Displacement PDF

The asymptotic form of $p(x)$ can be addressed through various approaches. Here, we opt for a straightforward method derived from a suitable variable change. We focus on determining the probability $p(x)$ for $\mid x \mid \gg b_0$. To achieve this, let's examine the probability $w(j)$ defined by the formula (7.6), where the variable j is expressed as b^j. It's crucial to recall that, based on the definition of Weierstrass flights (7.9), $b^j = \mid x \mid / b_0 \Leftrightarrow j = \frac{1}{\ln b}\ln\left(\frac{|x|}{b_0}\right)$. Utilizing probability invariance as a scalar, we can introduce the following equality:

$$w(j)dj = p(\mid x \mid)d \mid x \mid$$
$$\Leftrightarrow p(\mid x \mid) = \check{w}(\mid x \mid)\frac{dj(\mid x \mid)}{d \mid x \mid} = \frac{1 - \frac{1}{N}}{\ln b}\frac{b_0^{\beta}}{\mid x \mid^{1+\beta}}, \tag{7.10}$$

where we took advantage of the fact that after replacing the variable j with $\mid x \mid$, the probability $w(j)dj$ becomes the probability $p(\mid x \mid)d \mid x \mid$, wherein

$$\check{w}(\mid x \mid) = w(j(\mid x \mid)) = w\left(j = \frac{1}{\ln b}\ln\left(\frac{\mid x \mid}{b_0}\right)\right) = \left(1 - \frac{1}{N}\right)\frac{b_0^{\beta}}{\mid x \mid^{\beta}}, \tag{7.11}$$

as

$$\frac{1}{N^j} = \exp(-j \ln N) = \exp\left(-\beta \ln\left(\frac{|x|}{b_0}\right)\right) = \left(\frac{|x|}{b_0}\right)^{-\beta} \tag{7.12}$$

and $\beta = \ln N / \ln b > 0$. Here, we also took advantage of the possibility of extending the discrete variable j. This results from the necessity and sufficient condition, i.e.,

$$\frac{d\, j(|x|)}{d\, |x|} = \frac{1}{\ln b} \frac{1}{|x|} \ll 1 \Leftrightarrow |x| \gg \frac{1}{\ln b} \tag{7.13}$$

is easy to achieve. As evident, the asymptotic distribution (7.10) is a fat-tailed symmetric distribution with an unbounded positive value of the exponent $\beta > 0$. This stems directly from the boundless hierarchical nature of this wandering process. The Weierstrass walk facilitates exploring the connection between fractal and autocorrelated anomalous walks.

Let us add, for useful information, that in the case of a d-dimensional Weierstrass walk, the asymptotic one-dimensional PDF $p(x)$ extends into the form [4]

$$p(\mathbf{r}) \propto \frac{b_0^{\beta}}{r^{d+\beta}}, \tag{7.14}$$

where $r = |\mathbf{r}|$ is the length of the vector $\mathbf{r}$. We presented an analogous result for the three-dimensional spherical Weierstrass wander presented in Appendix A.12.

Recall that the fat/heavy tail above causes the distributions of individual/single test particle displacements to overlap, leading to their autocorrelation. The concept is that the power-law decay of the distributions for individual flights or walks generates, via the CTRW formalism, a power-law decay of the aggregate flights or walks. This leads to autocorrelations present in anomalous or even singular diffusion.

7.2.2 *Self-similar Structure of Weierstrass Flights*

Our deliberations are of crucial rationale. We began our exploration by analyzing a single displacement test particle. We will soon observe that Weierstrass flights exhibit self-similarity in the probabilistic sense. Therefore, it is reasonable to expect that even a basic solitary displacement would encapsulate significant information about the trajectory. This aspect mirrors the behavior observed in Brownian motion, where the self-diffusion coefficient, a macroscopically measurable quantity, could be expressed, on the one hand, by microscopic parameters characterizing only an individual molecule's displacement and, on the other hand, by its macroscopic/asymptotic characteristics.

The pattern interpretation (7.9) becomes exceptionally straightforward when the parameter N is a natural number greater than 1. Specifically, the formula (7.5) indicates that flights of length $b_0 b^{j+1}$ are N times less likely than flights of length $b_0 b^j$. Consequently, we can assert that, on average, before a molecule undertakes a flight of the order of $j+1$, it must undergo N flights of the order of j. Temporarily overlooking the inevitable fluctuations in the sequence of flights and the variations in the proportions between the numbers of flights of different lengths, this can be depicted schematically in graphical form. As a simple example, assume that $N = 3$ and $b = 2$.

The subsequent (two-part) Fig. 7.1 illustrates, for enhanced clarity, regularized (i.e., devoid of fluctuations) wandering in two-dimensional space. Importantly, this depiction in no way contradicts the overarching rules (7.5) and (7.7) that define Weierstrass flights.

A striking feature of such an ordered trajectory of a wandering particle is its self-similar character, which can be seen by simply comparing the Fig. 7.1a, b.

Figure 7.1a illustrates the trajectory in detail. To achieve a flight of length $b^1 = 2$, the molecule must undergo $N = 3$ flights of unit length (zero order). Similarly, a flight of length $b^2 = 4$ requires $N = 3$ flights of the first order, etc. Consequently, before starting a flight of $0 \leq j = 2$, the $N^{j=2} = 9$ units of length flights (of zero order, i.e., of length b^0) must be completed. Generally, the quantity N^j can be interpreted as the average number of zero-order flights required for a flight of length b^j to occur.

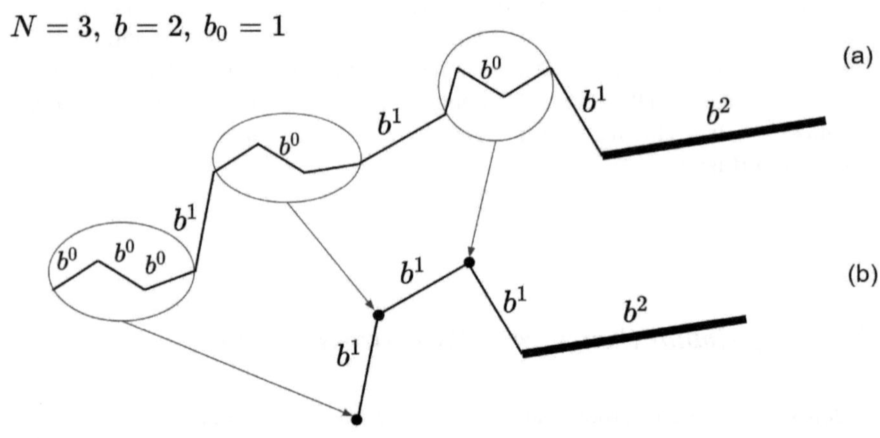

Fig. 7.1 **a** An initial part of the trajectory, depicted schematically and devoid of fluctuations through regularization, consists of hierarchically ordered single-step displacements. This trajectory is characterized by a discretized single-step PDF (7.9), illustrated here for the case of $N = 3$, $b = 2$, $b_0 = 1$, and $0 \leq j \leq 2$. **b** When observed from a considerable distance, the regions outlined as ovals in figure (**a**) condense into points (black dots) in this illustration. Notably, the self-similarity and scaling properties become apparent, accompanied by the representation of a rare event—a displacement of length b^2 (section bolded in black)

The single flight that initiates the jth generation in the flight hierarchy belongs to a set consisting of $(N^{j+1} - 1)/(N - 1)$, which equals $N^j + N^{j-1} + \cdots + N^1 + N^0$. This set defines the so-called "rare, extreme event" if it satisfies an additional condition introduced below. As we demonstrate, these events play a crucial role in "anomalous diffusion." The transformation of exponential relaxation into a power law is a consequence of these events, causing ordinary diffusion to deviate from its typical behavior.

In broad terms, a rare extreme event, particularly a rare extreme flyby, stands out as distinct from previously observed events and is at least an order of magnitude larger. However, as we demonstrate, this condition is necessary but insufficient. Under specific circumstances, a stochastic self-similar trajectory can generate rare, extreme events. Conversely, as we will elucidate, rare, extreme events consistently contribute to constructing a stochastic self-similar trajectory. Moving forward, we aim to provide a more precise definition of a rare, extreme event.

The quantity N^j can be formally regarded as the elementary "mass" of the Weierstrass wander, with the total "mass" defined as $(N^{j+1} - 1)/(N - 1)$, accumulated within the region characterized by the linear size b^j. We will demonstrate that the connection between the accumulated "mass" and the linear size gives rise to the presence of a universal exponent. This exponent, which we term the *self-similarity dimension*, is denoted as d_s.

In Fig. 7.1b, the trajectory is captured from a specific distance so that the resolution of the "photograph" prevents the discernment of specific details. This inability to distinguish finer details results in grouping each set of three individual displacements into a single point. Consequently, given this particular photo resolution, the smallest distinguishable group comprises three displacements of length b^1. As observed, the trajectory in Fig. 7.1b remains indistinguishable from the one in Fig. 7.1a, except for the transition

(1) the scale—it is drawn at the scale b times larger
(2) (possibly) accidental differences in the orientation of subsequent displacement vectors corresponding to each other in both plots.

A stochastic trajectory that satisfies the above properties is termed **self-similar stochastic trajectory**, which implies its **scaling-invariance** on average. Expanding our observation to even greater distances, descending to lower levels of image granularity, we would render groups comprising three displacements of length b^1 visible only as points. This iterative process does not significantly alter the original representation in Fig. 7.1a, etc. This procedure can be seamlessly extended without hindrance, bounded only by the size of the trajectory itself.

The above procedure can be formalized by writing for each granularity level $k(= 0, 1, 2, \ldots, j)$ the following entailment relation,

$$N^{j-k} \text{ of } b^k \text{ precedes } b^j. \tag{7.15}$$

That is, the k-th level of granularity entails treating a flight of length b^k as a singular entity, with all preceding flights of shorter lengths regarded (at this resolution) as mere points. Consequently, an average number of N^{j-k} displacements, each of length b^k, precedes displacement of length b^j, which is correspondingly longer (by a factor of b^{j-k}).

It can be demonstrated that the self-similar structure does not just characterize the entire trajectory; it also encompasses the ensemble of all turning points, denoted by solid circles in the figure. Additionally, the fractal dimension of this point set is denoted by $d_f = \beta$.

As indicated in the stochastic context here, the property of self-similarity is a fundamental characteristic of stochastic fractal structures, which will be explored further. Identification of the self-similarity property was facilitated by shedding the inevitable fluctuations in flight (both in quantity and order/shuffling). Essentially, we are dealing with a stochastic self-similar structure or, more broadly, a fractal structure. After undergoing the regularization (ordering) procedure, where fluctuations (disorder) are eliminated, the structure attains self-similarity in the deterministic sense. In a broader sense, it transforms into a deterministic fractal, at least within a substantial number of generations.

Conversely, a deterministic structure can transition into a stochastic one by introducing disorder, such as a specific type of fluctuation, ensuring that the self-similarity property is preserved on average. Consequently, for actual trajectories, it becomes essential to construct statistics on the length of flights, referencing the frequency of flights of particular lengths. Based on this data, the determination of the appropriate weight ratios becomes crucial. The primary objective of these considerations is to establish such statistics in a closed form rather than as an infinite sum.

Now, we can dive into a fundamental inquiry: how does the elementary/basic "mass" of the stochastic structure under consideration scale concerning the linear size of the extreme event? To address this query promptly, we can employ the relation for the order/generation j. Specifically, we express the equality as follows:

$$N^j = (b^j)^\beta, \tag{7.16}$$

from which the exponent β can be determined in a form independent of j,

$$\beta = \frac{\ln N}{\ln b} \tag{7.17}$$

This result directly follows from the inherent self-similar nature of the structure. The exponent β, the **self-similarity dimension**, essentially represents d_s. Consequently, the self-similarity dimension d_s becomes a distinctive characteristic of the self-similar structure. It is worth noting that this assertion holds not only for the specific case at hand but also for all types of fractal structures, whether stochastic or deterministic. In such structures, the fractal dimension is consistently expressed as the quotient of two linear logarithms related to the structure's characteristics.

In Sect. 7.6, we emphasize the threshold nature of wandering, elucidating the specific values of the exponent β that make the structure stable. The structure remains intact after numerous flights and does not blur beyond these critical values.

7.3 Structural Factor of Weierstrass Flights

The structure factor of Weierstrass flights, $\tilde{p}(k)$, also called the characteristic function of Weierstrass flights/PDF, is defined as the Fourier transform,

$$\tilde{p}(k) = \int_{-\infty}^{\infty} dx \exp(-ikx)p(x), \tag{7.18}$$

assuming (after substituting the formula (7.9) and integration) form of the sum of a geometric-trigonometric series,

$$\tilde{p}(k) = \left(1 - \frac{1}{N}\right) \sum_{j=0}^{\infty} \frac{1}{N^j} \cos(kb^j); \tag{7.19}$$

the structural factor of the Weierstrass flights given by the expression (7.19) is called the Weierstrass function or half Weierstrass-Mandelbrot function—its first few modes are presented in Fig. 7.2,

The basic properties of the Weierstrass function (7.19) are discussed below.

7.4 Spatial Scaling Equation

Does this raise the fundamental question of the conditions under which $\tilde{p}(k)$, expressed by the formula (7.19), can be represented (with a good approximation) in a closed form? The answer to this question is in two steps.

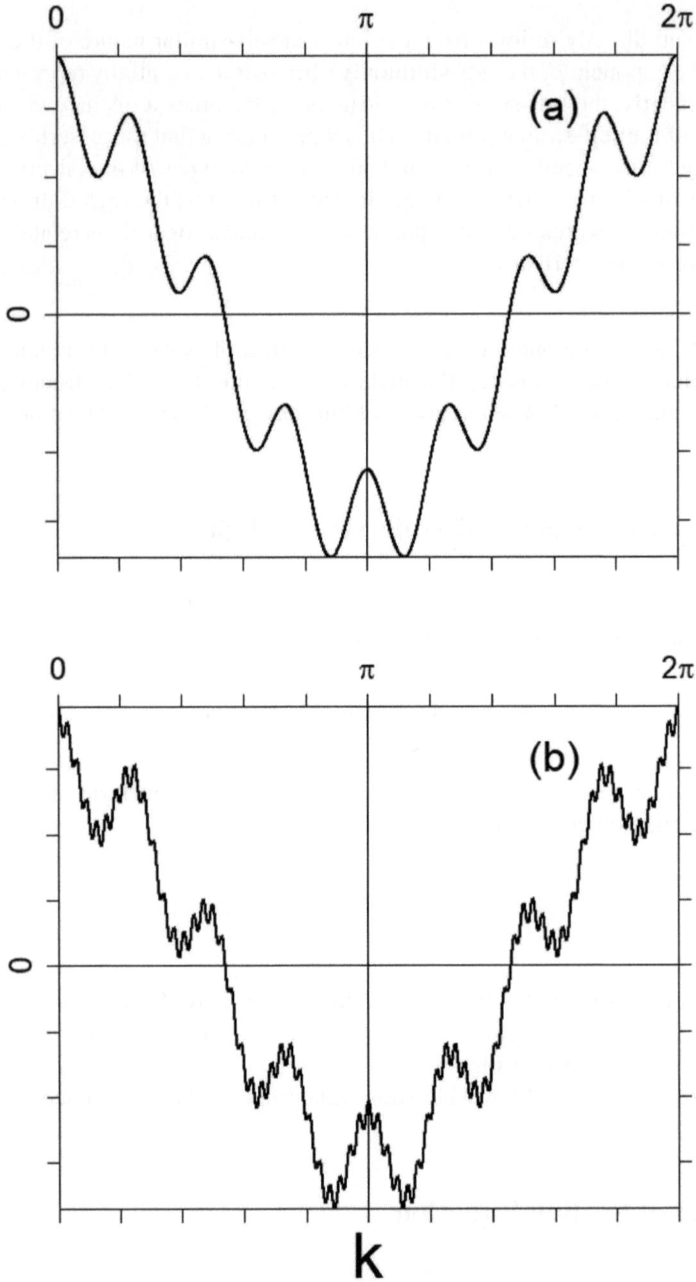

Fig. 7.2 Schematic depiction of various characteristic function components of the Weierstrass single-step PDF (7.19) for $N = 4$, $b = 8$ (yielding $\beta = \ln N/\ln b = 2/3$), and $b_0 = 1$. The upper plot (**a**) illustrates the sum of the first two components, while the lower one (**b**) displays the first three. It can be inferred that in the limit of an infinite sum of components, the continuity of this function persists. However, differentiability is forfeited due to the infinite hierarchy of progressively smaller, more pronounced bends in the characteristic function of the Weierstrass PDF, evident at each function turning point

First note that $\tilde{p}(k)$ satisfies the following non-uniform/non-homogeneous scaling equation,

$$\tilde{p}(bk) = N\tilde{p}(k) - (N-1)\cos(k), \tag{7.20}$$

which allows searching for its solution $\tilde{p}(k)$ in the form of the sum of the general solution (regular, normal), $\tilde{p}_n(k)$, non-homogeneous equation (7.20) and the special (singular) solution, $\tilde{p}_s(k)$, of the homogeneous equation,

$$\tilde{p}_s(bk) = N\tilde{p}_s(k). \tag{7.21}$$

The form of the general solution is inherently dictated by the heterogeneity of the Eq. (7.20), specifically $\cos(k)$. As a result, it takes the form of a series of powers that exclusively includes even powers of the variable k. The coefficients are determined conventionally by substituting this series into Eq. (7.20) and equating expressions with identical powers of k to derive the sought-after coefficients. Consequently, the general solution takes the following form:

$$\tilde{p}_n(k) \approx 1 - D'k^2, \tag{7.22}$$

where the so-called *generalized diffusion coefficient*

$$D' = \frac{1}{2}\frac{1-\frac{1}{N}}{1-\frac{b^2}{N}}. \tag{7.23}$$

More specifically, the discussion about the diffusion coefficient becomes pertinent when the average time required for an individual to complete a single flight has been explicitly defined—implicitly assumed for individuals in this context, a topic we will delve into more extensively later. The provided form (7.23) of the D' coefficient will be instrumental for subsequent analysis, particularly in the classification of diffusion types. Our focus in this context is restricted to the first two terms of the expansion. This limitation is intentional, as we are concerned with both regular and singular solutions exclusively in scenarios where $| k | \ll 1$, indicating our emphasis on solutions that primarily describe long flights.

Secondly, the singular solution $\tilde{p}_s(k)$ of the Eq. (7.21) can be designed in the form of a product of a slowly variable function and a power function (it was already mentioned in Sect. 3.2 in the context of comparison with empirical data)

$$\tilde{p}_s(k) \approx -Q\left(\frac{\ln | k |}{\ln b}\right) | k |^{\beta'}, \tag{7.24}$$

where the slowly changing pre-exponential function has the following property,

$$Q\left(\frac{\ln \mid bk \mid}{\ln b}\right) = Q\left(\frac{\ln \mid k \mid}{\ln b} + 1\right) = Q\left(\frac{\ln \mid k \mid}{\ln b}\right) \tag{7.25}$$

it is a periodic function with a period equal to 1. Therefore, it has the following Fourier expansion,

$$Q\left(\frac{\mid k \mid}{\ln b}\right) = \sum_{n=-\infty}^{\infty} A_n \exp\left(2\pi i n \frac{\ln \mid k \mid}{\ln b}\right), \tag{7.26}$$

called logarithmic periodicity (already discussed in Sect. 3.2). In the simplest case, it reduces to a constant, which we denote by $D'_f (= A_0)$—in this chapter, for simplicity of considerations, we limit ourselves only to this special case; the most general form of the function Q as well as the explicit form of the *fractal diffusion coefficient* D'_f we have derived in the Appendix A.13 using the Mellin transform.[2]

We find the exponent β' by substituting the expression (7.24) into the Eq. (7.21). In this way, we get that

$$\beta' = \frac{\ln N}{\ln b} = \beta. \tag{7.27}$$

Unfortunately, it is impossible to determine the fractal diffusion coefficient D'_f in this way.

It is characteristic that the exponent β introduced earlier is, in fact, responsible for the non-analytical nature of the singular solution. However, the appearance of this exponent was caused by the self-similar nature of Weierstrass flights, so its **self-similar property is the primary cause of the existence of the singular solution**. Therefore, the exponent β can be treated as primary in all analyses, including classifications.

Finally, combining (7.22) and (7.24) together with (7.27), we find the structural factor of Weierstrass flights as

$$\begin{aligned}\tilde{p}(k) &= \tilde{p}_s(k) + \tilde{p}_n(k) \approx 1 - D'k^2 - D'_f \mid k \mid^{\beta} \\ &\approx \exp(-D'k^2 - D'_f \mid k \mid^{\beta}).\end{aligned} \tag{7.28}$$

As one can see, the Gaussian and the singular components coexist—a significant conclusion of the Weierstrass wander analysis with a universal character. The role of these components depends critically on the exponent β value. Namely (remembering that $\mid k \mid \ll 1$) we get,

[2] See, Bateman and Erdéley [2], and methods of residuals for calculating integrals in the complex plane—see [3].

$$\tilde{p}(k) \approx \begin{cases} 1 - D'k^2 \approx \exp\left(-D'k^2\right), & \text{for } \beta > 2 \\ 1 - D'_f \mid k \mid^{\beta} \approx \exp\left(-D'_f \mid k \mid^{\beta}\right), & \text{for } \beta < 2. \end{cases} \tag{7.29}$$

The marginal case where $\beta = 2$ requires a different and more advanced approach (using the Mellin transform and integration by residuals), which we discuss later. The above form of the structural factor of Weierstrass flights is valid for any dimension of the Euclidean space where the flights occur. This form allows us to find (in the form of a function and not a distribution as in the formulas (7.1) and (7.9)) the asymptotic form of the distribution $p(x)$.

7.5 Renormalization Solution of the Spatial Scaling Equation

Now, let us define the inverse problem. Namely, if we have the scaling equation (7.20), we will find its solution using the renormalization method, i.e., systematically and not by "guessing." The renormalization method is multistep.

Step zero is the scaling equation itself, which we rewrite as:

$$\tilde{p}(k) = N^{-1}\tilde{p}(bk) + G(k) \tag{7.30}$$

where heterogeneity

$$G(k) \stackrel{\text{def.}}{=} \left(1 - \frac{1}{N}\right)\cos(k). \tag{7.31}$$

In the first step we discreetly renormalize (rescale) the independent variable k in the above equation using the real constant b. We get,

$$N^{-1}\tilde{p}(bk) = N^{-2}\tilde{p}(b^2k) + N^{-1}G(bk), \tag{7.32}$$

where we additionally divided the obtained equation by N. Substituting the Eq. (7.32) into the Eq. (7.30), we obtain after rearranging:

$$\tilde{p}(k) = N^{-2}\tilde{p}(b^2k) + N^{-1}G(bk) + G(k) \tag{7.33}$$

As one can see, we used the initial equation (7.30) in two ways, i.e., first, we renormalized it, and then we replaced the one-time quantity with the renormalized $\tilde{p}(bk)$. Of course, we proceed similarly in the next steps of the renormalization procedure.

In the second step, renormalizing the Eq. (7.33), we get,

$$N^{-1}\tilde{p}(bk) = N^{-3}\tilde{p}(b^3k) + N^{-2}G(b^2k) + N^{-1}G(bk) \tag{7.34}$$

and as a result, using the Eq. (7.30), we have,

$$\tilde{p}(k) = N^{-3}\tilde{p}(b^3k) + N^{-2}G(b^2k) + N^{-1}G(bk) + G(k). \tag{7.35}$$

From the above equation, one can easily deduce what form we will get in the lth step. Namely,

$$\tilde{p}(k) = N^{-l}\tilde{p}(b^lk) + \sum_{j=0}^{l-1} N^{-j}G(b^jk). \tag{7.36}$$

We continue to assume that in the limit $l \to \infty$ the size of $\tilde{p}\left(b^lk\right)$ is bounded. So finally, passing with $l \to \infty$, we get the formula,

$$\tilde{p}(k) = \left(1 - \frac{1}{N}\right)\sum_{j=0}^{\infty}\frac{1}{N^j}\cos(kb^j), \tag{7.37}$$

which is identical to the expression (7.19), as it was required.

7.6 The Weierstrass Diffusion Worlds

In the first step, we examine under what conditions the average length of a single flight is finite and in which it is not. Let's, for this purpose, analyze the expression

$$\langle| x |\rangle = \int_{-\infty}^{\infty} dxp(x) \mid x \mid= \left(1 - \frac{1}{N}\right)\sum_{j=0}^{\infty}\left(\frac{b}{N}\right)^j, \tag{7.38}$$

where we used the formula (7.9) and (7.18). We distinguish here two significantly different cases,

(A) $\langle| x |\rangle < \infty$
(B) $\langle| x |\rangle = \infty$.

The expression (7.38) shows that we deal with case (1) if and only if $b/N < 1$, i.e., when $\beta > 1$, while case (2) if and only when $b/N \geq 1$, i.e., when $\beta \leq 1$. Considering both situations in conjunction with the analysis of the properties of the squared average of a single flight $\langle x^2\rangle$ is interesting.

Let us note that

$$\langle x^2\rangle = \int_{-\infty}^{\infty} dxp(x)x^2\left(= -\frac{d^2\tilde{p}(k)}{dk^2}\mid_{k=0}\right) = \left(1 - \frac{1}{N}\right)\sum_{j=0}^{\infty}\left(\frac{b^2}{N}\right)^j, \tag{7.39}$$

where, as before, we used the definition (7.1) and the expression (7.18)). Similarly, as before, let us consider two fundamentally different situations.

(i) Normal, where $\langle x^2 \rangle < \infty$.
(ii) Anomalous, where $\langle x^2 \rangle = \infty$.

We deal with first situation (i) if and only if the sum of the geometric series in the expression (7.39) converges, i.e., when

$$\frac{b^2}{N} < 1 \Leftrightarrow \beta > 2. \tag{7.40}$$

As a result, we get that

$$\langle x^2 \rangle = \frac{1 - \frac{1}{N}}{1 - \frac{b^2}{N}} = 2D' < \infty. \tag{7.41}$$

The abovementioned relationship is crucial as it allows for a generalized diffusion coefficient to be expressed. This coefficient is a macroscopic quantity that utilizes microscopic sizes, representing the average of a single flight's square. This feasibility arises from the self-similar structure of flights, meaning that flights occurring at various scales form similar sets.

Let us note that the existence of (generic) inequality

$$\langle | \, x \, | \rangle^2 < \langle x^2 \rangle, \tag{7.42}$$

implies in this case that

$$\langle | \, x \, | \rangle < \infty. \tag{7.43}$$

It is crucial to recognize that the finiteness of the second moment $\langle x^2 \rangle$ (attributed to the finite value of the structural factor $\tilde{p}(k)$ for any vector k) serves as a sufficient and necessary condition for the existence of the Gaussian distribution $p(x)$ for asymptotically large values of $| \, x \, |$. Therefore, there is no requirement for finite higher moments for the random variable x. In this scenario, the Weierstrass process is equivalent to the Wiener process, signifying a description of a random walk known as Brownian motion.

The second situation (ii) takes place if and only if

$$\frac{b^2}{N} \geq 1 \Leftrightarrow \beta \leq 2 \tag{7.44}$$

and as a consequence

$$\langle x^2 \rangle = \infty. \tag{7.45}$$

The situation described above is entirely unrelated to the behavior of the first absolute moment; it can be either finite or infinite. However, if the first absolute moment is infinite (as implied by the inequality (7.42)), it necessarily follows that the second moment is also infinite. Further analysis of higher moments is irrelevant in this context.

7.7 Simulations of Two-Dimensional Weierstrass Flights

Figure 7.3 shows simulations of two-dimensional Weierstrass flights for two distinct situations [4].

The top plots for the Brownian Weierstrass flights (linear time-dependence of the MSD) concern the situation for $2 < \beta = \ln 13/\ln 2 = 3.70$, while the bottom plots for the Lévy Weierstrass flights (MSD doesn't exist) concern the situation for $1 < \beta = \ln 3/\ln 2 = 1.585 < 2$. These are two fundamentally different situations that represent the asymptotically Brownian and Gaussian process having mainly limited Weierstrass flights (because $\langle x^2 \rangle < \infty$) and mainly unlimited Weierstrass flights (since $\langle x^2 \rangle = \infty$), respectively. Generally speaking, we can say that the Weierstrass wandering covers both worlds of stochastic processes: from the normal to the singular.[3] It asymptotically depends only on the parameter values controlling the Weierstrass process, i.e., exponent β.

At this juncture, Figs. 7.3 and 7.4 answer an important question: How will the boundless nature of an elementary flight's length or its dispersion manifest in experiments (numerical, i.e., simulations, or in the real world)? This query has broader implications and extends beyond the elementary flight vector; it can be applied to any random variable. The response to this inquiry is relatively clear-cut today. However, it made its way into the physics community at a surprisingly gradual pace [5].

A consequence of the above properties is the singular behavior of total mean square displacement, $\langle R^2(t) \rangle$, and therefore its estimate $\overline{R^2(t)}_M$ versus t presented in Figs. 7.3 (bottom right plot) and 7.4. This estimate is calculated using a statistical

[3] The category of singular stochastic processes includes anomalous processes and processes whose variance does not exist or diverges.

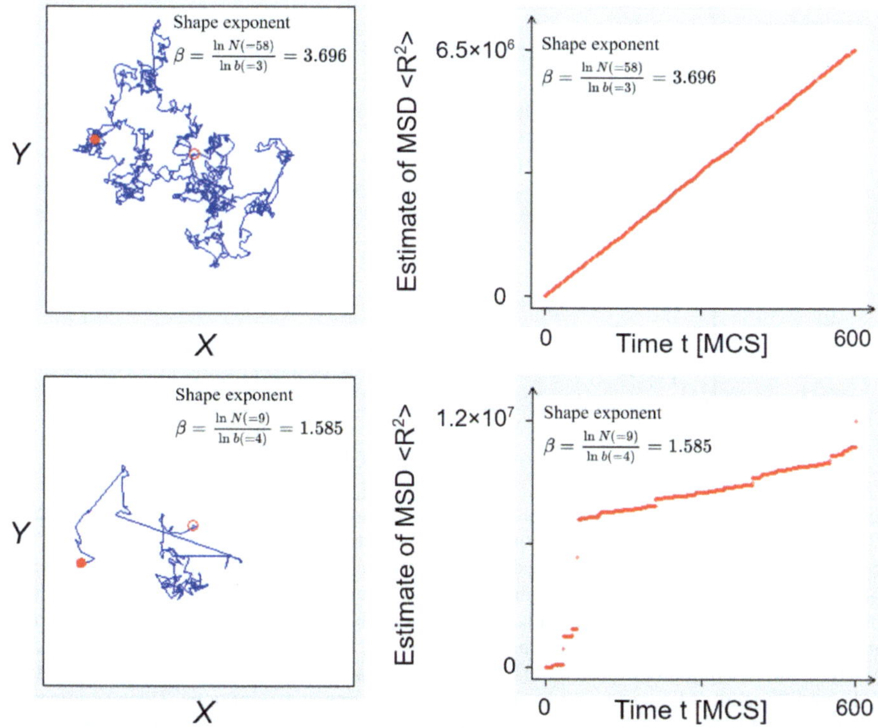

Fig. 7.3 Realizations obtained through Monte Carlo simulations (MCSs) depict two-dimensional Weierstrass flights. The shape exponent $\beta = 3.696$ governs the upper graphical representations, delineating an asymptotically Brownian and Gaussian domain synonymous with the Wiener process. In contrast, the shape exponent $\beta = 1.585$ influences the lower graphical representations, delineating an asymptotically non-Brownian and non-Gaussian domain characterized by a stable symmetric Lévy probability density function (PDF). In this latter context, the estimation of mean square displacement (MSD) reveals faults ("black swans"), which are conspicuously visible as trajectory elements or single-step displacements in the lower left graph. The left column illustrates instances of singular realizations of two-dimensional Weierstrass flight trajectories. The journey of the Weierstrass particle begins at a point marked by an empty circle and ends at a full one. The right column delineates the corresponding temporal mean square displacement estimations, calculated as an average across a statistical ensemble comprising $M = 10^4$ replicas. The results for larger ensemble sizes are elucidated in Fig. 7.4 below. The erratic and unstable behavior of the MSD estimates exemplifies its theoretical divergence within the Lévy PDF framework

ensemble of similar random-walk trajectories (i.e., stochastic replicas of the same random walk). Its value depends on the size M of the statistical ensemble. The bottom right plot in Fig. 7.3 shows that the count of $M = 10^5$ is insufficient to stabilize the estimate $\overline{R^2(t)}_{M=10^5}$.

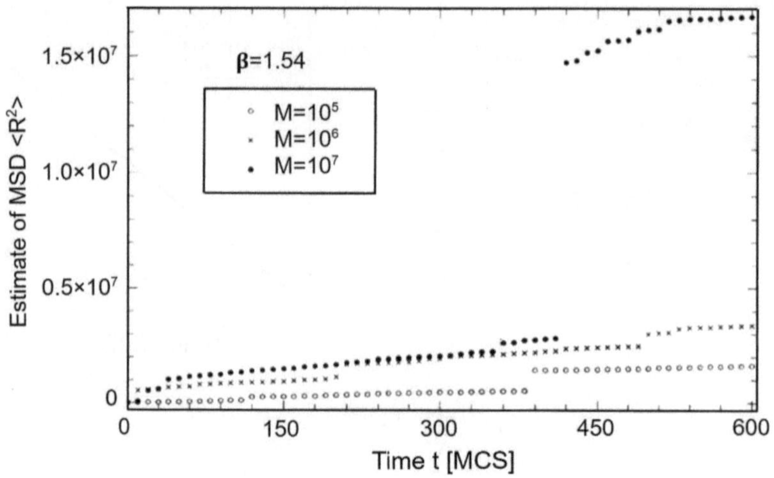

Fig. 7.4 The analysis of two-dimensional estimate, $\overline{R^2(t)_M}$, of mean-square displacement (MSD) $\langle R(t)^2\rangle$ against time t, derived from Monte Carlo simulations for scenario (ii), where $1 < \beta = 1.540 = \ln(1.158/1.10) < 2$, as discussed in our paper. We used three statistical ensembles (i.e., replicas of trajectories of Weierstrass flights) with increasing sizes $M = 10^5$ ($\circ$), 10^6 ($\times$), 10^7 ($\bullet$), representing a growing count of individual trajectory realizations until time $t = 6 \times 10^2$ Monte Carlo steps (MCS). It should be noted that expanding the size of the statistical ensemble causes an increase in the amplitudes of the faults. This indicates an instability or divergence, suggesting a failure of Bernoulli's Law of Large Numbers because of unmet assumptions. This phenomenon is attributed to the increased occurrences of prolonged flights of the test particle, i.e., rare extreme events or "black swans"—refer to Fig. 7.3 and Sect. 7.8. As the number of displacements of test particles increases, the duration of the process improves, and the probability of encountering such rare events increases. Adapted with permission from [4].

Figures 7.4 shows that a further increase in the number of M has the opposite effect to that intended: **as M increases, the amplitude of faults increases and not decreases**, as might be expected based on Bernoulli's Law of Large Numbers. As one can see, for singular random walks, Bernoulli's Law of Large Numbers stops working because its assumptions are not met.

7.8 Rare Extreme Events "Black Swans"

Applying the earlier considerations that led to formulas (7.16) and (7.17), we can now address a more profound question: How does the size (length), $\mid x_{\max} \mid$, of this rare and singular event (flight) scale with the total number of L flights (see Fig. 7.1 for details)? It's important to note that $\mid x_{max} \mid$ represents the maximum single flight value attained during these L flights. Therefore, for a substantial number of flights where $L \gg 1$, the relationship (i.e., the last approximate equality) is maintained,

$$L = \sum_{j=0}^{j_{max}} N^j = \frac{N^{j_{max}+1} - 1}{N - 1} \approx \frac{1}{1 - \frac{1}{N}} N^{j_{max}}, \tag{7.46}$$

because $N^{j_{max}} \gg 1$, where j_{max} is the highest value of the j-th generation that appeared during these L flights (in Fig. 7.1, $j_{max} = 2$ for simplicity), and

$$\mid x_{max} \mid = b_0 b^{j_{max}}, \tag{7.47}$$

of which, after eliminating the auxiliary quantity j_{max}, we obtain the desired relationship

$$\mid x_{max} \mid \approx A L^{1/\beta}, \tag{7.48}$$

where exponent $\beta = \ln N / \ln b$ and prefactor $A = b_0(1 - 1/N)^{1/\beta}$.

Let us note by the way that the formula (7.46) gives[4] that the probability, $w(x_{max})$, of a single flight of maximum length $\mid x_{max} \mid$ is, to a good approximation, $1/L$. Therefore, based on the above definition of the probability $w(x_{max})$ and the formula (7.48), we get that

$$w(x_{max}) \approx \frac{B}{\mid x_{max} \mid^{\beta}}, \tag{7.49}$$

where coefficient $B = A^{\beta}$.

It can be easily shown (using instead of Eq. (7.46) a simple expression $L_0 = N^{j_{max}}$ on the total number of elementary/shortest passes L_0 of length b_0) that $\mid x_{max} \mid$ scales with L_0 analogously to L (as given by Eq. (7.48)) where the proportionality coefficient is equal to unity. For this reason, sometimes L_0 and not L is called the "mass" of the random walk.

Let's focus on the two averages we tackled in this section. In the upcoming section, we'll delve into the third type. The first kind of average comprises (absolute) moments, denoted $\langle \mid x \mid^n \rangle_L$ for $n = 1, 2, \ldots$. These are calculated along the trajectory of the wandering molecule and depend on the total number of L flights, where $\langle \ldots \rangle_L$ represents the arithmetic mean over all L flights. The relationship holds in the scenario of an asymptotically large number of flights (for arbitrary natural n; extension to real numbers is straightforward),

$$\langle \mid x \mid^n \rangle_{L \to \infty} = \langle \mid x \mid^n \rangle = \int_{-\infty}^{\infty} \mid x \mid^n \, p(x) dx, \tag{7.50}$$

[4] A more subtle derivation, leading to a more precise formula (7.11), has been presented in Sect. 7.2.

where $\langle | x |^n \rangle$ is the second kind of average, the statistical average or weighted average (an expected value) for $n = 1$ and $n = 2$ was already discussed in the previous section. In general, the equality of averages

$$\langle \ldots \rangle_{L\to\infty} = \langle \ldots \rangle, \tag{7.51}$$

is nothing else than the kind of self-averaging property.[5]

This section will address a crucial question regarding the first type of averages: *How are the moments calculated along the trajectory scale with the number of flights L?*

To answer, we will employ an approximate relationship that holds for large values of L, allowing us to disregard both the fluctuations in the sequence of flights and the fluctuations in the proportions between the numbers of flights of different orders in a single sequence. The result is as follows:

$$\begin{aligned}\langle | x |^n \rangle_L &= \frac{(b^n)^0 N^{j_{max}} + (b^n)^1 N^{j_{max}-1} + (b^n)^2 N^{j_{max}-2} + \ldots + (b^n)^{j_{max}} N^0}{L} \\ &= \frac{N^{j_{max}}}{L} \sum_{j=0}^{j_{max}} \left(\frac{b^n}{N}\right)^j = \frac{N^{j_{max}}}{L} \frac{(\frac{b^n}{N})^{j_{max}+1} - 1}{\frac{b^n}{N} - 1} \\ &\approx \frac{N^{j_{max}}}{L} \frac{(\frac{b^n}{N})^{j_{max}+1}}{\frac{b^n}{N} - 1} \approx \frac{1}{L} \frac{1}{1 - \frac{N}{b^n}} (b^n)^{j_{max}}, \text{ for } \beta < n,\end{aligned} \tag{7.52}$$

where index j_{max} is the largest value among all indexes that appeared during L flights (see (7.46)). Furthermore, we assumed that $j_{max} \gg 1$ (which is a stronger assumption than $L \gg 1$). Let us note that the fraction $N^{j_{max}-j}/L,\ j = 0, \ldots, j_{max}$, estimates the probability of occurrence of the flight (among L of flights) that has length $b_0 b^j$.

Using the above formula and the expression (7.46), we obtain, after simple transformations,

$$\langle | x |^n \rangle_L \approx C_n L^{n/\beta - 1} \approx \frac{\hat{C}_n}{L} | x_{max} |^n, \text{ dla } \beta < n, \tag{7.53}$$

where the coefficient $C_n = A^n[1/(1 - N/b^n)]$ and $\hat{C} = 1/(1 - N/b^n)$; the case of $\beta = n$ is marginal, and therefore we do not address it here. It is worth realizing that we are primarily interested in the case of $n = 1$ and $n = 2$.

[5] Another type of average, considered in Sect. 7.9, is ensemble average $\langle \ldots \rangle_M$, where M denotes the size of this statistical ensemble. The equality $\langle \ldots \rangle_{L\to\infty} = \langle \ldots \rangle_{M\to\infty}$ is maintained, which is also a kind of self-averaging property. This should not be confused with ergodicity because we are not dealing here with phase space.

For thoroughness, let us explore the complementary scenario where the exponent is $\beta > n$. In this case, the inequality $b^n/N < 1$ holds, leading to a modified form of formula (7.52):

$$\langle | x |^n \rangle_L = \frac{N^{j_{max}}}{L} \frac{(\frac{b^n}{N})^{j_{max}+1} - 1}{\frac{b^n}{N} - 1} \approx \frac{N^{j_{max}}}{L} \frac{1}{1 - \frac{b^n}{N}} \approx \frac{1 - \frac{1}{N}}{1 - \frac{b^n}{N}} = 2\,\hat{D}. \tag{7.54}$$

Please note how eloquent Eq. (7.53) is. It says that an absolute moment of an arbitrary order n of a single displacement evolves as the n-th power of a single displacement of maximum length. This is a rare extreme event ("black swan") that governs the Weierstrass walk for $\beta < n$ for a trajectory consisting of L flights.

In the following, both relations (7.52) and (7.54) are used to determine, for example, the length of the path traveled by the particle and its total variance.

7.9 Average over the Statistical Ensemble

Let's delve into two profoundly significant implications of the scaling relation (7.53). Specifically, we aim to answer two questions:

(i) How does the average length of the trajectory traveled by the test particle (over L flights) scale with L?
(ii) How does the average of the square of the particle's displacement (as a result of its L flights) scale with L?

To address these questions, we must first clarify the type of averages involved. To achieve this, let's introduce a statistical ensemble consisting of a number $M \gg 1$ of similar trajectories, each resembling the others (i.e., stochastic replicas), with each trajectory encompassing L flights. When referring to averages in this context, we consider arithmetic averages across the statistical ensemble. Here is how we obtain them: for case (i), we calculate the length of the distance traveled due to L flights for the first trajectory, then the second, and so on, until the last trajectory numbered L. We then compute the arithmetic mean of these results. For $m \to \infty$, we denote this average over the ensemble as $\langle \ldots \rangle_{M \to \infty}$. This also applies to case (ii) below.

Similarly, for case (ii), we calculate the square of the resultant displacement for each trajectory and find the arithmetic mean. Crucially, the size of the statistical ensemble of trajectories aligns with the number of flights within each trajectory. This construction introduces a third type of average, and all three types prove immensely valuable in our considerations.

Consider case (i), denoting by $S(L)$ the length of a single molecule trajectory; the average mentioned above can be written in the form of the following scaling relation,

$$\begin{aligned}\langle S(L)\rangle_M &= \langle\mid x_1 \mid\rangle_M + \langle\mid x_2 \mid\rangle_M + \ldots + \langle\mid x_L \mid\rangle_M \\ &= L\langle\mid x \mid\rangle_M \approx C_{n=1} M^{1/\beta} \\ &= C_{n=1} L^{1/\beta} \approx \hat{C}_{n=1} \mid x_{max} \mid, \text{ for } \beta < 1; \end{aligned} \quad (7.55)$$

we assumed the trajectory's number of flights L equals the statistical ensemble's number/size M.

In the above derivation, we made use of the following:

(a) The definition of the trajectory length $S(L) =\mid x_1 \mid + \mid x_2 \mid + \cdots + \mid x_L \mid$, where $\mid x_l \mid$, $l = 1, 2, \ldots, L$, represents the lengths of individual flights, assumed to be proportional to powers of the similarity coefficient $b_0 b^j$. We also used the following assumptions in case (ii) below.
(b) We assumed (full) self-averaging, i.e., the equivalence of all three averages $\langle \ldots \rangle_{L\to\infty} = \langle \ldots \rangle_{M\to\infty} = \langle \ldots \rangle$.
(c) An almost self-evident relationship $\langle\mid x_l \mid\rangle = \langle\mid x_{l'} \mid\rangle$, where $l, l' = 1, 2, \ldots$, $L = M \gg 1$.
(d) The scaling relation (7.53).

Consider case (ii), denoted by $X(L) = x_1 + x_2 + \ldots + x_L$ the resulting displacement of the molecule as a result of L single flights. We will still use the assumption that individual flights are statistically independent, which leads (for $n = 2$ after using the expression (7.53)) to the following scaling relation,

$$\langle [X(L)]^2 \rangle_M = L \langle \mid x \mid^2 \rangle_M \approx \begin{cases} C_{n=2} L^{2/\beta} \approx \hat{C}_{n=2} (x_{max})^2, \text{ for } \beta < 2, \\ 2\,\hat{D}\,L, \text{ for } \beta > 2; \end{cases}$$

We remind that (as in case (i)), we assumed $M = L$.

Both cases reiterate a crucial insight: the significant role that the dimension of self-similarity β plays in Weierstrass flights. Moreover, and arguably more importantly, they underscore that beneath these scaling relations lies a singular phenomenon: the occurrence of rare, extreme events. This phenomenon not only emphasizes the relevance of statistical physics but also unveils a new realm of research and applications in the field.

7.10 Asymptotic of Pareto-Lévy Distribution Without Drift

We now prove the following key theorem

Lévy's theorem: Let be a characteristic function of the form (7.29) for an anomalous situation ($\beta < 2$) then,

$$p(x \to \pm\infty) \to\propto \frac{1}{\mid x \mid^{d+\beta}}, \tag{7.56}$$

where d is the Euclidean dimension of the space wanders occur. We continue to consider the situation $d = 1$, which does not (actually) change the generality of the proof.

The proof is in three parts for three different ranges of β. We will quote it here in full because of its central importance for our lecture and its highly characteristic calculation method. For the ranges mentioned, our goal is to calculate the Fourier transform of the form,

$$p(x \to \pm\infty) \approx \frac{1}{2\pi} \int_{-\infty}^{\infty} dk \exp(-ikx) \exp(-D'_f \mid k \mid^{\beta}). \tag{7.57}$$

It's worth noting that to derive the asymptotic form of $p(x)$, it suffices to leverage the information about the structural factor $\tilde{p}(k)$ for $D'_f \mid k \mid \ll 1$. This enables the formal extension of the integration limits from negative to positive infinity. In other words, what occurs in the far reaches of inverse space has negligible impact on the events unfolding in real space, simplifying the calculations significantly.

Part I: $\beta < 1$

We will gradually transform the right side of the initial expression (7.57)

$$\begin{aligned} p(x \to \pm\infty) &\approx \frac{1}{\pi} \int_0^{\infty} dk \cos(k \mid x \mid) \exp(-D'_f \mid k \mid^{\beta}) \\ &= \frac{1}{\pi} \frac{1}{\mid x \mid} \int_0^{\infty} dk \left(\frac{d}{dk} \sin(k \mid x \mid) \right) \exp(-D'_f \mid k \mid^{\beta}) \\ &= \frac{1}{\pi} \frac{D'_f \beta}{\mid x \mid} \int_0^{\infty} dk \mid k \mid^{\beta-1} \sin(k \mid x \mid) \exp(-D'_f \mid k \mid^{\beta}) \\ &= \frac{1}{\pi} \frac{D'_f \beta}{\mid x \mid^{1+\beta}} \int_0^{\infty} dy y^{\beta-1} \sin(y) \exp\left(-D'_f \left(\frac{y}{\mid x \mid}\right)^{\beta}\right). \end{aligned} \tag{7.58}$$

Along the way, we used the integration by parts theorem. To determine the sought asymptotic form of $p(x)$ we will use a known relationship

$$\int_0^{\infty} dy y^{\beta-1} \sin(y) = \mathbf{\Gamma}(\beta) \sin\left(\frac{\pi}{2}\beta\right). \tag{7.59}$$

So ultimately,

$$p(x \to \pm\infty) \approx \frac{1}{\pi} \frac{D'_f}{\mid x \mid^{1+\beta}} \mathbf{\Gamma}(1+\beta) \sin\left(\frac{\pi}{2}\beta\right) \tag{7.60}$$

thus taking the form of Pareto-Lévy. We took advantage of the fact that the exponential factor in Eq. (7.58) tends to 1 as $\mid x \mid \to \infty$.

Part II: $\beta = 1$

In this marginal case, the expression (7.57) takes on a simpler form,

$$p(x \to \pm\infty) \approx \frac{1}{2\pi} \int_{-\infty}^{\infty} dk \exp(-ikx) \exp(-D'_f \mid k \mid). \tag{7.61}$$

Then, thanks to the parity of the function $\exp(-D'_f \mid k \mid)$, we get

$$\begin{aligned} p(x \to \pm\infty) &\approx \frac{1}{2\pi} \int_0^{\infty} dk \exp(-ikx) \exp(-D'_f \mid k \mid) \\ &+ \frac{1}{2\pi} \int_0^{\infty} dk \exp(ikx) \exp(-D'_f \mid k \mid) \\ &= \frac{1}{\pi} \frac{D'_f}{(D'_f)^2 + x^2} \approx \frac{1}{\pi} \frac{D'_f}{x^2}, \text{ dla } \mid x \mid \gg (D'_f)^{1/2}. \end{aligned} \tag{7.62}$$

In this case, the distribution $p(x)$ takes (as expected) on asymptotically, the Lorentzian form is identical to the Pareto-Lévy PDF with exponent $\beta = 1$.

Part III: $1 < \beta < 2$

As in the previous two cases, it can be written,

$$p(x \to \pm\infty) \approx \frac{1}{\pi} \int_0^{\infty} dk \cos(k \mid x \mid) \exp(-D'_f \mid k \mid^{\beta}). \tag{7.63}$$

Still, after changing the variables, we get:

$$p(x \to \pm\infty) \approx \frac{1}{\pi} \int_0^{\infty} dy \cos(y) \exp(-D'_f v y^{\beta}), \tag{7.64}$$

where we substituted $y = \mid x \mid k$ and $v = \mid x \mid^{-\beta}$; note that $D'_f v$ is a small quantity allowing expansion into a Taylor series. Namely,

$$\begin{aligned} p(x \to \pm\infty) &\approx \frac{1}{\pi}\int_0^\infty dy \exp(-D'_f v y)\cos(y)\exp(-D'_f v(y^\beta - y)) \\ &= \frac{1}{\pi}\int_0^\infty dy \exp(-D'_f v y)\cos(y)[1 - D'_f v(y^\beta - y) \\ &+ \frac{1}{2}(D'_f v)^2 (y^\beta - y)^2 + \dots\,]\,; \end{aligned} \tag{7.65}$$

where we used the helpful relationship,

$$\int_0^\infty dy \exp(-D'_f v y)\cos(y) y^{\nu-1} = \frac{\mathbf{\Gamma}(\nu)}{((D'_f v)^2 + 1)^{\nu/2}} \cos\left(\nu \arctan\left(\frac{1}{D'_f v}\right)\right), \quad \text{for } \nu \geq 1, \tag{7.66}$$

which allows us to express the distribution of $p(x)$ in the following asymptotic form,

$$\begin{aligned} p(x \to \pm\infty) &\approx -\frac{1}{\pi}\frac{1}{\mid x \mid}\left[\frac{D'_f v \mathbf{\Gamma}(1+\beta)}{\left(\left(D'_f v\right)^2 + 1\right)^{(1+\beta)/2}} \cos\left((1+\beta)\frac{\pi}{2}\right) + \vartheta\left(\left(D'_f v\right)^2\right)\right] \\ &\approx \frac{1}{\pi}\frac{D'_f}{\mid x \mid^{1+\beta}}\mathbf{\Gamma}(1+\beta)\sin\left(\frac{\pi}{2}\beta\right), \end{aligned} \tag{7.67}$$

where we also used the equivalent form,

$$\int_0^\infty dy \exp(-D'_f v y)\cos(y) = \frac{D'_f v}{(D'_f v)^2 + 1}, \quad \text{for } \nu = 1. \tag{7.68}$$

The arctangent function can be approximated as $\pi/2$ when its argument approaches infinity (as it is in Eq. (7.66)). It can be calculated based on Eqs. (7.66) and (7.68) that in the second equality in Eq. (7.65) terms associated with 1 and y cancel out. We denoted the non-linear terms associated with subsequent powers of $y^\beta - y$ after integration in the first approximate equality in Eq. (7.67) by $\vartheta\left(\left(D'_f v\right)^2\right)$.

Summarizing this theorem, it is noteworthy that, for all ranges of β (with a later consideration for the particular case of $\beta = 2$), we have successfully derived an identical asymptotic form for the distribution $p(x)$. Additionally, beyond establishing that $p(x)$ takes the Pareto-Lévy form as $|x|$ becomes large, we've determined expressions for the pre-exponential coefficient. This coefficient is significant, especially when comparing theoretical predictions to experimental data.

To briefly summarize this chapter, we showed the influence of stochastic hierarchy/self-similarity on the properties of random walks, emphasizing their singular nature. We presented the analysis of time dependence in Appendix A.9 in the context of the Weierstrass waiting time distribution function.

Control and Review Questions and Exercises

7.1 Exercise and Question
Define Weierstrass's process/wandering and its interpretation. Derive its asymptotic form. For what values of the shape exponent is it asymptotically identical to the symmetric Lévy distribution?

7.2 Exercise
Explain self-similarity in the probabilistic sense of Weierstrass's wandering.

7.3 Exercise and Problem
Derive:
(a) Structural factor of Weierstrass's wandering.
(b) Explain why the structural factor is a continuous nowhere differentiable function.
(c) Weierstrass discrete walk scaling equation and its solution.
(d) Prove that the structural factor of the Wierstrass wander is asymptotically the sum of the regular and singular components. Indicate the situation in which the former dominates and the situations in which the latter dominates.

7.4 Exercise and Problem
Discuss the renormalization approach to the Weierstrass process. Indicate that the obtained solution (structure factor) for the Weierstrass walk is a special solution of the general solution. Discuss the role of the heterogeneity of the spatial scaling equation in this solution.

7.5 Exercise
Discuss the role of length and the role of variance of a single displacement/flight and indicate their impact on the Weierstrass wander. Provide conditions that violate CLT and BLLN.

7.6 Exercise
Explain how infinite variance manifests itself in Weierstrass's walk and why.

7.7 Exercise
Explain what the so-called "black swan" is.

7.8 Exercise
Derive the dependency of $\langle | x |^n \rangle, n = 1, 2, \ldots,$ on L. This relationship is governed by "black swans." Specify the conditions under which this takes place.

7.9 Questions
Is Weierstrass random walk asymptotically equivalent to Pareto-Lévy PDF? What is the asymptotic form of this distribution?

7.10 Exercise
Derive the formula (7.67) from the formula (7.63).

References

1. Montroll, E.W., Shlesinger, M.F.: On the wonderful world of random walks. In: Lebowitz, J.L., Montroll, E.W. (eds.) Nonequilibrium Phenomena II. From Stochastics to Hydrodynamics. Studies in Statistical Mechanics, vol. XI, pp. 1–121. North-Holland, Amsterdam (1984)
2. Bateman, H., Erdélyi, A.: Tables of Integral Transforms, vol. 1. McGraw-Hill, New York (1954)
3. Stein, E.M., Shakarchi, R.: Complex Analysis. Princeton Lectures in Analysis, vol. II. Princeton University, Princeton & Oxford (2003)
4. Kutner, R., Wysocki, K.: Application of statistical mechanics to non-Brownian random motion. Physica A **274**, 67–84 (1999). https://doi.org/10.1016/S0378-4371(99)00313-1
5. Mandelbrot, B.B.: The Paul Lévy I Knew. In: Shlesinger, M.F., Zaslavsky, G.M., Frisch, U. (eds.) Lévy Flights and Related Topics in Physics, LNP vol. 450, pp. VIII–XII. Springer, Berlin (1995). https://doi.org/10.1007/3-540-59222-9

Chapter 8
Valley Model of Multifractal Continuous-Time Random Wandering on Amorphous Substrates

Abstract This chapter introduces two implementations of the valley model for multifractal continuous-time random walks on amorphous substrates. The first implementation focuses on the Gaussian substrate (Sect. 8.2), while the second involves the stretched exponential amorphous substrate (Sects. 8.3 and 8.3.3). In both examples, we demonstrate that the (unconditional) q-moments of inter-event times, reliant on complex statistical frameworks, can be represented in a scale-invariant form governed by exponents that are nonlinear functions of the order q. In this chapter, we illustrate how to transition from a disordered exponential substrate that fails to exhibit multifractality to one that does.

8.1 Important Announcement

We illustrate the impact of the energy landscape, transitioning from exponential/Poissonian to nonexponential/non-Poissonian, leading to a significant alteration in the model's properties, shifting this model from fractal to multifractal representations. This provides an original perspective on the genesis of multifractality from amorphous/disordered substrates or (generally) from a statistical thermodynamic perspective.

Research on multifractality induced by time series consisting of price increments where the price is sampling with intertransaction (or inter-trade, interevent) time intervals fluctuating was shown by Kwapień and Drożdż [1]. Furthermore, it was shown [2, 3] that both stochastic processes (i.e., price and time intervals) considered separately are multifractal. In this chapter, we want to extract the multifractality of the latter process (i.e., series of time intervals) within the valley model CTRW operates.

M. Chorowski et al., *Anomalous Stochastics*, Understanding Complex Systems,
https://doi.org/10.1007/978-3-031-80392-5_8

8.2 Gaussian Amorphous Substrate

In this exposition, we delineate an advanced modification of the Continuous Time Random Walk (CTRW) framework, incorporating thermally activated transits that surmount a potential barrier atop a Gaussian/parabolically configured amorphous substrate. In prior analyses, a variant of CTRW was scrutinized, distinguished by its reliance on a Poissonian/exponentially amorphous substrate (see Chap. 6), as detailed in Sect. 6.4.1 and explicated in Eq. (6.58). It is imperative to emphasize that the average over-disorder within this framework does not culminate in multifractality. To elucidate, not all forms of averaging over disordered systems induce multifractality. Specifically, exponential and Gaussian substrates simulate the structural complexities of amorphous and glassy materials.

Our methodology is structured into two main phases. In the foundational phase (Sect. 8.2.1), we introduce a dynamic partition function predicated on waiting/interevent durations, underscoring its divergence from traditional thermodynamic principles. Progressing to the ensuing phase (see Appendix A.16), we elucidate a linear transformation that engenders a dynamic partition function predicated on frequencies (viz., the reciprocals of waiting times). This transformation facilitates the manifestation of a multifractal attribute that integrates coherently with the principles of thermodynamic formalism.

8.2.1 Non-thermodynamic Representation of the Parabolic Multifractality

Let us consider a scenario where we are exclusively addressing disorder arising from random local minima in the potential energy of the substrate. Assume these energies are sampled from a "narrow" Gaussian PDF

$$\rho(\mathcal{E}) = \frac{1}{\sqrt{2\pi\sigma^2}} \exp\left(-\frac{(\mathcal{E}-\bar{\mathcal{E}})^2}{2\sigma^2}\right), \tag{8.1}$$

centered at a point $\mathcal{E} = \bar{\mathcal{E}}$, where $\bar{\mathcal{E}} \geq 3\sigma$. This lower truncation allows us to replace (with a good approximation) the integral over the entire real axis with the integral over its positive semi-axis. That is, it allows multifractality to be controlled analytically. Therefore, the normalization condition takes the form

$$\int_{-\infty}^{\infty} \rho(\mathcal{E}) d\mathcal{E} \approx \int_{0}^{\infty} \rho(\mathcal{E}) d\mathcal{E} \approx 1. \tag{8.2}$$

Hence, the exponential distribution (6.58) found in the averaged waiting time distribution function (6.65) should be substituted with those mentioned above (8.1).

Consequently, the expression (6.67) transforms into the following superstatistics (i.e., composite/complex statistics).

$$\phi(t)=\int_0^{\infty}\frac{1}{\sqrt{2\pi\sigma^2}}\exp\left(-\frac{(\mathcal{E}-\bar{\mathcal{E}})^2}{2\sigma^2}\right)\phi_{\mathcal{E}}(t)d\mathcal{E}, \tag{8.3}$$

where

$$\phi_{\mathcal{E}}(t)=\gamma_0\exp\left(-\hat{\beta}\mathcal{E}\right)\exp\left(-\gamma_0\exp\left(-\hat{\beta}\mathcal{E}\right)t\right)=\gamma_0\gamma^{\mathcal{E}/\sigma}\exp\left(-\gamma_0\gamma^{\mathcal{E}/\sigma}t\right). \tag{8.4}$$

We used the notation (6.62) by putting $\Delta=\sigma$ and we remind that $\gamma=\exp(-\hat{\beta}\sigma)$. We emphasize that $\left(\gamma_0\gamma^{\mathcal{E}/\sigma}\right)^{-1}$ is the WTD conditional relaxation time, defining the physical time scale. In the text below, dimensionless γ is fundamental/universal (i.e., independent of the time unit) in scaling q-moments of inter-event times. The exponent $\hat{\beta}\sigma$ combines the thermodynamic conditions in which the sample is located (described by the dissipative parameter $\hat{\beta}$) with the statistical/fluctuational properties of the sample/substrate itself (described by the standard deviation σ). One can see here the direct impact of both types of disorder (thermodynamic and static) on the local stochastic dynamics locked in $\phi_{\mathcal{E}}(t)$.

Our initial objective is to compute the moment $\langle t^{q-1}\rangle$ of the order $q-1$, where q is currently an arbitrary actual number. We recall that the (fluctuating) time interval between subsequent events is a random variable. We will employ the provided expression (8.3). Therefore,

$$\langle t^{q-1}\rangle=\int_0^{\infty}dt\,t^{q-1}\phi(t)\approx\int_0^{\infty}d\mathcal{E}\rho(\mathcal{E})\langle t^{q-1}\rangle_{\mathcal{E}}, \tag{8.5}$$

where

$$\begin{aligned}\langle t^{q-1}\rangle_{\mathcal{E}}&=\int_0^{\infty}dt\,t^{q-1}\phi_{\mathcal{E}}(t)=\mathbf{\Gamma}(q)\cdot\left[\gamma_0\exp(-\hat{\beta}\mathcal{E})\right]^{-(q-1)}\\&=\mathbf{\Gamma}(q)\cdot\left(\gamma_0\gamma^{\mathcal{E}/\sigma}\right)^{-(q-1)},\end{aligned} \tag{8.6}$$

where $\mathbf{\Gamma}(q)$ is the gamma Euler function defined (we recall) as follows

$$\mathbf{\Gamma}(q)=\int_0^{\infty}dy\,y^{q-1}\exp(-y). \tag{8.7}$$

Let us note that the following key assignment can be adopted (based on the axiomatic definition of the partition function given at the end of Sect. 5.1.1)

$$Z_q\propto\langle t^{q-1}\rangle \tag{8.8}$$

In essence, the moment of waiting times of order $q - 1$ may be formally conceptualized as a quantity proportional to the partition function Z_q.

The proportionality coefficient is determined by excluding the gamma Euler function, which manifests undesirable singularities at $q = -n$, where $n = 0, 1, 2, \ldots$. This is accomplished while preserving the fundamental characteristics of the partition function.

By substituting the expression (8.6) into (8.5) and then using (8.8) we get

$$Z_q = \frac{\langle(\gamma_0 t)^{q-1}\rangle}{\Gamma(q)} = \int_0^\infty \rho(\mathcal{E}) \left(\frac{1}{\gamma}\right)^{(q-1)\frac{\mathcal{E}}{\sigma}} d\mathcal{E} = \int_0^\infty \exp(G(\mathcal{E}))d\mathcal{E}, \tag{8.9}$$

and

$$G(\mathcal{E}) = -\frac{(\mathcal{E} - \bar{\mathcal{E}})^2}{2\sigma^2} + (q-1)\hat{\beta}\mathcal{E} - \frac{1}{2}\ln(2\pi\sigma^2); \tag{8.10}$$

the function/exponent $G(\mathcal{E})$ can be easily transformed into a more convenient parabolic form (for integrating over an independent random variable $\mathcal{E}$) with a visible peak at the center point $\mathcal{E} = \mathcal{E}^\star$,

$$\begin{aligned} G(\mathcal{E}) &= -\frac{1}{2\sigma^2}(\mathcal{E} - \mathcal{E}^\star)^2 - \frac{1}{2}\ln(2\pi\sigma^2) \\ &+ (q-1)\hat{\beta}\bar{\mathcal{E}} + \frac{1}{2}(q-1)^2\left(\hat{\beta}\sigma\right)^2, \end{aligned} \tag{8.11}$$

where

$$3\sigma \leq \mathcal{E}^\star = \bar{\mathcal{E}} + (q-1)\hat{\beta}\sigma^2, \tag{8.12}$$

thereby imposing the following constraint $q \geq 1$ (as from the beginning $\bar{\mathcal{E}} \geq 3\sigma$).

In other words, we assumed that the Gaussian integral in the final equality in (8.9) peaks sharply at $\mathcal{E}^\star$, aligning with the assumption of introducing a narrow PDF (8.1). For simplicity, we have omitted the indexing of the exponent $G(\mathcal{E})$ with order/generation q. It is noteworthy that the narrow parabolic dependence on $\mathcal{E}$ in Eq. (8.1) transitions to the parabolic dependence of the global exponent $\tau(q)$ on q in Eq. (8.14) below. This holds crucial implications for characterizing this CTRW variant.

The weak inequality (8.12) places constraint on exponent q, as we said above,

$$q \geq q_- = 1 + \frac{1}{\hat{\beta}\sigma}\left(3 - \frac{\bar{\mathcal{E}}}{\sigma}\right) \tag{8.13}$$

where q_- defines the lower bond of q, respectively (notably, the upper bond doesn't exist).

Substituting Eqs. (8.11) to (8.9) and using Eq. (8.12), we obtain the key form of the generalized partition function,

$$Z_q = \left(\frac{1}{\gamma}\right)^{\tau(q)}, \tag{8.14}$$

where global multiscaling exponent[1] obeys inequality,

$$\tau(q) = (q-1)D(q) = q\eta(q) - f(\eta(q)) \geq \tau_- = \tau(q_-) = 0, \tag{8.15}$$

while

$$D(q) = \frac{\bar{\mathcal{E}}}{\sigma} + \frac{1}{2}\hat{\beta}\sigma(q-1) \geq= D_- = D(q_-) = \frac{1}{2}\left(3 + \frac{\bar{\mathcal{E}}}{\sigma}\right) \tag{8.16}$$

These are the so-called Rényi dimensions, a linear function of $q - 1$ in the parabolic multifractality.

Based on the above considerations, we define the generalized Hurst exponent, $h(q)$, in the form of the relation,

$$\tau(q) = q\; h(q) - h(q = 1), \tag{8.17}$$

where $h(q = 1) = -\tau(q = 0) = D(q = 0)$ and $h_- = h(q_-) = \frac{h(q=1)}{q_-}$. We can obtain the value of $h(q = 1)$ from empirical data, e.g., from the course of the fluctuation function for $q = 0$ (as long as we operate within MF-DFA).

Moreover, we introduce the key quantities for our considerations, moving at the same time from the q variable to the variable $\eta = \frac{d\tau(q)}{dq}$,

$$\begin{aligned} \eta(q) &= \frac{\bar{\mathcal{E}}}{\sigma} + (q-1)\hat{\beta}\sigma \geq \eta_- = \eta(q_-) = 3, \\ f(\eta_-) &= f_- = q_-\eta_- - \tau_- = 3q_- = 3 + \frac{3}{\hat{\beta}\sigma}\left(3 - \frac{\bar{\mathcal{E}}}{\sigma}\right), \\ f(\eta(q)) &= q\eta(q) - \tau(q) = \eta + \frac{1}{2}\frac{1}{\hat{\beta}\sigma}\left(\eta - \frac{\bar{\mathcal{E}}}{\sigma}\right)^2 \end{aligned} \tag{8.18}$$

The formulation delineated in expression (8.15) encapsulates, among other things, the application of the Legendre transformation, previously employed in Sect. (5.1.6),

[1] More precisely, it is a family of global multiscaling exponents indexed by q values.

to amalgamate the global multiscale exponent $\tau(q)$ with the singularity spectrum[2] $f(\eta(q))$, where

$$\frac{d\tau(q)}{dq} = \eta(q),$$
$$\frac{df(\eta)}{d\eta} = q(\eta). \tag{8.19}$$

Let's add that the quantity η is often called the local scaling exponent.

It is worth highlighting that $\tau(q)$ and the spectrum of local fractal dimensions $f(\eta)$ convey identical information about the system, as they are connected through the Legendre transformation. The benefit of employing the $f(\eta)$ spectrum lies in its "physical" interpretation of how the density of states $\rho(\eta)$ scales. Specifically, by utilizing (8.1) and (8.18), we obtain:

$$\rho(\eta) = \frac{1}{\sqrt{2\pi\sigma^2}} \left(\frac{1}{\gamma}\right)^{-f(\eta)}. \tag{8.20}$$

As one can see, the spectrum $f(\eta)$ (here the parabolic function of η) plays the role of a fractal exponent scaling the density measure of the set of points having the same feature, namely, identical local fractal dimension η.

Using (8.20), one can express (8.14) in a form that justifies treating η as a local scaling exponent

$$Z_q = \sqrt{2\pi\sigma^2}\, \rho(\eta) \left(\frac{1}{\gamma}\right)^{q\eta(q)}. \tag{8.21}$$

We won't conduct a more in-depth, microscopic analysis of this exponent here[3]; instead, we will focus mainly on analyzing the relationship between multifractality and thermodynamic formalism.

> This section's discussion of multifractality suggests its presence in the random walk process continuous in time on a Gaussian substrate. This contrasts with the previously examined walk on a monofractal Poisson substrate.

The approach presented here, namely the multifractal representation, is inadequate to illustrate its connection with thermodynamic formalism. We establish such a relationship in the Appendix A.16 only after the linear transformation of

[2] Also referred to as the 'spectrum of local dimensions.'

[3] We remind that the global/generalized scaling exponent $\tau(q) \overset{\text{def.}}{=} q\eta(q) - f(\eta(q))$ is determined with precision to the scale (i.e., the multiplicative constant) and the basis of the power function with the corresponding precision to the inverse of this scale.

the exponent q. Following this transformation, as we will observe, the spectrum of singularities/dimensions becomes a convex function, with its arms pointing downwards—contrary to the current scenario where they point upwards.

8.2.2 Spectrum of Singularities' Analysis

After performing simple algebraic transformations,

$$\begin{aligned} q &\Rightarrow \tilde{q} = 2 - q, \\ \tau(q) &\Rightarrow \tilde{\tau}(\tilde{q}) = -\tau(q(\tilde{q})) \end{aligned} \tag{8.22}$$

used in the Appendix A.16, we obtained there $\tilde{f}(\tilde{\eta})$ (shown also in Table 8.1) and properties characterizing Fig. 8.1,

(1) from Eq. (8.13) and the first equation in (8.22), we get $\tilde{q} \le \tilde{q}_- = 1 - \frac{1}{\hat{\beta}\sigma}\left(3 - \frac{\bar{\mathcal{E}}}{\sigma}\right)$;
(2) local scaling exponent $\tilde{\eta}(\tilde{q}) = \frac{d\tilde{\tau}(\tilde{q})}{d\tilde{q}} = \tilde{D}(\tilde{q}) - (\tilde{q} - 1)\frac{d\tilde{D}(\tilde{q})}{d\tilde{q}}$, where $\tilde{\tau}(\tilde{q}) = (\tilde{q} - 1)\tilde{D}(\tilde{q})$, where $\tilde{D}(\tilde{q}) = \frac{\bar{\mathcal{E}}}{\sigma} - \frac{1}{2}\hat{\beta}\sigma\,(\tilde{q} - 1)$ was found in Appendix A.16; hence and from item (1) we get $\tilde{\eta}(\tilde{q}_-) = 3$;
(3) using the above and the Legendre transformation (second equality in (A.147)), the singularity spectrum can be expressed as follows $\tilde{f}(\tilde{\eta}) = \tilde{D}(\tilde{q}(\tilde{\eta})) + \tilde{q}(\tilde{\eta})(\tilde{q}(\tilde{\eta}) - 1)\tilde{D}'(\tilde{q}(\tilde{\eta}))$ (where $\ldots'$ means here the derivative with respect to $\tilde{q}$);
(4) the singularity spectrum reaches its maximum $\tilde{f}(\tilde{\eta}(\tilde{q} = 0)) = \tilde{D}(\tilde{q} = 0) = \frac{\bar{\mathcal{E}}}{\sigma} + \frac{1}{2}\hat{\beta}\sigma$ for $\tilde{\eta}(\tilde{q} = 0) = \frac{\bar{\mathcal{E}}}{\sigma} + \hat{\beta}\sigma$;
(5) for $\tilde{\eta}(\tilde{q} = 1)$ spectrum of singularities $\tilde{f}(\tilde{\eta}(\tilde{q} = 1) = \tilde{\eta}(\tilde{q} = 1)) = \tilde{D}(\tilde{q} = 1) = \frac{\bar{\mathcal{E}}}{\sigma}$; moreover, $\frac{d\tilde{f}(\tilde{\eta})}{d\tilde{\eta}}|_{\tilde{\eta}(\tilde{q}=1)} = 1$;

Table 8.1 Assignment of multifractal specific characteristics to thermodynamic specific quantities

Parabolic multifractality	Thermodynamics
$\tilde{q}$	β
$\hat{\beta}\sigma$	V
$\tilde{\eta}(\tilde{q}) = \frac{\bar{\mathcal{E}}}{\sigma} - \hat{\beta}\sigma\,(\tilde{q} - 1)$	$\frac{U(\beta)}{V}$
$\tilde{\tau}(\tilde{q}) = (\tilde{q} - 1)\left(\frac{\bar{\mathcal{E}}}{\sigma} - \frac{1}{2}\hat{\beta}\sigma\,(\tilde{q} - 1)\right)$	$\frac{\beta F(\beta)}{V}$
$\tilde{f}(\tilde{\eta}(\tilde{q})) = -\frac{1}{2}\frac{1}{\hat{\beta}\sigma}\left(\tilde{\eta} - \frac{\bar{\mathcal{E}}}{\sigma}\right)^2 + \tilde{\eta}$	$\frac{S(\beta)}{V}$
$c_{\hat{\beta}\sigma}(\tilde{q}) = -\tilde{q}^2\frac{d\tilde{\eta}}{d\tilde{q}} = \hat{\beta}\sigma\tilde{q}^2$	$c_V(\beta)$

(6) spectrum of singularities has two different roots $\tilde{\eta}^0_{\mp} = \tilde{\eta}(\tilde{q}^0_{\pm}) = \frac{\bar{\mathcal{E}}}{\sigma} + \hat{\beta}\sigma\left(1 - q^0_{\pm}\right) = \frac{\bar{\mathcal{E}}}{\sigma} + \hat{\beta}\sigma \mp \hat{\beta}\sigma\sqrt{1 + 2\frac{1}{\hat{\beta}\sigma}\frac{\bar{\mathcal{E}}}{\sigma}}$, where $\tilde{q}^0_{\pm} = \pm\sqrt{1 + 2\frac{1}{\hat{\beta}\sigma}\frac{\bar{\mathcal{E}}}{\sigma}}$. Hence, the full span of the singularity spectrum $\Theta^0 = \tilde{\eta}^0_{+} - \tilde{\eta}^0_{-} = \tilde{\eta}(\tilde{q}^0_{-}) - \tilde{\eta}(\tilde{q}^0_{+}) = 2\hat{\beta}\sigma\sqrt{1 + 2\frac{1}{\hat{\beta}\sigma}\frac{\bar{\mathcal{E}}}{\sigma}}$ denoted on the zero level (i.e., at level $\tilde{f}(\tilde{\eta}) = 0$); similarly, the allowable span $\Theta = \tilde{\eta}^0_{+} - \tilde{\eta}_{-} = \hat{\beta}\sigma\left(1 - q^0_{-}\right)$;

(7) derivatives of the singularity spectrum at extreme points: $0 < \frac{d\tilde{f}}{d\tilde{\eta}}|_{\tilde{\eta}^0_{-}} = \tilde{q}^0_{+} < +\infty$ and $-\infty < \frac{d\tilde{f}}{d\tilde{\eta}}|_{\tilde{\eta}^0_{+}} = \tilde{q}^0_{-} < 0$ do not achieve values $\pm\infty$, respectively. This results directly from the parabolic relationship $\tilde{f}(\tilde{\eta})$ versus $\tilde{\eta}$.

Figure 8.1 shows a plot (for example, for $\frac{\bar{\mathcal{E}}}{\sigma} = 3$ and $\hat{\beta}\sigma = 1/2$) of the relationship (A.149) with the marked characteristic points.

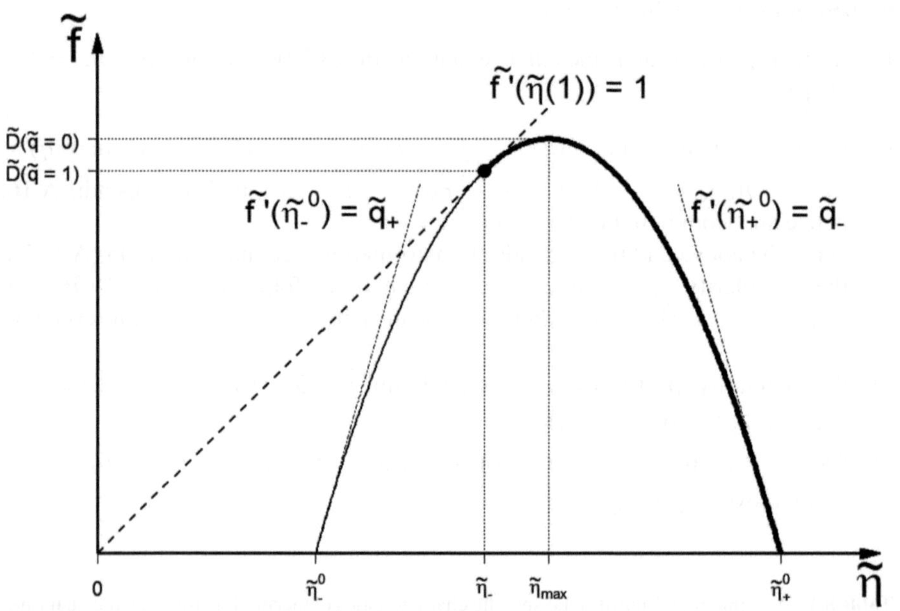

Fig. 8.1 Plot of the parabolic spectrum of singularities $\tilde{f}(\tilde{\eta}) = -\frac{1}{2}\frac{1}{\hat{\beta}\sigma}\left(\tilde{\eta} - \frac{\bar{\mathcal{E}}}{\sigma}\right)^2 + \tilde{\eta}$ versus the local scaling exponent $\tilde{\eta}$ is presented. For example, for $\frac{\bar{\mathcal{E}}}{\sigma} = 3$ and $\hat{\beta}\sigma = 1/2$. This plot is validated using the expressions specified in Table 8.1 and quantities in Sect. 8.2.2. The allowable range of $\tilde{\eta}$ extends from $\tilde{\eta}_{-} = \tilde{f}(\tilde{\eta}(\tilde{q} = 1)) = \tilde{D}(\tilde{q} = 1) = 3.0$ to $\tilde{\eta}^0_{+} = 5.303$, with the entire range of $\tilde{\eta}^0_{-} = 1.697$ to $\tilde{\eta}^0_{+} = 5.303$. The permissible section of the $\tilde{f}(\tilde{\eta})$ versus $\tilde{\eta}$ is the parabola highlighted in black (bold curve). The maximum of the parabola occurs at $\tilde{\eta}_{max} = \tilde{\eta}(\tilde{q} = 0) = 3.5$ yielding $\tilde{f}(\tilde{\eta}_{max}) = \tilde{D}(\tilde{q} = 0) = 3.25$. Furthermore, the derivatives are $\tilde{f}'(\tilde{\eta}^0_{\mp}) = \pm 3.606$ instead of $\pm\infty$ as for fully developed spectrum of dimensions

8.2.3 *The Relation of the Parabolic Multifractality with Thermodynamics*

This subsection explores thermally activated over-barrier hopping on a Gaussian substrate. We have shown that this stochastic process exhibits characteristics of a parabolic multifractal.

Specifically, the dimension/singularity spectrum, $\tilde{f}(\tilde{\eta})$, appears as a downward-facing parabola concerning the variable $\tilde{\eta}$ (refer to Table 8.1). The exponent $\tilde{\tau}$ is expressed in a quadratic form in terms of the variable/order $\tilde{q}$. As a result, the relationship between the variables $\tilde{q}$ and $\tilde{\eta}$, connected through the Legendre transformation, forms a linear function restricted to a narrow range of the independent variable $\tilde{q} \in (\tilde{q}_-^0, \tilde{q}_+^0)$. This range defines the orders for the generalized partition function.

Following the analytical approach outlined in Sect. 5.1.6 and the associated Table 5.2, which maps the connections between multifractal and thermodynamic variables, we have constructed Table 8.1 in this section, drawing on Appendix A.16.

Table 8.1 presents the distinct forms of multifractal characteristics for the parabolic multifractal continuous-time random walk on a Gaussian substrate. In particular, two specific temperatures are discussed: $\frac{1}{\tilde{\beta}}$ and $\frac{1}{\tilde{q}}$. The first is a fixed external parameter within the theory, while the second denotes the actual multifractal temperature. This parabolic behavior mirrors patterns observed in specific heat versus temperature plots, similar to those found in phenomena like paramagnetism [4] or in amorphous substances of independent, low-energy two-level systems [5].

It should be noted after this section that parabolic multifractality does not represent the fully developed form of multifractality. This is because both derivatives $\tilde{f}'(\tilde{\eta}_-^0) < \infty$ and $\tilde{f}'(\tilde{\eta}_+^0) > -\infty$, meaning they are bounded. Nevertheless, it serves as a useful illustrative example. This utility arises because all its multifractal features are expressed in explicit analytical forms.

8.3 Stretched Exponential Amorphous Substrate

The Scher-Montroll valley model,[4] considered in Chap. 6, introduces a sharp conditional probability density of time intervals (also called pausing, waiting, or inter-event) times, Δt, denoted $\psi(\Delta t|\varepsilon)$.[5] This density describes a carrier trapped within a potential well of a random depth ε for a time interval Δt. After this interval, the carrier jumps to another potential valley. The energy ε is drawn from a density function $\rho(\varepsilon)$ [8, 9]. Thus, we have a superstatistics framework with the sharp unconditional pausing-time density $\psi(\Delta t)$, as defined by the equation,

$$\psi(\Delta t) = \int_{-\infty}^{\infty} \psi(\Delta t|\varepsilon)\rho(\varepsilon)d\varepsilon. \tag{8.23}$$

The density function $\rho(\varepsilon)$ is centered/peaked around a large value of ε. As a result, only a tiny part of the integrand function is likely to have negative values within the domain of ε.

The significance of a sharp waiting-time distribution $\psi(\Delta t)$ for the multifractality is that, as a basic distribution, it allows the construction of the partition function. This is general—it is presented (supplementarily) in Appendix A.17.

The sharp conditional probability density, $\psi(\Delta t|\varepsilon)$, is assumed to be of the simple exponential (Poisson) form,

$$\psi(\Delta t|\varepsilon) = \frac{1}{\tau(\varepsilon)} \exp\left(-\frac{\Delta t}{\tau(\varepsilon)}\right). \tag{8.24}$$

This choice is reasonable since the emerging statistics are homogeneous for a given ε because all occurrences have an identical origin. Consequently, they enjoy an identical characteristic time scale $\tau(\varepsilon)$. Scher and Montroll also assume that the relationship between the random energy ε and the characteristic time of the distribution is given by the simple exponential form:

$$\tau(\varepsilon) = \tau_0 e^{\beta\varepsilon} \qquad (\tau_0, \beta > 0), \tag{8.25}$$

where $\tau_0 = \tau(0)$, and β^{-1} a fundamental constant of the model is measured in units of energy. In Scher-Montroll's approach, $\beta^{-1} = k_B T$ is the thermal energy of the environment at absolute temperature T (k_B is the Boltzmann constant).

[4] This subsection discusses our most significant results included in the publications [6, 7], which we present in the sufficiently popular form.

[5] We hope that replacing t with Δt in this definition does not confuse; we intend to make this section more readily accessible to our publications listed in an earlier footnote.

We note that the valley model is consistent with the most fundamental properties of a queuing process recently addressed by Vázquez et al. [10]. A set of incoming messages or tasks arrives randomly in that process. A particular priority labeled ε is attached to these messages. The execution time of a given task with priority ε is described by the conditional density $\psi(\Delta t|\varepsilon)$. In the most general setting, ε is a random variable with a density $\rho(\varepsilon)$. Thus, we are faced again with the "superstatistics" mentioned above, since the timing of the outgoing tasks is governed by the unconditional $\psi(\Delta t)$ given by Eq. (8.23).

8.3.1 Properties of the Unconditional Waiting-Time Distribution

The unconditional $\psi(\Delta t)$ provides maximal information about the timing between successive events. Nonetheless, other quantities can more easily unveil hidden but relevant process features, such as multifractality. This is precisely the case of the pausing-time moments:

$$\langle \Delta t^q \rangle = \int_0^\infty \Delta t^q \psi(\Delta t) d\Delta t.$$

Within our framework based on superstatistics with some weight function $\rho(\varepsilon)$, q-moments are written as

$$\langle \Delta t^q \rangle = \int_{-\infty}^\infty \langle \Delta t^q |\varepsilon\rangle \rho(\varepsilon) d\varepsilon. \tag{8.26}$$

where $\langle \Delta t^q |\varepsilon\rangle$ are the conditional pausing-time moments defined by

$$\langle \Delta t^q |\varepsilon\rangle = \int_0^\infty \Delta t^q \psi(\Delta t|\varepsilon) d\Delta t.$$

From Eqs. (8.24) and (8.25) we get

$$\langle \Delta t^q |\varepsilon\rangle = \mathbf{\Gamma}(1+q)\tau_0^q e^{q\beta\varepsilon}, \tag{8.27}$$

and the moments read,

$$\langle \Delta t^q \rangle = \mathbf{\Gamma}(1+q)\tau_0^q \int_{-\infty}^\infty e^{q\beta\varepsilon} \rho(\varepsilon) d\varepsilon. \tag{8.28}$$

We remark that exponent q may be negative. However, for the Poissonian density (8.24), the conditional moment $\langle \Delta t^q |\varepsilon\rangle$ and, hence, $\langle \Delta t^q \rangle$, will exist as long as

$q > -1$ (which results from the properties of the Euler gamma function). Moreover, for the moments to exist, $\rho(\varepsilon)$ must decay faster than $e^{q\beta\varepsilon}$ as $\varepsilon \to \pm\infty$.

Before proceeding ahead, let us briefly recall that the timing process will show multifractality if q-moments, $\langle \Delta t^q \rangle$, have a part/factor, which behaves as $L^{f(q)}$, where $f(q)$ is a nonlinear function of q, and L is some suitable scale. When $f(q)$ is linear, the process is termed monofractal [11].

8.3.2 Choice of Weight Function for a Substrate

We will now explore the possible emergence of multifractality depending on the choice of weight function $\rho(\varepsilon)$. First, we need to consider the simplest assumption as a reference case $\rho(\varepsilon) = \delta(\varepsilon - \mu)$. No heterogeneity exists; all the "valleys" have identical depth μ. In such a case, we readily obtain the monofractal behavior

$$\langle \Delta t^q \rangle = \boldsymbol{\Gamma}(1+q)\tau_0^q l^q, \tag{8.29}$$

where

$$l \stackrel{\text{def.}}{=} \exp(\beta\mu). \tag{8.30}$$

To get multifractality, we should go one step further in the formalism and consider, for instance, a stretched exponential:

$$\rho(\varepsilon) = \frac{1}{2^{1+1/\alpha}\sigma\boldsymbol{\Gamma}(1+1/\alpha)} \exp\left\{-\frac{1}{2}\left|\frac{\varepsilon-\mu}{\sigma}\right|^\alpha\right\}, \tag{8.31}$$

($\alpha > 0$ and $\sigma > 0$). As one can see, Eq. (8.31) becomes the Gauss PDF used in the previous section when the exponent $\alpha = 2$.

Substituting Eq. (8.31) into Eq. (8.28), we get, after a simple change of variables, the unconditional moment in the intermediate form,

$$\langle \Delta t^q \rangle = \frac{\boldsymbol{\Gamma}(1+q)}{2^{1+1/\alpha}\boldsymbol{\Gamma}(1+1/\alpha)} \tau_0^q \, l^q \, I(q), \tag{8.32}$$

where l is defined in Eq. (8.30) and

$$I(q) = \int_{-\infty}^{\infty} \exp\left(-\frac{1}{2}|y|^\alpha + q\sigma\beta y\right) dy. \tag{8.33}$$

Looking at this integral, we see at once that the convergence is assured as long as $\alpha > 1$. In other words, the unconditional moment $\langle \Delta t^q \rangle$ exists if $\alpha > 1$ and $q > -1$ (see above).

When $\alpha = 2$ (that is, for a Gaussian ε variable), the integral in Eq. (8.33) can be done precisely in closed form with the result:

$$\langle \Delta t^q \rangle = \Gamma(1+q)\, \tau_0^q\, l^q\, \exp\left((\sigma\beta)^2 q^2/4\right),$$

which, after defining

$$L \stackrel{\text{def.}}{=} \exp\left((\sigma\beta)^2/4\right)$$

clearly shows for the Gaussian ε the parabolic multifractal behavior of the q-moment:

$$\langle \Delta t^q \rangle = \Gamma(1+q)\, \tau_0^q\, l^q\, L^{q^2}. \tag{8.34}$$

When $\alpha \neq 2$, the integral (8.33) cannot be done exactly in a closed form, and to elucidate any possible multifractal behavior of the q-moments, we have to resort to approximations. To this end we first define in Eq. (8.33) a new integration variable x by the change of scale $y = (\beta\sigma)^{1/(\alpha-1)}x$. We have

$$I(q) = \lambda^{1/\alpha} \int_{-\infty}^{\infty} \exp\left\{-\lambda\left[\frac{1}{2}|x|^\alpha - qx\right]\right\} dx, \tag{8.35}$$

where

$$\lambda \stackrel{\text{def.}}{=} (\beta\sigma)^{\alpha/(\alpha-1)}. \tag{8.36}$$

Let us suppose that the fluctuations of the random variable ε, represented by σ, are larger than "the thermal energy of the environment" β^{-1}, that is, $\beta\sigma > 1$. In such a case, the dimensionless parameter λ is large, and we can safely use the saddle-point approximation [12] to evaluate $I(q)$. This is done in the Appendix A.18 with the result

$$I(q) \simeq A(q) \exp\left\{b|q|^{\alpha/(\alpha-1)}\right\}, \tag{8.37}$$

where $A(q)$ is given in Eq. (A.172) of the Appendix A.18, and

$$b = 2^{1/(\alpha-1)}(\alpha-1)\,(\beta\sigma/\alpha)^{\alpha/(\alpha-1)} > 0. \tag{8.38}$$

Substituting Eq. (8.37) into Eq. (8.32), we get an approximate analytical expression for the q-moment of the time intervals between events. We write such an expression in a form that enhances its multifractal character:

$$\langle \Delta t^q \rangle \simeq \Gamma(1+q)\tau_0^q l^q L^{|q|^{\alpha/(\alpha-1)}}, \tag{8.39}$$

where, as before, we have introduced two different scales l and L. The former $l = e^{\mu\beta}$ and given by Eq. (8.30) is related to dissipation because it depends on the dissipative term μ of the weight density (8.31). On the other hand, the scale

$$L \stackrel{\text{def.}}{=} \exp(b) \Leftrightarrow l = L^{\frac{\beta\mu}{b}} \tag{8.40}$$

providing the multifractal behavior is responsible for fluctuations since b defined in Eq. (8.38) depends on σ which, in turn, is related to the variance of $\rho(\varepsilon)$; the latter is given by $[\Gamma(3/\alpha)/\Gamma(1/\alpha)]\sigma^2$. Note that all q-moments considered above obey the normalization condition, i.e., they are equal to 1 for $q = 0$. Just Eq. (8.39) introduced a multifractal variant of CTRW, which we called the Multifractal Continuous-Time Random Walk (MF-CTRW).

The dissipative and fluctuating scales l and L merge into a single scale when $b = \mu\beta$. This equality means that dissipation μ and fluctuation σ are linked by

$$\mu = k\sigma^{\alpha/(\alpha-1)}, \tag{8.41}$$

where $k = (1 - 1/\alpha)(\beta/\alpha)^{1/(\alpha-1)}$. For $\alpha = 2$ (the Gaussian case), this relation reads

$$\mu = (\sigma/2)^2,$$

which is the analog of the usual fluctuation-dissipation relation. This leads us to look at Eq. (8.41) as the fractional version of the fluctuation-dissipation theorem resulting from the present approach.

Note that using Eqs. (8.30) and (8.39), we rewrite Eq. (8.40) in a form that is slightly more convenient for use in numerical applications,

$$\frac{\langle \Delta t^q \rangle}{\Gamma(1+q)} \simeq L^{\frac{1}{b}(\ln \tau_0 + \beta\mu)q + |q|^{\alpha/(\alpha-1)}}. \tag{8.42}$$

Let us finally observe that when the fractional fluctuation-dissipation relation holds the monofractal and multifractal parts of the q-moment are both governed by the same scale $l = L$ hence,

$$\frac{\langle \Delta t^q \rangle}{\Gamma(1+q)} \simeq L^{\left(\frac{\ln \tau_0}{b}+1\right)q + |q|^{\alpha/(\alpha-1)}}. \tag{8.43}$$

8.3.2.1 Comparison with Empirical Data

We compared the prediction of the formula (8.42) with the empirical data presented in Fig. 8.2 coming (for example) from financial markets.

The heuristic formula we propose is:

$$\frac{\langle \Delta t^q \rangle}{\Gamma(1+q)} = L^{\frac{b_1}{b}(\ln \tau_0 + \beta\mu)q + \varphi(q)}, \tag{8.44}$$

where

$$\varphi(q) = \left[1 - \exp\left(-b_1 \mid q \mid^{1/(\alpha-1)}\right)\right] q. \tag{8.45}$$

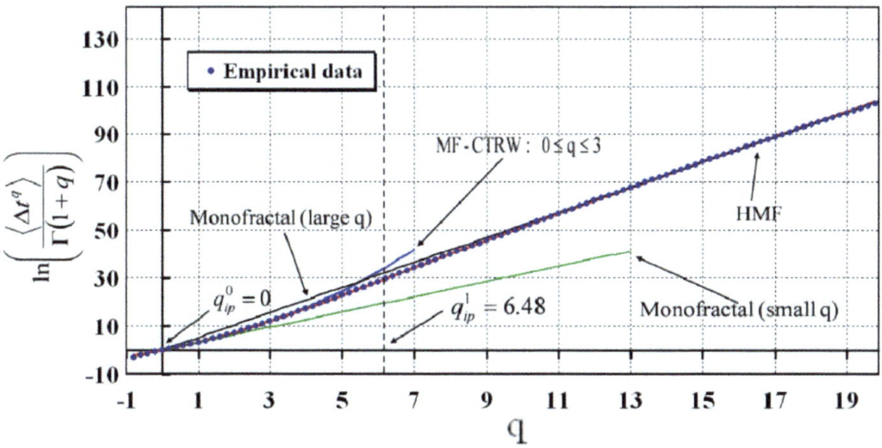

Fig. 8.2 Semi-logarithmic plot comparing the prediction of the theoretical q-moment $\langle \Delta\tau^q \rangle / \Gamma(1+q)$ versus order q (different solid curves) with empirical data/quotation concerning futures at USDM rate on the Forex (marked by dots; quotation taken from year 2008). The solid curve marked by MF-CTRW (Multifractal Continuous-Time Random Walk): $0 \le q \le 3$ is the formula (8.42) prediction. The solid curve, which fits the empirical data in the whole range of q, predicts the heuristic multifractal formula (HMF) (8.44). As can be seen, the multifractal curve is located for $q \ge 0$ between two straight lines defining monofractal curves (for q in the vicinity of zero and asymptotically large q). In addition, we marked the inflection points of this multifractal curve with q^0_{ip} and q^1_{ip}. Adapted with permission from [7]. © 2010, Springer Nature. All rights reserved

We have added a fourth parameter, $b_1 > 0$, and changed L to new $L = \exp(b/b_1)$. Equation (8.44) satisfies the normalization condition. That is, both sides are equal to 1 for $q = 0$.

Moreover, for $\mid q \mid$ small enough, we have $1 - \exp\left(-b_1 \mid q \mid^{1/(\alpha-1)}\right) \approx b_1 \mid q \mid^{1/(\alpha-1)}$ and Eq. (8.44) gives

$$\frac{\langle \Delta t^q \rangle}{\Gamma(1+q)} = L^{\frac{b_1}{b}(\ln \tau_0 + \beta\mu)q + b_1 |q|^{\alpha/(\alpha-1)} sgn(q)}, \tag{8.46}$$

which recover Eq. (8.42) for $q \ge 0$.

For $q \gg 1$ we get $1 - \exp\left(-b_1 \mid q \mid^{1/(\alpha-1)}\right) \approx 1$ and Eq. (8.44) tends to the monofractal form:

$$\frac{\langle \Delta t^q \rangle}{\Gamma(1+q)} \simeq L^{\frac{b_1}{b}(\ln \tau_0 + \beta\mu)q + q}.$$

Figure 8.2 illustrates (solid curves) the fitting of empirical data for futures at the USDM rate on Forex by the MF-CTRW and HMF formulas. The corresponding parameter values obtained by fitting the theoretical expressions to the empirical data are compared in Table 8.2.

Table 8.2 Fit parameters of the MF-CTRW (top line) and the HMF (bottom line) of formulas (8.42) and (8.44), respectively

Fit q range	α	$\ln \tau_0 + \mu\beta$	b	b_1
$0 - 3.0$	1.754 ± 0.080	3.008 ± 0.011	0.227 ± 0.009	–
$-0.8 - 19.8$	1.676 ± 0.018	3.074 ± 0.063	0.228 ± 0.028	0.106 ± 0.010

It is the existence of a nonlinear dependence of $\ln \frac{\langle \Delta t^q \rangle}{\Gamma(1+q)}$ on the degree/order q in empirical data that is the basis for the considerations further in this chapter.

8.3.3 Multiscaling Representation of Partition Function

Since integral $J(q')$ and therefore the partition function $Z(q')$ was already calculated in Appendix A.19, again using the saddle-point approximation/method (cf. Eq. (A.192)), one readily finds that

$$Z(q') \approx B(q')\, L^{-\tau(q')}, \tag{8.47}$$

where

$$B(q') = M^{1-q'} q'^{(1-2\alpha)/2(\alpha-1)} \exp((bc - \beta\mu)(q' - 1)), \tag{8.48}$$

and $q' = q + 1$, while c is the constant that should be chosen to make prefactor $B(q')$ independent of the scale L as expected, therefore

$$c = \frac{\beta\mu}{b}; \tag{8.49}$$

the general scaling exponent

$$\begin{aligned} \tau(q') &= -c\,(1 - q') - \mid 1 - q' \mid \left(\frac{\mid 1 - q' \mid}{q' + q_0'} \right)^{1/(\alpha-1)} \\ &= (q' - 1) D(q'), \end{aligned} \tag{8.50}$$

where q_0' is the ad hoc correction constant, which could be estimated by the self-consistency of our procedure (cf. relation (i) below). Here, the (positive) Rényi dimensions

$$D(q') = \frac{\tau(q')}{q' - 1} = c + sgn(1 - q') \left(\frac{\mid 1 - q' \mid}{q' + q_0'} \right)^{1/(\alpha-1)} \tag{8.51}$$

obey the following relations:

(i) $\lim_{q'\to 0} D(q') = c + \frac{1}{q_0'^{1/(\alpha-1)}} \equiv q_0' = \frac{1}{[D(0)-c]^{\alpha-1}} \geq 0$
(ii) $\lim_{q'\to\infty} D(q') = c - 1 \geq 0$
(iii) $\lim_{q'\to 1} D(q') = c > 0$
(iv) the first derivative $D(q')$ over q'

$$D'(q') = -\frac{1}{\alpha-1}\left(\frac{|\,1-q'\,|}{q'+q_0'}\right)^{(2-\alpha)/(\alpha-1)} \times \\ \times \frac{1+q_0'}{(q'+q_0')^2} \leq 0, \tag{8.52}$$

which for $q' = \infty$ is equal to 0, while it vanishes at $q' = 1$ for $\alpha < 2$ otherwise (for $\alpha > 2$) it diverges.

Of course, if we could know the Hausdorff dimension, we could determine q_0' from relation (i).

For large q' we obtain, from Eq. (8.50) and relation (ii), the expression concerning the general exponent

$$\tau(q') \approx (c-1)q' = \eta_{min} q', \tag{8.53}$$

which is important for the analysis of the spectrum of singularities.

In Fig. 8.3, we plotted the general scaling exponent $\tau(q')$ versus q' for two different examples $q_0' = 0$ (lower curve) and $q_0' = 1$ (upper curve). As seen, there is an essential difference between them since, for the former one, this exponent diverges for $q' \to 0$ (according to the power low) in contrast to the latter case (where it is finite). For both examples, for $q' \to \infty$, the general scaling exponent diverges linearly while it vanishes for $q' = 1$. Nevertheless, limited Rényi dimensions are proper quantities, which are the most essential (multiplicative) part of the general scaling exponent describing different exciting properties of the system. They are ready for comparison with the corresponding empirical data.

Property (ii) follows directly from Eq. (8.51). Therefore, as usual (see paper [13] for explanation), the Rényi dimensions are positive monotonically decreasing functions of q'.

8.3.4 Rescaled Spectrum of Local Dimensions

Since, in general, the scale can be determined up to some arbitrary power ω, the general scaling exponent and the Rényi dimensions are rescaled by some multiplicative factor $1/\omega$. Thus, the question arises: Which scale definition is most useful for our situation?

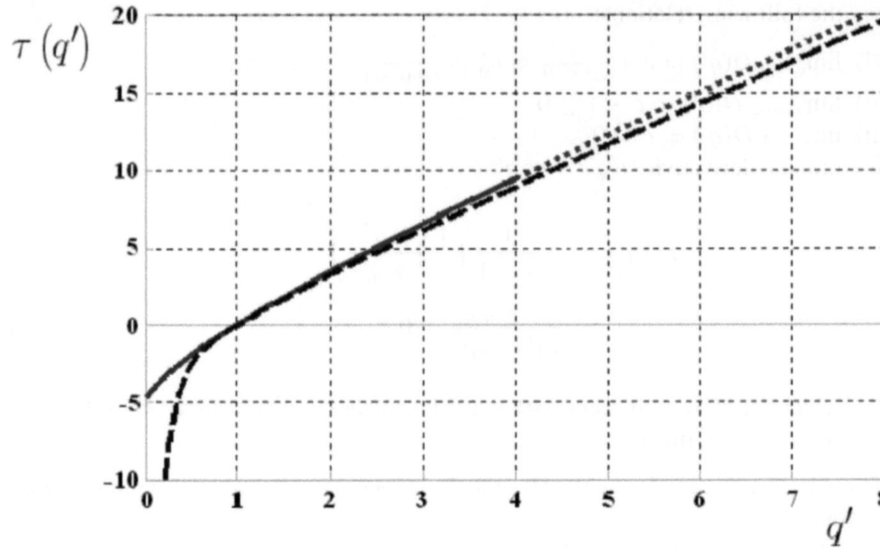

Fig. 8.3 The plot of the general scaling exponent $\tau(q')$ versus order $q'(\geq 0)$ at $\alpha = 1.754$ and both for $q'_0 = 1.0$ (the upper, solid curve into the dotted curve) and $q'_0 = 0$ (the lower, dashed curve), defined only for the non-negative values of q'. The solid part of the upper curve corresponds to the range of q' values where the MF-CTRW formula (8.50) fits the empirical data satisfactorily. In contrast, the dotted part corresponds to the unsatisfactory fit range (cf. Fig. 8.2). Both curves vanish (as they should) for $q' = 1$. Their non-linear segments are visible compared to the linear ones. Adapted with permission from [7].

8.3.4.1 Spectrum of Local Dimensions

In this section, in addition to discussing further essential properties of Rényi dimensions, we consider the spectrum of local dimensions to construct the self-consistent multifractal procedure. Since Legendre or contact transformation can be well defined here, we can obtain the spectrum of local dimensions $f(\eta)$ from the global scaling exponent $\tau(q')$ given by Eq. (8.50)

$$\tau(q') = q'\eta - f(\eta), \tag{8.54}$$

where variables

$$\eta(q') = \frac{d\tau(q')}{dq'},$$
$$q' = \frac{df(\eta)}{d\eta}. \tag{8.55}$$

By combining Eqs. (8.50)–(8.52) and (8.54), we derive an explicit dependence of the spectrum of local dimensions f and variable η on the independent variable q',

$$\begin{aligned}\eta(q') &= D(q') + (q'-1)D'(q') \\ &= c + \left(\frac{\mid 1-q' \mid}{q'+q'_0}\right)^{\nu}\left(1+\nu\,\frac{1+q'_0}{q'+q'_0}\right) sgn(1-q'),\end{aligned} \tag{8.56}$$

and

$$\begin{aligned}f(\eta(q')) &= D(q') + q'(q'-1)D'(q') \\ &= c + \left(\frac{\mid 1-q' \mid}{q'+q'_0}\right)^{\nu}\left(1+\nu\,q'\,\frac{1+q'_0}{q'+q'_0}\right) sgn(1-q'),\end{aligned} \tag{8.57}$$

where $\nu \overset{\text{def}}{=} 1/(\alpha-1)$. Although we obtained an explicit dependence of f on q', the analogous dependence of f on η requires the solution of the entangled transcendental equation. This equation was derived by combining Eqs. (8.56) and (8.57),

$$\mid \eta - c \mid = \omega_0\,(1+q'_0)\,\frac{\left(1-\frac{f-c}{\eta-c}\right)^{\nu}}{\left(q'_0+\frac{f-c}{\eta-c}\right)^{\nu+1}},\ \eta \neq c, \tag{8.58}$$

where $sgn(f-c) = sgn(\eta-c) = sgn(1-q')$. Equation (8.58) gives solution η (by using the method of successive iterations) as a convex function of f, i.e., we have $f \leq \eta$, while equality $\eta = f$ is valid only for a single point $q' = 1$; additionally, for this point $\eta = c$. For the solution of Eq. (8.58) (among others), the concrete value of the correction constant q'_0 must be known. For the (pathological) case of the disappearance of the correction constant, this equation reduces to the simplified form:

$$\eta = f + \omega_0^{-(\alpha-1)} \mid f - c \mid^{\alpha}. \tag{8.59}$$

8.3.4.2 Useful Rescaling of Variables

In fact, in Fig. 8.4, the plot $\tilde{f}$ versus $\tilde{\eta}$ was presented, where any quantity $\tilde{y} \overset{\text{def.}}{=} \omega_0^{-1}\, y$.

In this representation Eq. (8.58) takes the simpler form (which is consistent with Eq. (8.54), and both relations in (8.55)), where ω_0 is absent (and η, c and f are simply replaced by $\tilde{\eta}$, $\tilde{c}$ and $\tilde{f}$, respectively). Moreover, in this representation, both q' and q'_0 remain unchanged. Hence, Eq. (8.59) reduces to a simpler rescaled form:

$$\tilde{\eta} = \tilde{f} + \mid \tilde{f} - \tilde{c} \mid^{\alpha}. \tag{8.60}$$

The width of the singularity spectrum corresponds to the range $-1 < q < 3$ (illustrated in Fig. 8.2), where the prediction of the MF-CTRW model closely aligns with the empirical curve $\ln\left(\langle \Delta t^q \rangle / \boldsymbol{\Gamma}(1+q)\right)$. Fortunately, this range is crucial for investigating multifractality, encompassing the nonlinear dependence on q. Thus,

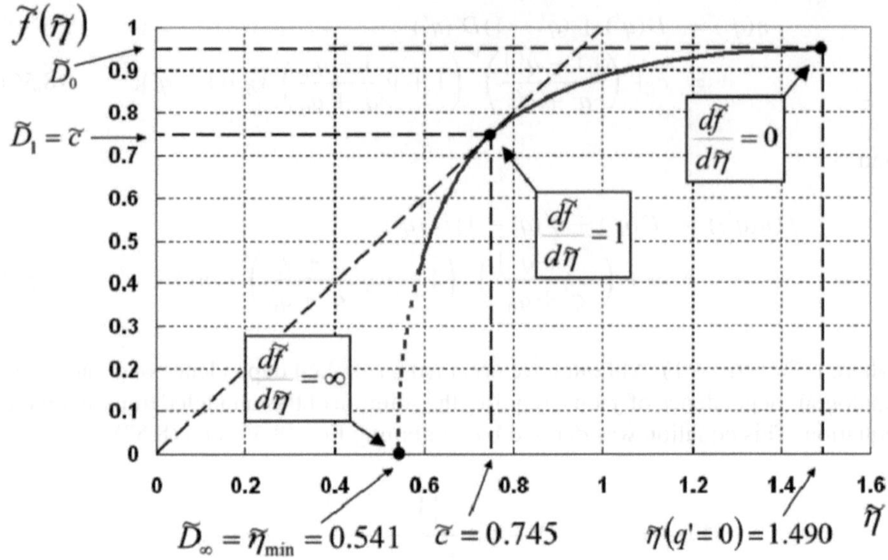

Fig. 8.4 The left-sided non-matured multifractality is examined through the comparison of the rescaled spectrum of local dimensions, $\tilde{f} = \omega_0^{-1} f$, with the rescaled local dimension $\tilde{\eta} = \omega_0^{-1}\eta$, for an exponent $\alpha = 1.754$, specifically addressing futures trading at the USD/DM exchange rate in Forex, with $q'_0 = 1.0$. Equations (8.57) and (8.56) provide expressions for f and η, respectively. This plot represents only the positive range of q', thus truncated at $\tilde{\eta}(q' = 0) = 1.490$, where $q' = 0$ means approaching 0 from the positive side. This spectrum, analyzed within the MF-CTRW model, remains valid approximately within the range of $(\tilde{\eta}, \tilde{f})$ from point $(\tilde{\eta}(q' = 4) = 0.586, \tilde{f}(\tilde{\eta}(q' = 4) = 0.421))$, which is the beginning of the solid curve, to point $(\tilde{\eta}(q' = 0) = 1.490, \tilde{f}(\tilde{\eta}(q' = 0) = 0.949)$, which is the end of the solid curve. This curve exhibits a sufficiently broad spectrum relative to the minimum $\tilde{\eta}_{min} = 0.541$. Adapted with permission from [7].

only within the range $\tilde{\eta}(q' = 4) < \tilde{\eta} < \tilde{\eta}(q' = 0)$ can the MF-CTRW model accurately describe empirical data. Despite this constraint, the spectrum's width remains sufficiently broad. This justifies the model's efficacy in examining the multifractal properties of transaction interarrival times generated by market fluctuations.

Certain left-sided multifractals, characterized by spectra of singularities, $f(\eta)$, resembling the left half of $\bigcap$, have been previously explored [14, 15]. This exploration delved into the mathematical aspects, examining spectra of singularities peaks at $\eta = \infty$.

> In this study, the left-sided multifractality was observed based on empirical evidence. Both examples of multifractality provided in this chapter exhibit incomplete multifractality. It is important to note that real-world multifractality often manifests as incomplete multifractals.

Control and Review Questions and Exercises

8.1 Questions

What is complex statistics? In the context of multifractality, why do we consider a substrate drawn from a narrow Gaussian PDF and not an arbitrary Gaussian PDF? Can exponential disorder/substrate lead to multifractality?

8.2 Question

Why does the (formal) definition of partition function (8.8) include the exponent $q - 1$ and not q?

8.3 Exercise and Problem

Provide the estimated allowable span of the exponent domain q. On this basis, determine the allowable range of

(a) global scaling exponent $\tau(q)$
(b) Rényi dimensions $D(q)$
(c) exponent $\eta(q)$
(d) the permissible span of the spectrum of dimensions $f(\eta(q))$.

8.4 Problem

Prove that the spectrum of dimensions, $f(\eta)$, is a concave function of η a $\tilde{f}(\tilde{\eta})$ convex function of $\tilde{\eta}$.

8.5 Problem

Prove that the derivative of $\frac{d\tilde{f}}{d\tilde{\eta}}$ is less than ∞ at the point $\tilde{\eta}_-$ and is greater than $-\infty$ at the point $\tilde{\eta}_+$

8.6 Exercise

Prove that replacing the variable η with the variable $\tilde{\eta}$ and the function τ with $\tilde{\tau}$ performs concave multifractality in convex multifractality.

8.7 Questions

What do we call incomplete multifractality? Why are the multifractals presented in this chapter of this type?

8.8 Questions

Is it possible to consistently scale the variables and functions characterizing multifractality so that the spectrum of dimensions for the vanishing order of fluctuations q equals 1? If so, how does it, and what is this scaling factor?

8.9 Problem

Solve numerically the Eq. (8.60) to obtain the dependence of $\tilde{f}$ on $\tilde{\eta}$ to compare very different three values of $\tilde{c}$ and three values of the exponent α.

References

1. Kwapień, J., Stanisław Drożdż. S.: Physical approach to complex systems. Phys. Rep. **515**, 115–226 (2012). https://doi.org/10.1016/j.physrep.2012.01.007
2. Paweł Oświęcimka, P., Kwapień, J., Stanisław Drożdż, S.: Multifractality in the stock market: price increments versus waiting times. Phys. A **337**, 626–638 (2005). https://doi.org/10.1016/j.physa.2004.08.025
3. Eisler, Z., Kértesz, J.: Size matters: some stylized facts of the stock market revisited. Eur. Phys. J. B **51**, 145–154 (2006). https://doi.org/10.1140/epjb/e2006-00189-6
4. Baierlein, R.: Thermal Physics. Cambridge University Press, Cambridge UK (1999). https://doi.org/10.1017/CBO9780511840227
5. Kittel, Ch.: Introduction to Solid State Physics, Eight Wiley, New York (2021)
6. Perelló, J., Masoliver, J., Kasprzak, A., Kutner, R.: Model for interevent times with long tails and multifractality in human communications: an application to financial trading. Phys. Rev. E **78**, 036108 (2008). https://doi.org/10.1103/PhysRevE.78.036108
7. Kasprzak, A., Kutner, R., Perelló, J., Masoliver, J.: Higher-order phase transitions on financial markets. Eur. Phys. J. B **76**, 513–527 (2010). https://doi.org/10.1140/epjb/e2010-00064-y
8. Scher, H., Montroll, E.W.: Anomalous transient-time dispersion in amorphous solids. Phys. Rev. B **15**, 2455–2477 (1975). https://doi.org/10.1103/PhysRevB.12.2455
9. Kehr, K.W., Kutner, R., Binder, K.: Diffusion in concentrated lattice gases. Self-diffusion of noninteracting particles in three-dimensional lattices. Phys. Rev. B **23**(10), 4931–4945 (1981). https://doi.org/10.1103/PhysRevB.23.4931
10. Vásquez, A., Oliveira, J.G., Dezsö, Z., Goh, K.-I., Kondor, I.: Modeling bursts and heavy tails in human dynamics. Phys. Rev. E **73**, 036127 (2006). https://doi.org/10.1103/PhysRevE.73.036127
11. Bunde, A., Havlin, S. (eds.): Fractals and Disordered Systems, Second Revised and Enlarged Edition. Springer, Berlin (1996)
12. Erdelyi, A.: Asymptotic Expansions. Dover, New York (1956)
13. Beck, C., Schlögl, F.: Thermodynamics of Chaotic Systems. An Introduction. Cambridge Nonlinear Science Series, Series Number 4. Cambridge University Press, Cambridge (1995)
14. Mandelbrot, B.B., Evertsz, C.J.G., Hayakawa, Y.: Exactly self-similar left-sided multifractal measures. Phys. Rev. A **42**(8), 4528–4536 (1990). https://doi.org/10.1103/PhysRevA.42.4528
15. Mandelbrot, B.B., Evertsz, C.J.G.: Exactly Self-similar Left-sided Multifractals. In: Bunde, A., Havlin, S. (eds.) Fractals and Disordered Systems, Second Revised and Enlarged. Springer, Berlin (1996)

Chapter 9
Statistics of Extremes

Abstract The central values and typical fluctuations are insufficient to fully characterize natural systems that exhibit rare but extreme events, which can often dominate the long-term behavior of time series generated by various random walks. Consequently, the statistics of extrema represent a classical subject of significant interest in mathematics, physics, and the social and economic sciences [1–5]. In physics, extreme events have been extensively studied across multiple disciplines [6] (and references therein), including self-organized fluctuations, critical phenomena, material fracture, disordered systems, and turbulence. Understanding the statistics of extreme events is fundamentally important for predicting and assessing risks associated with various natural and man-made phenomena, such as earthquakes, climate change, floods, and substantial movements in financial markets. A relatively new area where extreme statistics has gained attention is complex networks [6]. In this chapter, we present the key aspects of Extreme Value Theory (EVT) articulated by physicists. This framework facilitates the direct application of EVT in the analysis of risk and random walks in amorphous materials.

9.1 Limit Theorem for Maxims

The cornerstone of the Statistical Theory of Extreme Events, often called Extreme Value Theory (EVT), is the seminal Maximum Limit Theorem by Fisher, Tippett, and Gnedenko [7, 8].

In this context, we present the theorem for a continuous random variable, deferring the proof for later sections while emphasizing its practical significance. Subsequently, we will approach this theorem from a perspective that enhances the physical interpretation of the parameters associated with the distributions featured in this theorem.

M. Chorowski et al., *Anomalous Stochastics*, Understanding Complex Systems,
https://doi.org/10.1007/978-3-031-80392-5_9

The importance of the Maximum Limit Theorem (MLT) for extreme events can be likened to that of the Central Limit Theorem for limit theorems. We discuss the MLT in detail (albeit without proof) below [1, 4, 5, 8].

Theorem 9.1.1 (Maximum Limit Theorem) *Consider a sequence of statistically independent continuous random variables, i.e., the sequence* $(x_1, x_2, \ldots, x_n)$ *with an identical distribution, and let* $x_n^{\max} = \max(x_1, x_2, \ldots, x_n)$*. If there exists a convergent sequence of triplets of real numbers* $(a_n(>0), b_n, \gamma_n) \to (a(>0), b, \gamma)$*, representing parameters for standardization/scaling, centralization/location, and shape, respectively, and if a non-degenerate limit cumulative distribution function (CDF),*

$$\lim_{n\to\infty} P\left(x_n^{\max} \leq x\right) = \lim_{n\to\infty} H_{\gamma_n}\left(\frac{x-b_n}{a_n}\right) = H_\gamma\left(\frac{x-b}{a}\right) = H_\gamma^{Acr}(x),$$

is well defined/established for every x*, where acronim index* $Acr = Fr, Gu, Wei$ *(see complementary notation directly below). Then, this distribution function can take only one of the following three limiting forms:*

(i) Fréchet

$$H_{\gamma=\alpha}^{Fr}(x) = \begin{cases} 0, & \text{for } x < 0 \\ \exp\left(-\left(\frac{x-b}{a}\right)^{-\alpha}\right), & \text{for } x \geq b,\ \alpha > 0. \end{cases} \tag{9.1}$$

(ii) Gumbel

$$H_{\gamma=1}^{Gu}(x) = \exp\left(-\exp\left(-\left(\frac{x-b}{a}\right)\right)\right), \text{ for } x \in \mathcal{R}. \tag{9.2}$$

(iii) Weibull[1]

$$H_{\gamma=\alpha}^{Wei}(x) = \begin{cases} \exp\left(-\left(-\frac{x-b}{a}\right)^{-\alpha}\right), & \text{for } x \leq b < 0,\ \alpha < 0 \\ 0, & \text{for } x > 0. \end{cases} \tag{9.3}$$

Of course, Fréchet, Gumbel, and Weibull PDFs are obtained as derivatives of the appropriate CDFs given above in their upper limits, i.e., on the variable x. Accordingly,

[1] More precisely, it is the so-called inverse Weibull distribution. The Weibull distribution describes the statistics of minimum values.

$$P^{Fr}(x) = \frac{\alpha}{a}\left(\frac{x-b}{a}\right)^{-\alpha-1} \exp\left(-\left(\frac{x-b}{a}\right)^{-\alpha}\right),$$

$$P^{Gu}(x) = \frac{1}{a}\exp\left(-\left(\frac{x-b}{a} + \exp\left(-\frac{x-b}{a}\right)\right)\right),$$

$$P^{Wei}(x) = \frac{|\,\alpha\,|}{a}\left(-\frac{x-b}{a}\right)^{-\alpha-1} \exp\left(-\left(-\frac{x-b}{a}\right)^{-\alpha}\right), \quad (9.4)$$

where we limit ourselves only to those parts of the domains where the distributions do not disappear. Let us add that the default parameter values are $a = 1$ and $b = 0$.

It is worth noting that the (inverse) Weibull and Fréchet distributions are complementary because all their variability is located in complementary domains. However, their shapes differ significantly (despite their formal similarity) because the shape parameters defining them are of the opposite sign.

It can be shown that the above three distribution functions can be expressed using the Generalized Extreme Value Distribution (GEVD).

$$H_\gamma(x) = \exp\left(-\left(1+\gamma\cdot\frac{x-b}{a}\right)_+^{-1/\gamma}\right), \; a > 0, \quad (9.5)$$

where $(\ldots)_+ = max[0, (\ldots)]$ and argument $1+\gamma\cdot\frac{x-b}{a} \geq 0$. The domain of this generalized CDF depends on the parameters a, b and the shape parameter[2] γ as follows:

(1) when $\gamma > 0$ then $x \in]-a/\gamma - b, \infty]$ and then we are dealing with the CDF (shifted and scaled) of the Fréchet distribution (see the formula (9.1)), where $1/\gamma = \alpha$;
(2) in the case of $\gamma \to 0$ the domain of x is not limited—we are then dealing with the CDF of the Gumbel distribution (see the formula (9.2));
(3) for $\gamma < 0$ we have $x \in [-\infty, -a/\gamma + b[$ and then we are dealing with the CDF (shifted and scaled) of the Weibull distribution (see the formula (9.3)), where $1/\gamma = \alpha$.

[2] Shape parameter $\gamma \propto \frac{1}{\alpha}$, where α is the exponent defining the Fréchet distribution and the Weibull distribution in Theorem 9.1.1.

The most crucial element of the limit theorem is the statistical independence of the sequence elements referred to in the theorem. As in CLT, the introduction of long-term dependencies/autocorrelation between sequence elements brings the analysis into the area of anomalous diffusion; in this case, we move to the area defined by extreme value statistics (EVS) [9]. This is a "hot" area of interest for current research. Therefore, we encourage readers to conduct independent research in this field.

In Sect. 9.2 we present a complementary approach to the statistics of extreme events based on base distributions from which the sequences of independent random variables $(x_1, x_2, \ldots, x_n)$ mentioned in Theorem 9.1.1 are drawn. This allowed us to physically interpret the positional parameters by linking them to the mentioned underlying distribution.

It should be emphasized that determining the positional limit parameters a and b, especially the shape parameter γ, from empirical data is the basic, pragmatic task of the Theory of Extreme Events.

The Maximum Limit Theorem (MLT), as defined in Theorem 9.1.1, is theoretically and practically significant, falling within the extensive category of limit theorems. Unlike other fundamental limit theorems, such as the Central Limit Theorem and Bernoulli's Law of Large Numbers, the MLT is not based on moments but on quantiles. One notable advantage of the MLT is its lack of limitations associated with finite process variance values. It is worth noting that, in this textbook, we primarily focus (except the MLT) on the Central Limit Theorem, Bernoulli's Law of Large Numbers, and the Generalized Central Limit Theorem of Lévy–Khintchine.

9.1.1 Value at Risk: A Prominent Application of the GEVD

One of the noteworthy achievements of extreme value theory (EVT) is the identification of the 'Value at Risk' (VaR), representing the maximum acceptable loss, through the use of Generalized Extreme Value Distribution (GEVD). In this subsection, we will determine this value.

So let's assume we have set a confidence level of $1-\alpha$. This type of notation[3] is of the form,

$$F(VaR_{1-\alpha}) = \mathcal{P}(x < VaR_{1-\alpha}) = 1-\alpha. \tag{9.6}$$

We can now easily determine the GEVD, remembering that by assumption, we only consider the random values of the x variable smaller than the threshold value $VaR_{1-\alpha}$. Namely, for sufficiently large values of n, by Theorem 9.1.1, holds with a good approximation,

$$H_{\gamma}\left(\frac{VaR_{1-\alpha}-b}{a}\right) \approx \mathcal{P}(x_n^{\max} < VaR_{1-\alpha}) = \left[\mathcal{P}(x < VaR_{1-\alpha})\right]^n = (1-\alpha)^n, \tag{9.7}$$

where the nth power appearing on the right side comes from the fact that each of the $n \gg 1$ random values of the sequence $(x_1, \ldots, x_n)$ of i.i.d. must be lower than the adopted threshold value $VaR_{1-\alpha}$.

Now, substituting (9.5) into the left side (9.7) we finally obtain, after simple transformations,

$$VaR_{1-\alpha} \approx b + \frac{a}{\gamma}\left[(-n\ln(1-\alpha))^{-\gamma} - 1\right]. \tag{9.8}$$

This expression rationalizes the acceptable loss threshold for a given confidence level—a factor of utmost practical importance. Nevertheless, it's crucial to emphasize that a more comprehensive analysis necessitates addressing how rapidly the parameters (a_n, b_n, γ_n) converge to their limit values, denoted as (a, b, γ), respectively. This consideration leads to answering the question of the rate at which the right side of the Eq. (9.8) reaches a plateau[4] with an increase in n (unless stated otherwise).

9.2 Distribution of Maxims and the Base Distribution—General Formula

Suppose one observes a series of L independent realizations of the same random phenomenon (or its stochastic replica). In that case, the central question of Extreme Value Theory (EVT) is how to characterize the maximum observed value of random

[3] The parameter α, as commonly used in this context, has nothing in common with the shape exponent present in power-law distributions. Let us add that the distribution function (9.6) is said to be a quantile of the order $1-\alpha$.

[4] This matter becomes more intricate. Specifically, a rigorous approach would entail utilizing the pre-limit distribution function H_{γ_n} for $n \gg 1$, as outlined in Theorem 9.1.1, the explicit form of which remains unknown.

independent and identically distributed (i.i.d.) variables[5] $x_{max} \stackrel{\text{def.}}{=} \max\{x_l\}_{l=1,\ldots,L}$. For example, the maximum value could represent the deepest trap encountered by a walker in a disordered medium/substrate (in this case, $x \equiv \varepsilon$, where ε is the energetic depth of the trap) or the longest mean residence time (also called the sojourn time of the walker) in such a trap (in this case, $x \equiv \tau$, where τ is the mean residence time).

The main objective of EVT is to characterize x_{max} by determining the probability distribution, $P(x_{max} = \Lambda)$, of the maximal value x_{max}, where Λ is an arbitrary threshold. In the case of dispersive transport and diffusion, we apply EVT to characterize the two related stochastic variables (ε and τ).

First, we calculate the cumulative probability distribution $\mathcal{P}(x_{max} < \Lambda)$ of the random variable x_{max} by noting that if the maximum x_{max} is smaller than Λ, all x_l's are also smaller than this threshold and vice versa. As these random variables are i.i.d., we can employ

$$\mathcal{P}(x_{max} < \Lambda) = [\rho_<(\Lambda)]^L = [1 - \rho_\geq(\Lambda)]^L, \tag{9.9}$$

by assuming the cumulative probability distribution of random variable x

$$\rho_<(\Lambda) = \int_{x_{down}}^{\Lambda} \rho(x)dx, \tag{9.10}$$

where $\rho(x)dx$ is the basic probability that the random variable x can be found in the interval $x, x + dx$, and x_{down} is the lowest value which this variable can assume. Of course, the second equality in expression (9.9) comes from the normalization of the probability density function $\rho(\ldots)$ where

$$\rho_\geq(\Lambda) = \int_{\Lambda}^{x_{up}} \rho(x)dx, \tag{9.11}$$

here x_{up} is the largest value which the variable x can assume. We set here $x_{down} \ll \Lambda \leq x_{up}$ so that the strong inequality $\rho_\geq(\Lambda) \ll 1$ is obeyed. Therefore, the second equality in expression (9.9) takes, with a good approximation, the useful form,

$$\mathcal{P}(x_{max} < \Lambda) \approx \exp(-L \cdot \rho_\geq(\Lambda)). \tag{9.12}$$

In this way, we reached our second step, namely the intermediate formula useful for further transformations

$$P(x_{max} = \Lambda) = \frac{d\mathcal{P}(x_{max} < \Lambda)}{d\Lambda} \approx L \cdot \rho(\Lambda) \cdot \exp(-L \cdot \rho_\geq(\Lambda)), \tag{9.13}$$

where the notation $\rho(\Lambda) = \rho(x = \Lambda)$ and definition (9.11) have been introduced.

[5] We developed the Extreme Value Theory by considering continuous variables and assuming simplified notation x_{max} instead of x_n^{max} used in Sect. 9.1, where $n \equiv L$.

In the third step, we relate the number of observations, L, to the rare event. Bernoulli's Law of Large Numbers tells us that one can expect to observe such events with a probability at least equal to $1/L$. Therefore, it would be surprising to encounter an event with a probability smaller than $1/L$. The largest single event of Λ_{max}, observed whenever in a single long series of $L \gg 1$ the i.i.d. random variables, which we call indeed the rare one,[6] is thus given by the relation

$$\rho_{\geq}(\Lambda_{max}) = \frac{1}{L}. \tag{9.14}$$

We can call the above definition of the rare event the weak one; the stronger definition (which seems to be even easier to interpret) could have the form,

$$\rho(\Lambda_{max}) = \frac{1}{L}, \tag{9.15}$$

which is, however, less convenient (from the technical point of view of the general approach).[7] Since we now operate with two types of 'max' variables, we aim to find their probabilistic relation.

By combining Eqs. (9.12), (9.13) and (9.14), we finally find the general formula for the searched distribution

$$P(x_{max} = \Lambda) \approx \frac{\rho(\Lambda)}{\rho_{\geq}(\Lambda_{max})} \cdot \exp\left(-\frac{\rho_{\geq}(\Lambda)}{\rho_{\geq}(\Lambda_{max})}\right). \tag{9.16}$$

We utilize the formula above to derive the Gumbel and Fréchet distributions and establish a relationship between them.

It is worth adding for completeness that the distribution function (9.12) now takes the general form,

$$\mathcal{P}(x_{max} < \Lambda) \approx \exp\left(-\frac{\rho_{\geq}(\Lambda)}{\rho_{\geq}(\Lambda_{max})}\right) \tag{9.17}$$

not as specified as the GEVD given in Sect. 9.1, although it can be shown that it follows from it (which we leave to the interested reader).

[6] Please do not confuse x_{max} with Λ_{max}. The variable x_{max} refers to a single time series of L observations. In contrast, Λ_{max} is the maximum value in a statistical ensemble consisting of a large number $J \gg 1$ of such time series. That is, $\Lambda_{max} = \max(x^j_{max})_{j=1,\dots,J}$. To put it more figuratively, Λ_{max} is the global top. So, remember that we are operating with three different quantities here: x_{max}, Λ, Λ_{max}.

[7] Note that for most cases analytically solvable, both definitions give (as expected) identical shapes of distributions of random variables, which require only rescaling by additive and multiplicative constants.

The approach in this section started from the base distribution $\rho(x)$ from which we randomly selected elements of the series $\{x_l\}_{l=1,2,\ldots,L}$. It is more intuitive—physical than that presented in Sect. 9.1. However, not as general.[8]

9.2.1 The Gumbel Distribution Versus the Fréchet One

We assume the disordered substrate/medium is characterized by the random trap or valley model defined on a regular lattice. Therefore, all valleys are equally spaced but have different depths, $\varepsilon(> 0)$, while the mountain peaks are all at the same energy level. It is assumed that the distribution of depths is exponential

$$\rho(\varepsilon) = \frac{1}{\langle\varepsilon\rangle} \exp\left(-\frac{\varepsilon}{\langle\varepsilon\rangle}\right) \tag{9.18}$$

which has been adopted by many authors. The conspicuous aspect of the random-trap model is its symmetry, where, in the absence of a bias, there is no tendency for the carrier to drift from any trap configuration. Hence, the carrier hops in any possible direction have an equal probability, and the different hops between valleys are uncorrelated (or independent). We use the distribution provided above as a foundation for further considerations.

9.2.1.1 The Gumbel Probability Distribution Function

As mentioned, we can identify the random variables $x \iff \varepsilon$. Moreover, from expression (9.18) we find

$$\begin{aligned} \rho(\Lambda) &= \frac{1}{\langle\varepsilon\rangle} \exp\left(-\frac{\Lambda}{\langle\varepsilon\rangle}\right), \\ \rho_{\geq}(\Lambda) &= \exp\left(-\frac{\Lambda}{\langle\varepsilon\rangle}\right), \ \rho_{\geq}(\Lambda_{\max}) = \exp\left(-\frac{\Lambda_{\max}}{\langle\varepsilon\rangle}\right), \end{aligned} \tag{9.19}$$

which enables to express formula (9.16) in an explicit form. Note that the third equality in (9.19) together with (9.15) gives an explicit, unique relation between the value of the rare event $\Lambda_{\max}$ and the number of observations L

$$\frac{\Lambda_{\max}}{\langle\varepsilon\rangle} = \ln L, \tag{9.20}$$

[8] The considerations presented in Sect. 9.2 are based on the corresponding ones presented by Bouchaud and Potters [1] and Kozłowska and Kutner [3].

which points to a slow (logarithmic) growth[9] with increasing L.

By using (9.19), formula (9.16) takes an intermediate form,

$$P(\varepsilon_{\max} = \Lambda) \approx \frac{1}{\langle \varepsilon \rangle} \exp\left(-\frac{\Lambda - \Lambda_{\max}}{\langle \varepsilon \rangle}\right) \exp\left(-\exp\left(-\frac{\Lambda - \Lambda_{\max}}{\langle \varepsilon \rangle}\right)\right), \quad (9.21)$$

which fully corresponds to the second equation in (9.4), since an assignment can be allowed $a = \langle \varepsilon \rangle$ and $b = \Lambda_{\max}$.

To obtain the searched distribution in a closed, standardized form, the following transformation of variable $\varepsilon_{\max}$ or Λ should be made;

$$u \stackrel{\text{def.}}{=} \frac{\varepsilon_{\max} - \Lambda_{\max}}{\langle \varepsilon \rangle} = \frac{\Lambda - \Lambda_{\max}}{\langle \varepsilon \rangle} \Rightarrow du = \frac{\Lambda}{\langle \varepsilon \rangle}; \quad (9.22)$$

hence and by expression (9.21) we finally obtain the well known Gumbel distribution,

$$P(u) = \exp(-u) \exp(-\exp(-u)) \quad (9.23)$$

of the u random variable, where we tacitly use the invariance of the probability under the monotonic transformation of a random variable (invariant measure); thus, we used the (positively oriented) equality,

$$P(\varepsilon_{\max} = \Lambda) d\Lambda = P(u) du. \quad (9.24)$$

Note that the distribution (9.23) can be obtained, as needed, as the derivative of the Gumbel cumulative distribution function given by (9.2), with $u \equiv x$.

Note that the most probable value of the distribution (9.23) is $u = 0$, which shows that, paradoxically, the rare event $\Lambda_{\max}$ is the most probable value among $\varepsilon_{\max}$'s. On the other hand, when $u \to \infty$ the Gumbell distribution $P(u \to \infty) \to \exp(-u)$. Hence, the distribution of random variable ε and the analogous (although asymptotic) one variable $\varepsilon_{\max}$ are exponential. We can say that the exponential distribution is asymptotically stable concerning the 'max' operation.

9.2.1.2 The Fréchet Probability Distribution Function

Now, we are ready to address the inquiry regarding the distribution of sojourn times of the walker in traps and determine (utilizing the formula (9.16)) the probability density function (PDF) of its longest values present within a given series of observations. Then (as previously stated in this section), we assume the random variable $x \equiv \tau$.

Accordingly, as the first step, we perform the transformation

$$\varepsilon \Rightarrow \tau(\varepsilon) = \tau_0 \exp(\hat{\beta}\varepsilon) = \tau_0 \cdot \hat{\tau}^{\,\varepsilon/\Delta},$$

---

[9] For the stronger definition of the rare event (9.15) we obtain $\Lambda_{\max}/\langle \varepsilon \rangle = \ln(L/\langle \varepsilon \rangle)$ while the Gumbel distribution (9.23) of variable u defined by (9.22) is unaffected.

$$\rho(\varepsilon) \Rightarrow \hat{\rho}(\tau(\varepsilon)) = \frac{1}{\tau_0} \frac{\alpha}{(\tau/\tau_0)^{\alpha+1}} \tag{9.25}$$

where we set $\hat{\tau} = exp(\hat{\beta} \cdot \Delta)$, as we consider over-barrier jumps of a carrier (here Δ denotes the energy unit), and the exponent

$$\alpha = \frac{\ln N}{\ln \hat{\tau}} = \frac{1}{\hat{\beta} \cdot \langle \varepsilon \rangle}. \tag{9.26}$$

To derive the second equality in (9.25), we used the invariance of the probability under the monotonic transformation of a random variable (as given by the first equation of (9.25)), i.e., we used the positively oriented equality

$$\hat{\rho}(\tau) d\tau = \rho(\varepsilon) d\varepsilon. \tag{9.27}$$

Note that the exponential transformation of a random variable leads to the transformation of its (invariant) PDF from the exponential to the power law. Conversely, the logarithmic transformation of random variables leads to the transformation of their PDF from the power law to exponential ones.

From the second relation in (9.25) and definition (9.11) we can easily calculate the probability,

$$\hat{\rho}_{\geq}(\Lambda) = \frac{1}{(\Lambda/\tau_0)^{\alpha}}. \tag{9.28}$$

and hence

$$\hat{\rho}_{\geq}(\Lambda_{\max}) = \frac{1}{(\Lambda_{\max}/\tau_0)^{\alpha}}. \tag{9.29}$$

necessary to obtain probability distribution (9.16) in an explicit form.[10] Note that by using Eq. (9.14) we obtain $\Lambda_{\max}$ as a power-law function of L[11]

$$\frac{\Lambda_{\max}}{\tau_0} = L^{1/\alpha}. \tag{9.30}$$

It should be noted that the same result is obtained if we use the rare event of energy depth of traps (9.20) as a power (divided by Δ) of τ', which gives self-consistency of the approach.

By substituting formulae (9.28) and (9.29) into (9.16), and performing straightforward calculations,

---

[10] The Λ variable used here relates to τ and not ε one.

[11] By using the stronger definition (9.15) of the rare event one gets a related scaling law $\Lambda_{\max}/\tau_0 = (L/(\tau_0 \cdot \alpha))^{1/(\alpha+1)}$.

$$P(\tau_{\max} = \Lambda) = \frac{1}{\Lambda_{\max}} \cdot \frac{\alpha}{(\Lambda/\Lambda_{\max})^{\alpha+1}} \exp(-1/(\Lambda/\Lambda_{\max})^{\alpha}), \tag{9.31}$$

where the correspondence to the first equation in (9.4) reduces to the position $a = \Lambda_{\max}$ and $b = 0$. Hence, we finally obtain the Fréchet distribution in the standardized form,

$$P(u) = \frac{\alpha}{u^{\alpha+1}} \exp\left(-\frac{1}{u^{\alpha}}\right) \tag{9.32}$$

of $u \stackrel{\text{def.}}{=} \Lambda/\Lambda_{\max}$ variable, where as usual, we used the invariance of the probability under the monotonic transformation of a random variable, i.e., we used the (positively oriented) equality

$$P(\tau_{\max} = \Lambda)d\Lambda = P(u)du. \tag{9.33}$$

Note that this distribution can be obtained, as needed, as the derivative of the Fréchet probability distribution function (9.1), where $u \iff x$.

It can be easily found that the most probable value of $\tau_{\max}$ is proportional to the value of the rare event $\Lambda_{\max}$.[12]

As it is seen, for $u \gg 1$, the Fréchet PDF is the power-law of the exponent $1 + \alpha$ with the power-law correction to the scaling,

$$P(u) \approx \frac{\alpha}{u^{\alpha+1}} \left(1 - \frac{1}{u^{\alpha}}\right). \tag{9.34}$$

Analogously to the Gumbel distribution, the power-law tail is asymptotically stable to the 'max' operation.

9.2.1.3 Relation Between the Gumbel and Fréchet Distributions

The above considerations demonstrate that transforming the random variable ε to its exponential form $\tau(\varepsilon)$ (see the first relation in (9.25)) results in the transformation of the Gumbel PDF into the Fréchet PDF. In other words, the Gumbel distribution describes an additive stochastic process, whereas the Fréchet distribution characterizes a multiplicative stochastic process. The logarithm operation defines the relationship between these two processes.

We leave the production of the Weibull PDF similarly (i.e., based on Eqs. (9.16) or (9.17)) to the interested reader.

---

[12] More precisely, $\tau_{\max} = (\alpha/(1+\alpha))^{1/\alpha} \cdot \Lambda_{\max}$, and only for $\alpha \to \infty$ variable $\tau_{\max} = \Lambda_{\max}$.

9.3 Ranking Within the Extreme Value Theory

In this section, we ask a more general question than in Sect. 9.2 although we consider again a series of L independent observations of identically distributed random variables. We can rank variables $x_l,\ l = 1, 2, \ldots, L$, in decreasing order of their amplitude. We denote by Λ_n the value encountered n^{th} among these random variables. Hence, $\Lambda_1 = x_{\max}$ and $\Lambda_L = x_{\min}$ (i.e., the minimal value of variables x_l).

As the first step, we are interested in the probability distribution function $P_n(\Lambda_n) = P_n(x = \Lambda_n)$ of the random variable x. We can write the exact formula

$$P_n(\Lambda_n) = L \cdot C_{L-1}^{n-1}\, \rho(x = \Lambda_n)[\rho_>(\Lambda_n)]^{n-1}[\rho_<(\Lambda_n)]^{L-n}, \tag{9.35}$$

where $C_{L-1}^{n-1} = \frac{(L-1)!}{(n-1)!(L-n)!}$ denotes the combinatorial (or Newton binomial) factor. The product $L \cdot C_{L-1}^{n-1}$ gives the total number of ways to set Λ_n within all possible configurations of $L - 1$ elements that remain random variables of the series. Note that for $n = 1$, the above formula simplifies to expression (9.13), as it is expected to be for $L \gg 1$. We continue to write briefly $\rho(\Lambda_n) = \rho(x = \Lambda_n)$

In the second step, we find the most probable value of $\Lambda_n^\star$ (for a given rank n). By differentiating probability distribution (9.35) and setting it equal to zero, we obtain the formula,

$$\frac{1}{L} \cdot \frac{d\rho(\Lambda_n)}{d\Lambda_n} \cdot \rho_>(\Lambda_n) \cdot \rho_<(\Lambda_n) - \frac{n-1}{L} \cdot [\rho(\Lambda_n)]^2 \cdot \rho_<(\Lambda_n) + (1 - \frac{n}{L}) \cdot [\rho(\Lambda_n)]^2 \cdot \rho_>(\Lambda_n) = 0 \tag{9.36}$$

useful for further considerations particularly when $n,\ L \to \infty$ with fixed ratio n/L. Then the first term in (9.36) vanishes and we obtain the formula

$$\rho_>(\Lambda_n^\star) = \rho_\geq(\Lambda_n^\star) \approx \frac{n}{L}. \tag{9.37}$$

which generalizes (9.14). We used here the normalization condition $\rho_<(\Lambda_n) = 1 - \rho_>(\Lambda_n)$ and equality $\rho_>(\Lambda_n) = \rho_\geq(\Lambda_n)$ which are valid for the continuous random variable x.

To complete information on the distribution $P_n(\Lambda_n)$ in the vicinity of $\Lambda_n^\star$ we, as our third step, calculate its width σ_n. We find σ_n by using the saddle-point (or Gaussian) approximation from the second derivative of $\ln P_n(\Lambda_n)$ calculated at $\Lambda_n^\star$ since in this approximation, one can use the general equation,

$$\frac{d^2}{d\Lambda_n^2} \ln P_n(\Lambda_n) \mid_{\Lambda_n^\star} = -\frac{1}{\sigma_n^2}. \tag{9.38}$$

Hence and from (9.35), we obtain the width of the probability distribution $P_n(\Lambda_n)$ in the form

$$\sigma_n \approx \frac{1}{\sqrt{L}} \cdot \frac{\sqrt{\frac{n}{L} \cdot (1 - \frac{n}{L})}}{\rho(\Lambda_n^\star)} \tag{9.39}$$

which is more and more sharply peaked around its most probable value $\Lambda_n^\star$ as L tends to infinity (with fixed ratio n/L).

Two useful cases. Let us assume the case of the exponential tail (given in Sect. 9.2.1 by Eq. (9.18)). Apply the second relation of Eq. (9.19) to Eq. (9.36), we obtain

$$\Lambda_n^\star \approx \langle \varepsilon \rangle \cdot \ln\left(\frac{L}{n}\right). \tag{9.40}$$

In the case of the power-law tail (given again in Sect. 9.2.1 by the second equation in (9.25)), we obtain,

$$\Lambda_n^\star \approx \tau_0 \cdot \left(\frac{L}{n}\right)^{1/\alpha}, \tag{9.41}$$

which was already derived in Sect. 6.4.10 by the simplified approach (of course, $\Lambda_n^\star$ present in the above formula is equivalent to $\tau(n)$ in formula (6.143)).

In Fig. 9.1, we show both the above-derived results in the log-log plot (where we applied $L = 9841$ and $\alpha = 1.50$). We used the notation,

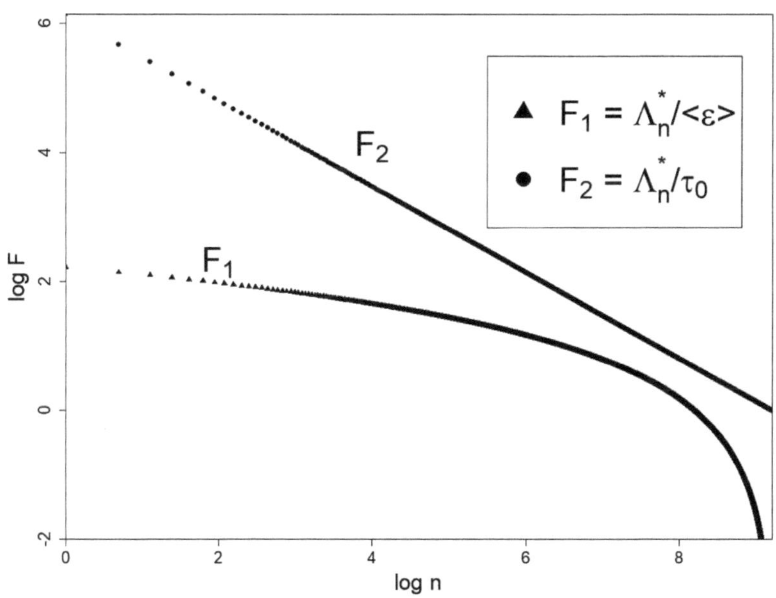

Fig. 9.1 The rank ordering of residence times and depths of traps in log-log scale described by the logarithmic function $F_1(n)$, where $\Lambda_n^\star$ is given by Eq. (9.40), and the power-law function $F_2(n)$, where $\Lambda_n^\star$ is given by Eq. (9.41). Adapted with permission from [3]. ©1995, Elsevier. All rights reserved

$$F(n) = \begin{cases} F_1(n) = \Lambda_n^\star/\langle\varepsilon\rangle, & \text{from Eq. (9.40)} \\ F_2(n) = \Lambda_n^\star/\tau_0, & \text{from Eq. (9.41).} \end{cases} \tag{9.42}$$

We observe the trend in the slope of the exponential distribution (represented by $F_1(n)$ vs. n), which decreases as the rank variable n increases. However, it's essential to note that in the case of a power law (represented by $F_2(n)$ vs. n), the slope remains constant when plotted on a log-log scale.

Control & Review Questions and Exercises

9.1 Problem Develop the Maximum Limit Theorem. Extract the Probability Density Functions (PDFs) of Fréchet, Gumbel, and Weibull from their Cumulative Distribution Functions (CDFs). Moreover, validate that the expression (9.5) is congruent to the expressions (9.1), (9.2), and (9.3).

9.2 Exercise
Deduce the expression (9.8) using the expression (9.5).

9.3 Exercise and Question
Obtain the universal formula (9.13) and the Eq. (9.16). Use this latter equation for the base exponential and power law distributions. What is the relationship between $\Lambda_{\max}$ and L in these instances?

9.4 Exercise and Question
Demonstrate that the Gumbel PDF and Fréchet PDF are asymptotically stable concerning the 'max' operation.

9.5 Question
Which nonlinear transformation results in a bijective relationship between the Gumbel and Fréchet PDFs, and what is the reason for this?

9.6 Exercise and Question
Obtain an exact formula for the PDF, $P_n(\Lambda_n) = P_n(x = \Lambda_n)$, of the stochastic variable x, which represents the nth element in size of the L-element time series. What is the primary presupposition in this derivation?

9.7 Problem
Derive a universal, estimated dependence of the nth largest element of a time series on the number of its elements L for the exponential and power-law probability distributions.

References

1. Bouchaud, J.-P., Potters, M.: Theory of Financial Risks and Derivative Pricing. From Statistical Physics to Risk Management. Cambridge University Press, Cambridge (2003)
2. Sornette, D.: Critical Phenomena in Natural Sciences. Chaos, Fractals, Self-Organization, and Disorder: Concepts and Tools, 2nd edn. Springer Series in Synergetics. Springer Complexity. Springer, Berlin (2006). https://doi.org/10.1007/3-540-33182-4
3. Kozłowska, M., Kutner, R.: Anomalous transport and diffusion versus extreme value theory. Phys. A **357**, 282–304 (2005). https://doi.org/10.1016/j.physa.2005.06.030
4. Albeverio, S., Jentsch, V., Kantz, H. (eds.): Extreme Events in Nature and Society. (The Frontiers Collections). Springer, Berlin (2006). https://doi.org/10.1007/s10955-007-9389-7
5. Embrechts, P., Klüppelberg, C., Mikosh, T.: Modelling Extremal Events for Insurence and Finance. Springer, Berlin (1997)
6. Moreira, A.A., Andrade, J.S., Jr., Amaral, L.A.N.: Extremum statistics in scale-free network models. Phys. Rev. Lett. **89**(26), 268703-1-4 (2002). https://doi.org/10.1103/PhysRevLett.89.268703
7. Albeverio, S., Piterbarg, V.: Mathematical methods and concepts for the analysis of extreme events. In: Albeverio, S., Jentsch, V., Kantz, H. (eds.) Extreme Events in Nature and Society, pp. 47–68. Springer, Berlin (2006)
8. Malevergne, Y., Sornette, D.: Extreme Financial Risks. From Dependence to Risk Management. Springer, Berlin (2006). https://doi.org/10.1007/b138841
9. Klinger, J., Voituriez, R., Bénichou, O.: Extreme value statistics of jump processes. Phys. Rev. E **109**(105), L052101 (2024). https://doi.org/10.1103/PhysRevE.109.L052101

Chapter 10
Limit Theorems on the Stock Market

Abstract As is well-documented, time series data of stock quotes can be regarded as the realization of a trajectory associated with a one-dimensional random walk. This chapter utilizes empirical data from the S&P 500 index. The probability density function (PDF) derived from this data exemplifies heavy-tailed distributions, characteristic of real-world phenomena, particularly in the context of the stock market. The analysis introduces the concept of moral hazard, where the potential for rare but substantial losses accompanies significant profits. This phenomenon necessitates a modernized understanding of risk, especially in investment paradigms. Lévy processes are well-suited for describing such behavior, in notable contrast to Gaussian processes, characterized by narrower distributions. Crucially, significant information regarding the statistical properties of the system resides in the tails of the distribution function. Consequently, this leads to infinite variance and kurtosis, rendering traditional risk assessments that rely on variability ineffective. Therefore, it is evident that a robust market risk analysis demands a reconceptualization of risk. More generally, non-Brownian and non-Gaussian stochastic processes observed in financial markets have contributed to a new/contemporary perspective on market risk in the broad sense.

10.1 Remarks Concerning Generalized Central Limit Theorem on NYSE

As a highly instructive example, we pose a pivotal question for traditional stock dynamics analysis. Do CLT predictions manifest on the stock exchange or not? The answer is complex and depends on factors such as the specific security or index, the time horizon, and the period under consideration. Specifically, for the Standard & Poor 500 (S&P500) index listed on the New York Stock Exchange (NYSE, one of the largest stock exchanges in the world), empirical data, scaled by the factor $(\Delta t)^{-1/\alpha}$, converges, with good approximation, to a stable symmetric Lévy PDF across time horizons ranging from (high-frequency empirical data) $\Delta t = 1$ [min] to

M. Chorowski et al., *Anomalous Stochastics*, Understanding Complex Systems,
https://doi.org/10.1007/978-3-031-80392-5_10

a (daily empirical data) $\Delta t = 1$ [td].[1] Here, α represents the index (shape exponent) of the Lévy PDF. This property is the Generalized Central Limit Theorem (GCLT) or the Lévy–Khintchine limit theorem (LKLT). Physicists Rosario Mantegna and Eugene Stanley derived these findings [1–4]. Wang and Hui observed similar results for the Hang Seng index on the Hong Kong stock market [5].

Of course, there are also stock market quotations whose daily changes are subject to a Gaussian distribution and, therefore, are standardized. This allows for a Gaussian data collapse that is fundamentally different from the abovementioned one. In the following, we compare the conclusions drawn from analyzing both types of limit statements using empirical data from financial markets.

10.1.1 The Approach

It must be emphasized that Mantegna and Stanley were among the first to observe departures from the assumptions of the Central Limit Theorem in financial markets, leading to the development of the Generalized Central Limit Theorem. They analyzed the S&P500 index, focusing on the high-frequency financial data across various time horizons. The findings are illustrated in Fig. 10.1 using appropriate probability density functions.

As anticipated, the initial value of the distribution decreases with an increase in the time horizon. The notable increase in flattening indicated an increase in excess kurtosis (leptokurtosis) as the time horizon increased. For example, the curve for $\Delta t = 1000$ [min] exhibits the highest leptokurtosis for larger values of Z variable, while the curve for $\Delta t = 1$ [min] is the least leptokurtic.

This contradicts the Central Limit Theorem (CLT) prediction, which suggests that as the time horizon extends, the distribution of the aggregated random variable tends to approach a Gaussian distribution, resulting in a reduction of leptokurtosis towards zero.[2] This unexpected result prompted physicists to analyze stock market quotations and marked the onset of econophysics, signifying a notable increase in physicist interest in financial time series.

Figure 10.2 shows the probability distribution function of the standardized change in the index/variable Z/σ, where $Z(t) \equiv Z_{\Delta t}(t) = Y(t + \Delta t) - Y(t)$, with $Y(t + \Delta t)$ and $Y(t)$ representing the values of the S&P500 index at times $t + \Delta t$ and t, respectively, and $\sigma \equiv \sigma(\Delta t)$ denotes the estimate of dispersion calculated based on the empirical data presented in the figure for the selected time horizon $\Delta t = 1$ [min].

---

[1] The abbreviation "td" stands for "trading day."

[2] It is worth noting that if the distribution is Gaussian, excess kurtosis is always zero regardless of the time horizon Δt.

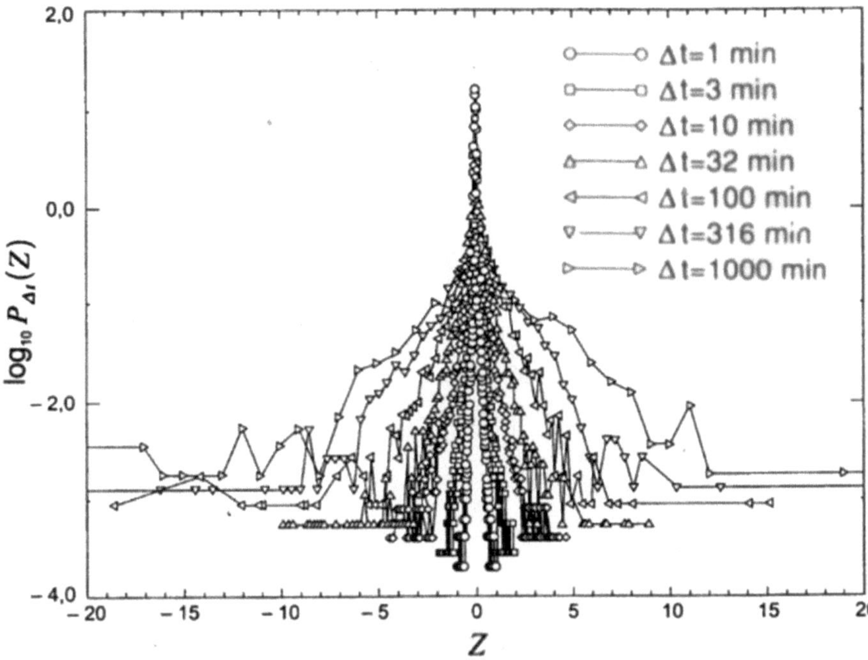

Fig. 10.1 Comparison of probability distribution functions derived from high-frequency empirical data on the S&P500 index is conducted for various time horizons, specifically $\Delta t = 1, 3, 10, 32, 100, 316, 1000$ [min]. The analysis involves the variable $Z(t) \equiv Z_{\Delta t}(t) = Y(t + \Delta t) - Y(t)$, where $Y(t + \Delta t)$ and $Y(t)$ represent the S&P 500 index at times $t + \Delta t$ and t, respectively. It's important to note that the excess kurtosis of the distribution increases for the large Z variable values when it goes to a larger time horizon. This results in a slight decrease and flattening of the peak at the central part of the PDF (although this may not be visible on the adopted figure scale). Additionally, the arms of the distribution widen and the maximum decreases as the time horizon Δt extends, which means that the distribution is narrowest for $\Delta t = 1$ [min] and widest for $\Delta t = 1000$ [min]. Adapted with permission from [4].

Figure 10.3 displays the scaled statistics/histogram of the S&P500 index shown at the beginning (in Fig. 10.1), depending on the scaled random variable, i.e.,

$$\tilde{P}(\tilde{Z}) = (\gamma \Delta t)^{1/\alpha} \cdot P_{\Delta t}(Z(\tilde{Z})), \ \tilde{Z}(Z) = \frac{Z}{(\gamma \Delta t)^{1/\alpha}} \iff Z(\tilde{Z}) = (\gamma \Delta t)^{1/\alpha} \tilde{Z}, \tag{10.1}$$

where the shape exponent α and the coefficient γ, common to all histograms $P_{\Delta t}(Z)$, were determined from their values for $Z = 0$, i.e., from the probability density function of the process returning to the origin[3] $P_{\Delta t}(Z = 0)$ or, more precisely, from the

[3] Such a probability distribution function describes trajectories returning to the starting point $Z = 0$ after time Δt or waiting at that point for that time. We've already discussed this in Sect. 3.7.1.

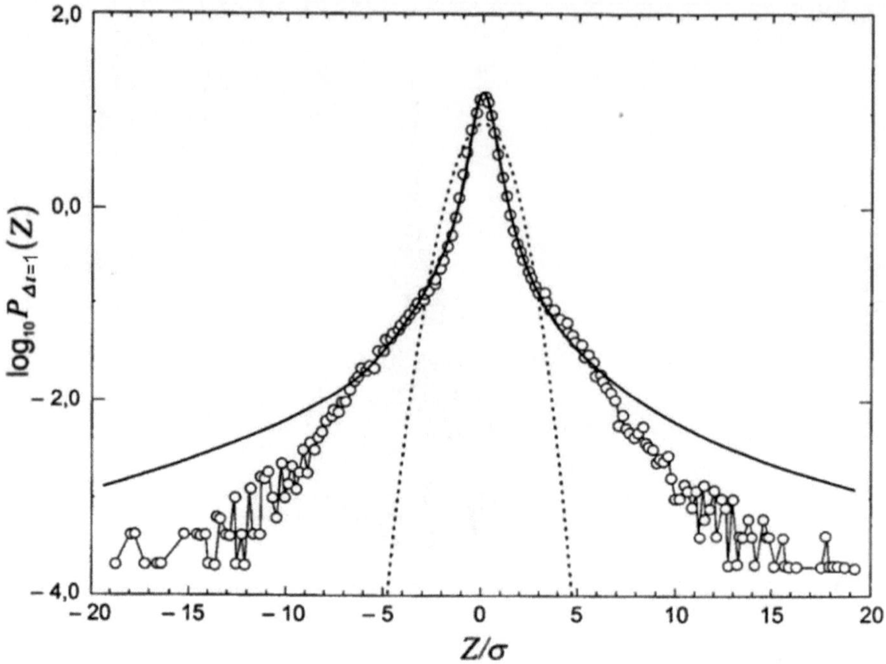

Fig. 10.2 Exemplary presentation of the logarithm of the probability distribution function $\log P_{\Delta t}(Z)$ obtained for high-frequency empirical data (open circles connected by line segments; in Fig. 10.1 it is the narrowest curve) regarding the S&P500 index for the time horizon $\Delta t = 1$ [min], versus $Z(t)/\sigma \iff [Z_{\Delta t}(t) = Y(t + \Delta t) - Y(t)]/\sigma$, where $Y(t + \Delta t)$ and $Y(t)$ represent the values of the S&P500 index at times $t + \Delta t$ and t. At the same time, $\sigma = 0.0508$ is the dispersion calculated based on all empirical data presented in the figure (which is always finite). The solid line depicts the Lévy PDF with a shape exponent $\alpha = 1.40$ and a scaling factor $\gamma = 0.00375$. In contrast, the dotted line represents a Gaussian PDF centered at $Z/\sigma = 0$ and parameterized only by the dispersion σ mentioned above. Adapted with permission from [4].

slope of the line $P_{\Delta t}(Z = 0)$ (on the $log - log$ scale) as a function of Δt shown in Fig. 10.4. When deriving the formulas (10.1), we used the invariance of probabilities,

$$\tilde{P}(\tilde{Z})d\tilde{Z} \overset{\text{orient.}}{=} P_{\Delta t}(Z)dZ, \tag{10.2}$$

as scalars.[4]

[4] Equality (10.2) is co-oriented, i.e., $\tilde{P}^L(\tilde{Z}) = P^L_{\Delta t}(Z(\tilde{Z})) \mid \frac{d\tilde{Z}(Z)}{dZ} \mid^{-1}$. The superscript "$L$" means that we are considering the PDF given by Lévy PDF.

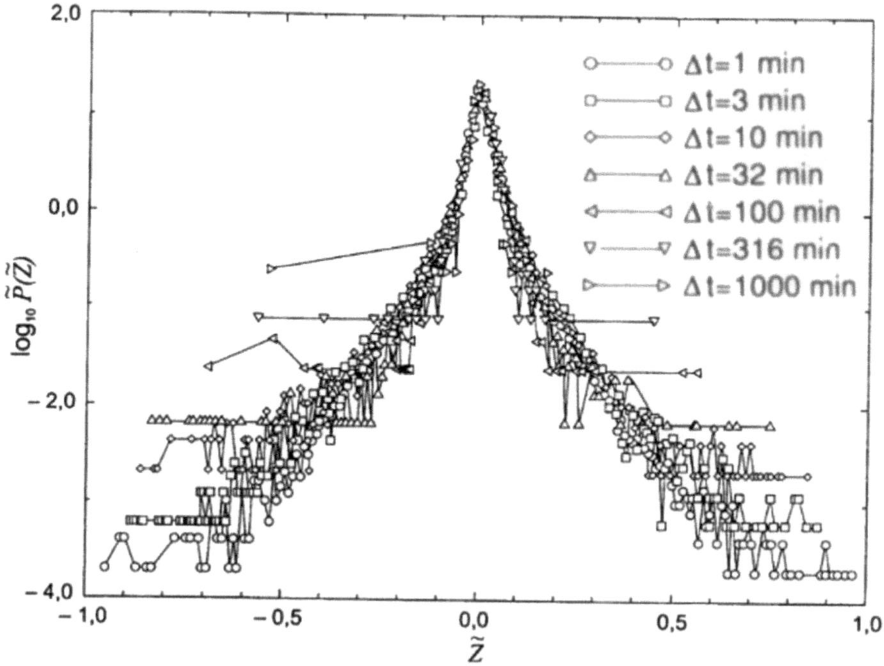

Fig. 10.3 Comparison of appropriately standardized probability distribution functions, $\tilde{P}$, obtained for empirical data on the S&P500 index for a time horizon $\Delta t = 1, 3, 10, 32, 100, 316, 1000$ [min], where $\tilde{Z}(t)$ is a standardized random variable. As observed, there is (to a good approximation) a data collapse. Adapted with permission from [4].

10.2 Stable Symmetric Lévy Probability Distribution Function

Data collapse obtained through scaling (10.1) (see also Fig. 10.3), where the exponent α was obtained by analyzing the probability of returning to the origin $P_{\Delta t}(Z = 0)$ (see Fig. 10.4), indicates that the statistics for the discussed time horizons can be expressed using a stable symmetric Lévy PDF controlled by exponent α

$$P^L_{\Delta t}(Z) = \frac{1}{\pi} \int_0^{\infty} \chi_{\Delta t}(q) \cos(qZ) dq, \tag{10.3}$$

where

$$\chi_{\Delta t}(q) \overset{\text{def.}}{=} \exp(-\gamma \Delta t \mid q \mid^{\alpha}), \ \gamma = \frac{\gamma_0}{\tau}, \quad \Delta t = n\tau, \tag{10.4}$$

is its characteristic function. In the case of $\alpha = 2$ and $\gamma_0 = \sigma^2/2$, we obtain a one-dimensional Gaussian PDF. From the expressions (10.3) and (10.4), we obtain (after

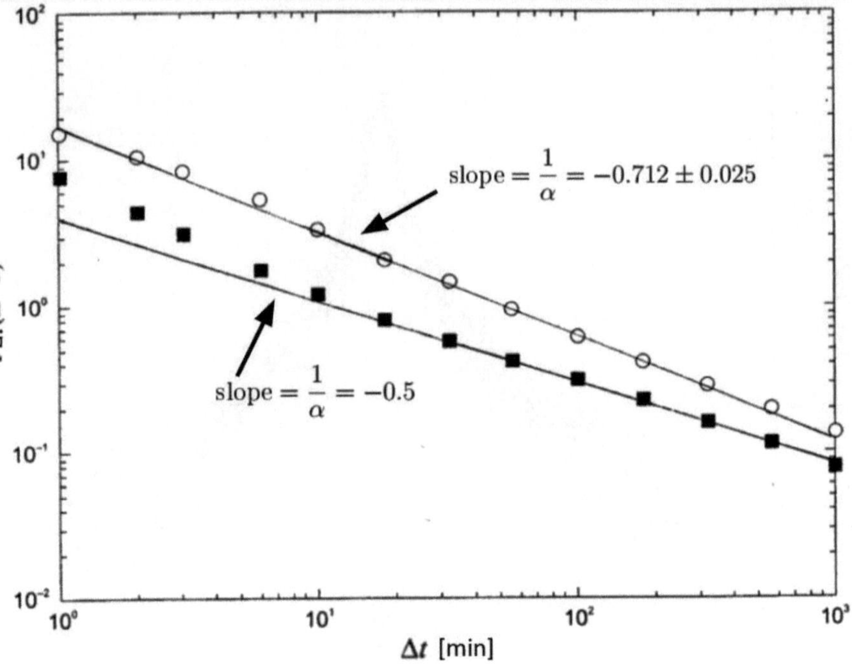

Fig. 10.4 The Lévy PDF of the process returning to the origin $P_{\Delta t}(Z=0) = P^L_{\Delta t}(Z=0)$ (open circles) as a function of time horizon Δt (counted here in trading minutes; the fit was done using Eq. (10.5)). We identify the slope of the straight solid line (on the log − log scale) with the exponent $1/\alpha = 0.712 \pm 0.025$, which gives $\alpha = 1.40 \mp 0.05$. The results are compared for the analogous probability $P_{\Delta t}(Z=0) = P^G_{\Delta t}(Z=0)$ (black squares) obtained from the Gaussian PDF (i.e., Eq. (10.5) where $\alpha = 2.0$ was placed). The variances of this process were determined from the empirical data for each value of Δt separately. The more significant the difference in the slopes of the two lines, the more the distribution described here by open circles differs from the Gauss PDF (cf. Fig. 10.2). As can be seen, the Gauss PDF cannot fit the empirical data over the entire range of Δt, unlike the Lévy PDF. Adapted with permission from [1].

a simple change of variables $q \Rightarrow \tilde{q} = (\gamma\Delta t)^{1/\alpha} q)$ that the probability distribution function of returning to the origin is:

$$P^L_{\Delta t}(Z=0) = \frac{1}{\pi}\Gamma\left(\frac{1}{\alpha}\right) \cdot \frac{1}{\alpha \cdot (\gamma\Delta t)^{1/\alpha}} = \frac{1}{\pi}\Gamma\left(1+\frac{1}{\alpha}\right) \cdot (\gamma\Delta t)^{-1/\alpha}, \quad (10.5)$$

because

$$\frac{1}{\alpha}\Gamma\left(\frac{1}{\alpha}\right) = \int_0^\infty \exp(-\mid \tilde{q} \mid^\alpha) d\tilde{q} = \frac{1}{\alpha}\int_0^\infty \exp(-y) \cdot y^{1/\alpha - 1} dy, \quad (10.6)$$

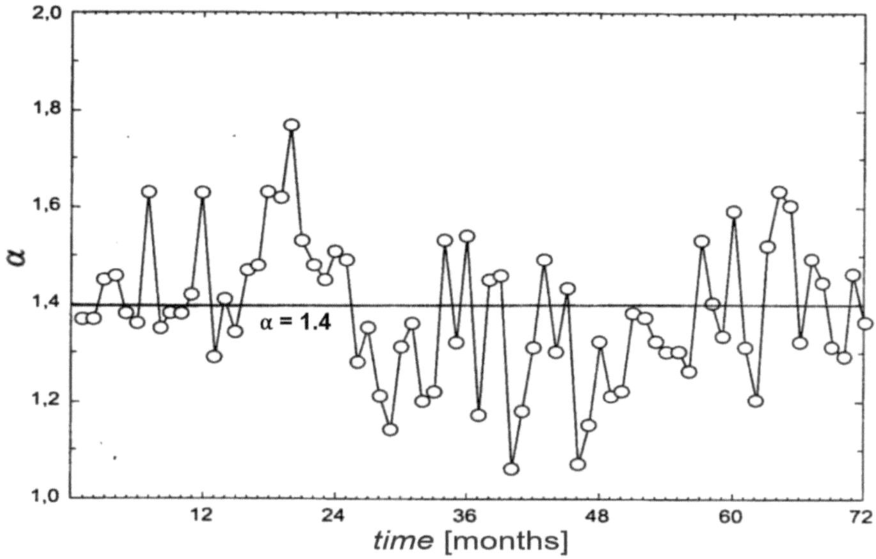

Fig. 10.5 The time dependence of the exponent α is determined based on the dependence of the probability of returning to the origin on time (calculated in months). A solid horizontal line is drawn for $\alpha = 1.40$ obtained from the fit presented in Fig. 10.4. Adapted with permission from [1]. ©1999, Cambridge University Press. All rights reserved

where we replaced the variable $y =| \tilde{q} |^{\alpha}$. Given a specific value of α, we determine the scaling factor γ from the expression (10.5) and the shift of the straight line (fitted to white circles) on the plot Fig. 10.4.

Note that the exponent α and the scaling factor γ change monthly. These monthly fluctuations are shown in Figs. 10.5 and 10.6. As can be seen, even the most considerable value of α is much smaller than 2 (the typical error of a single value of the exponent α for a given month is given in the caption of Fig. 10.4).

Therefore, the empirical data suggest that these parameters controlling the Lévy distribution can be treated as random variables. This is the main argument favoring the averaging procedure used in Sect. 3.7.1 leading to multifractality.

The consequence of replacing q with $\tilde{q}$ is to replace the variable Z in the output integral (10.3) with $\tilde{Z} = Z/(\gamma \Delta t)^{1/\alpha}$—thanks to this we get the necessary expression,

$$\tilde{P}^L(\tilde{Z}) = \frac{1}{\pi} \int_0^{\infty} \exp(- \mid \tilde{q} \mid^{\alpha}) \cos(\tilde{q} \tilde{Z}) d\tilde{q} \tag{10.7}$$

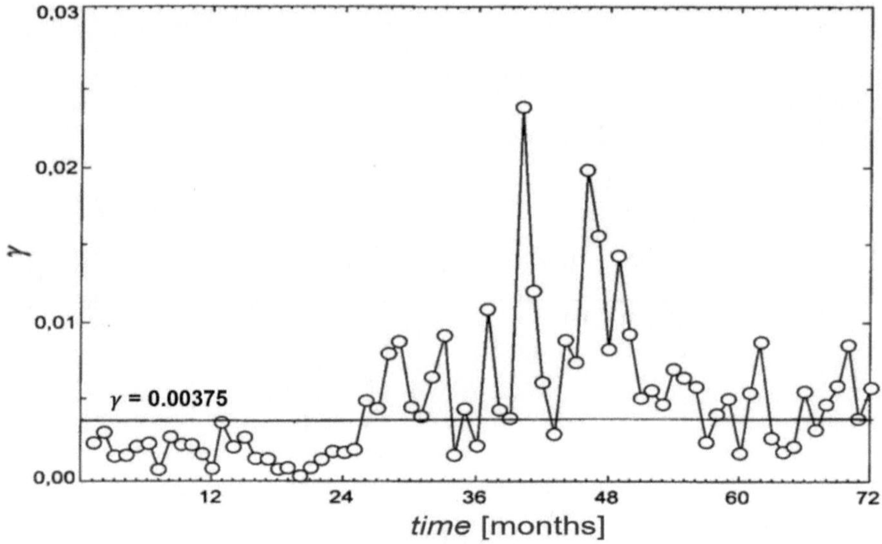

Fig. 10.6 Time dependence of the scaling factor γ determined based on the dependence of the probability of returning to the origin on time (calculated in months). A horizontal solid line is drawn for $\gamma = 0.00375$ obtained from the fit presented in Fig. 10.4. Adapted with permission from [1]. ©1999, Cambridge University Press. All rights reserved

independent of Δt, and it is related to the output probability $P_{\Delta t}(Z)$ by the first relation in (10.1).

Therefore, the description of the main part of the empirical data using the Lévy distribution is proven, although the problem of describing the shape of the "tails" themselves, as well as the significant spread of the data on the "tails" (see Fig. 10.3 for $|\tilde{Z}| \gtrsim 0.5$) is still an open issue of great interest [6, 7].

It is worth emphasizing that, for example, the probability of returning to the origin for the Gaussian PDF scales with the exponent $\alpha = 2$, which defines a completely different (normal as opposed to anomalous) "statistical world," in which there is no room for rare events.

Control & Review Questions and Exercises

10.1 Exercise

Derive the scaling of the symmetric Lévy PDF leading to data collapse.

10.2 Problem

What parameters of the Lévy PDF can be obtained with the return-to-origin probability distribution? Can they be treated as random variables?

10.3 Questions

Give the formula for a symmetrical, stable Lévy PDF. Demonstrate that it can be scaled to a form leading to data collapses.

10.4 Question

Why can Lévy distributions observed in the real world lead to multifractality?

References

1. Mantegna, R.N., Stanley, H.E.: An Introduction to Econophysics. Correlations and Complexity in Finance. Cambridge University Press, Cambridge UK (1999)
2. Bouchaud, J.-P., Potters, M.: Theory of Financial Risks. From Statistical Physics to Risk Management, Second reprint. Cambridge University Press, Cambridge (2001)
3. Sornette, D.: Critical Phenomena in Natural Sciences. Chaos, Fractals, Self-Organization, and Disorder: Concepts and Tools, 2nd edn. Springer Series in Synergetics. Springer Complexity. Springer, Berlin (2006). https://doi.org/10.1007/3-540-33182-4
4. Mantegna, R.N., Stanley, H.E.: Scaling behavior in the dynamics of an economic index. Nature **376**(6535), 46–49 (1995). https://doi.org/10.1038/376046a0
5. Wang, B.H., Hui, P.M.: The distribution and scaling of fluctuations for Hang Seng index in Hong Kong stock market. Eur. Phys. J. B **20**(4), 573–579 (2001). https://doi.org/10.1007/PL00022987
6. Mantegna, R.N., Stanley, H.E.: Turbulence and financial markets. Nature **383**(6601), 587–588 (1996). https://doi.org/10.1038/383587a0
7. Mantegna, R.N., Stanley, H.E.: Physics Investigation of Financial Markets. In: Mallamace, F., Stanley, H.E. (eds.) The Physics of Complex Systems. Proceedings of the International School of Physics "Enrico Fermi," Course CXXXIV. IOS Press, Amsterdam (1997)

Chapter 11
Comprehensive Partition Function: A Universal Tool in Multifractality

Abstract In this chapter, we examine the evolution of the free market characterized by competing firms. We explore its relationship to random walks, specifically positing that the time-dependent market shares of firms can be modeled as instances of random walks. Our approach complements the multifractal detrended fluctuation analysis (MF-DFA) [2, 6]. More precisely, we propose utilizing market shares for multifractal market analysis among competing firms. This innovative method begins with a partition function directly expressed in terms of market shares, thus eliminating the laborious preparatory work typically associated with traditional MF-DFA. Unlike conventional methodologies that rely on a fluctuation function derived from time series data, our approach streamlines the analysis by directly incorporating market share information. We successfully applied our method to publicly available empirical data. A key outcome of this chapter is the sought-after classification of the spectrum of dimensions, which greatly enhances the applicability of a multifractal description in this context.

11.1 Initiation

In the last two decades, multifractal properties have been a focal point of extensive research in various scientific domains, even far from canonical physics [1–14]. The ongoing trend of exploring new realms of multifractality persists. The spectrum of dimensions (singularities), including their shape, location, and spread, the primary indicator of multifractality, offers valuable information on structural organization. Our methodology, based on the formalism [15], characterizes systems in a state of statistical equilibrium and stationary states. Moreover, we emphasize that this can readily be extended to transient states (which, however, is beyond the scope of the textbook).

The chapter is divided into two parts. The first part encompasses Sect. 11.1: 'Initiation' and Sect. 11.2: 'Theory,' where we explain the general multifractal approach through a comprehensive partition function representation. The second part applies this multifractal approach to the real market of the S&P500 index. Furthermore, this

M. Chorowski et al., *Anomalous Stochastics*, Understanding Complex Systems,
https://doi.org/10.1007/978-3-031-80392-5_11

section conducts a comparative analysis between the results obtained from the actual market and those derived, for example, from the four-group market model.

11.2 Theory

11.2.1 Definition of the Comprehensive Partition Function

We recall that multifractality can be read from the power-law behavior of the comprehensive partition function (CPF)[1] defined with coarse-grained measure [15] as follows:

$$Z_\beta(N) = \sum_{n=1}^{N} (p_n)^\beta, \tag{11.1}$$

where the coarse-grained probability (measure) p_n, $n = 1, 2, \ldots, N$, is attributed to the disjoint boxes/cells of equal size (numbered with n) with which the domain of this measure is entirely covered in the number N. In the context of diffusion, one can look at p_n as the probability of finding a diffusing molecule in the box/cell with the number n. A non-uniformly distributed mass means the mass accumulated in the box n or, in the case of deterministic chaos, the concentration of points on the strange set. Information theory defines this probability using a bit-number (related to the crowding index).

However, this chapter considers another fresh example with some universal features. That is, it is the multifractal behavior of the market of competing firms. This concept of multifractality was founded on the normalized market shares, ω_n, where we put $p_n \stackrel{\text{def.}}{=} \omega_n$, for each firm $n = 1, 2, \ldots, N$, where N is the total number of firms in the market. These values, obtained from simulations, empirical data, or theoretical studies, are given in advance.

[1] Please do not confuse the comprehensive partition function given by the formula (11.1) with the more broadly defined generalized partition function introduced for cells of variable size in the paper Halsey–Jensen–Kadanoff–Procaccia–Shraiman [16].

11.2.2 Limitations of CPF

The CPF in the form given by Eq. (11.1) is ready to study the multiscale nature of the ω distribution. This multiscale nature comes from the hierarchical distribution of firms' sizes.

The partition function, $Z_\beta(N)$, obeys two basic properties,

$$Z_{\beta=0}(N) = N, \tag{11.2}$$

and

$$Z_{\beta=1}(N) = 1. \tag{11.3}$$

Of course, Eq. (11.2) describes the size of the multifractal substrate or company market, while Eq. (11.3) comes from the normalization condition of the shares.

Moreover, using the share limitation from below and above, we get,

$$Z_{\beta\to\mp\infty}(N) \approx \begin{cases} \left(\omega^{\min}\right)^\beta, & \beta < 0 \\ \left(\omega^{\max}\right)^\beta, & \beta > 0 \end{cases} \tag{11.4}$$

where $\omega^{\min}$ and $\omega^{\max}$ determine the marginal values of the companies' market shares.

11.2.3 Scaling Relations

We continue to show that the partition function $Z_\beta(N)$ takes the form of a power law,

$$Z_\beta(N) = N^{-\tau(\beta)} \Leftrightarrow \tau(\beta) = -\frac{\ln Z_\beta(N)}{\ln N}, \tag{11.5}$$

where $\tau(\beta)$ is the generalized scaling exponent. Let us add that the scale denoted by s in the second equality in (5.90) in Sect. 5.3.2, corresponds to $1/N$ in the above definition. Having the partition function at our disposal, we can build a thermodynamic formalism on this basis.

11.2.3.1 Properties of the Generalized Scaling Exponent

To say more about the partition function given by the first equality Eq. (11.5), we use two crucial generalized scaling exponent properties,

$$\tau(\beta) = (\beta - 1)D(\beta) \tag{11.6}$$

where $D(\beta) \geq 0$ is the Rényi dimensions and

$$\tau(\beta) = \beta h(\beta) - D(\beta = 0), \tag{11.7}$$

here, $h(\beta)$ is a generalized Hurst exponent, and $D(\beta = 0)$ is the Hausdorff dimension of the substrate/market, which we can put to 1 for our case. Notably, from Eqs. (11.6) and (11.7) we obtain $h(\beta = 1) = D(\beta = 0)$.

For $\beta \to 1$, the Rényi information approaches the Shannon information (which results directly from the definition (11.1)). That is, it becomes the information dimension,

$$D(\beta = 1) = -\frac{1}{\ln N} \sum_{n=1}^{N} \omega_n \ln \omega_n . \tag{11.8}$$

For $\beta \to 2$ the partition function (11.1) reduces to the well known correlation integral $C(N)$ of Grossberger and Procaccia [17], i.e.,

$$D(\beta = 2) = -\frac{\ln C(N)}{\ln N} . \tag{11.9}$$

Let us also note that always $D(\beta') \leq D(\beta)$ for $\beta < \beta'$; we are still accepting $D(\beta = 0) = 1$.

The above result, in combination with the scaling Eq. (11.5), allows us to present the borders of the scaling exponent in an explicit asymptotic form,

$$\tau(\beta \to \mp\infty) \approx \begin{cases} -\beta \frac{\ln \omega^{\min}}{\ln N}, & \beta < 0 \\ -\beta \frac{\ln \omega^{\max}}{\ln N}, & \beta > 0. \end{cases} \tag{11.10}$$

As one can see, the exponent β is (in both cases) multiplied by a factor with a fractal exponent's structure. The generalized scaling exponent $\tau(\beta)$ is bounded by two diagonal asymptotes defined by Eq. (11.10).

We consider the next two extreme cases. The first is when all but one of the company shares disappear (the case of a monopolized market). Then, with Eqs. (11.1), and the first equality in (11.5), we directly find that $\tau(\beta)$ vanishes independent of β.

The second case is when all shares are equal (the case of the egalitarian market), i.e., $\omega_n = \frac{1}{N}$, $n = 1, 2, \ldots, N$. Then, with Eqs. (11.1), (11.3), and (11.5), we get

$$\tau(\beta) = \beta - 1, \tag{11.11}$$

i.e., the scaling exponent is a linear function of β. We continue to deal mainly with cases distant from the above extreme cases.

We assume that company shares, ω_n, create the nonuniform/multiscale function ω_n versus n, a multifractal structure. In other words, here we are dealing with multifractality, the source of which is the heterogeneous distribution of company shares.

11.2.4 Rényi Dimensions and Generalized Hurst Exponent

Horizontal asymptotes limit also $D(q)$ and $h(q)$ quantities:

$$\begin{aligned} & D(\beta \to \mp\infty) = h(\beta \to \mp\infty) \\ = & \begin{cases} D^{\max} = h^{\max} = -\frac{\ln \omega^{\min}}{\ln N}, & \beta < 0, \\ D^{\min} = h^{\min} = -\frac{\ln \omega^{\max}}{\ln N}, & \beta > 0, \end{cases} \end{aligned} \tag{11.12}$$

while

$$\begin{aligned} & \Delta D(\beta \to \infty) = D^{\max} - D^{\min} \\ = & \ \Delta h(\beta \to \infty) = h^{\max} - h^{\min} \\ = & \ \ln\left(\frac{\omega^{\max}}{\omega^{\min}}\right). \end{aligned} \tag{11.13}$$

The asymptotic scaling exponent properties, as expressed in Eqs. (11.10), (11.6) and (11.7), respectively, directly yield Eqs. (11.12) and (11.13).

11.2.5 Spectrum of Dimensions

We now designate the most crucial multifractality signature, i.e., the spectrum of dimensions (singularities), f, given by the Legendre transformation,

$$f(\alpha) = \beta(\alpha)\alpha - \tau(\beta(\alpha)), \tag{11.14}$$

where the local dimension (singularity or Hölder exponent) is,

$$\alpha(\beta) = \frac{d\tau(\beta)}{d\beta} = -\frac{1}{\ln N} \frac{\sum_{n=1}^{N} \omega_n^{\beta} \ln \omega_n}{\sum_{n=1}^{N} \omega_n^{\beta}}. \tag{11.15}$$

Hence, we obtain a helpful equality locating the maximum of the spectrum of dimensions $f(\alpha(\beta = 0))$,

$$\alpha(\beta = 0) = -\frac{1}{N \ln N} \sum_{n=1}^{N} \ln \omega_n. \tag{11.16}$$

and we get, analogously as in Eq. (11.12),

$$\alpha(\beta \to \mp\infty) \approx \begin{cases} \alpha^{\max} = -\frac{\ln \omega^{\min}}{\ln N}, \\ \alpha^{\min} = -\frac{\ln \omega^{\max}}{\ln N}. \end{cases} \tag{11.17}$$

As one can see from Eq. (11.12), the quantities $D,\ h$, and α have the same lower and upper bounds.

Furthermore, from Eqs. (11.14) and (11.15) we get,

$$\beta(\alpha) = \frac{df(\alpha)}{d\alpha}. \tag{11.18}$$

We present below useful quantities, which characterize the spectrum of singularities,

(i) $f^0 = f(\alpha(\beta = 0) = \alpha^0) = D^0 = D(\beta = 0)$, which results from Eqs. (11.7), (11.14), and (11.18), and moreover we get $\frac{df(\alpha)}{d\alpha}|_{\beta=0} = 0$;
(ii) for $\beta = 1$ we immediately get from Eq. (11.18) $\frac{df(\alpha)}{d\alpha}|_{\beta=1} = 1$, hence $f^1 = f(\alpha(\beta = 1)) = \alpha(\beta = 1) = \alpha^1$;
(iii) for $\beta \to -\infty$ we get from Eqs. (11.14) and (11.15), that $f^{\max} = f(\alpha = \alpha^{\max} = -\frac{\ln \omega^{\min}}{\ln N})$ and $\frac{df(\alpha)}{d\alpha}|_{\beta\to-\infty} = -\infty$; similarly for $\beta \to \infty$ we get $f^{\min} = f\left(\alpha = \alpha^{\min} = -\frac{\ln \omega^{\max}}{\ln N}\right)$ and $\frac{df(\alpha)}{d\alpha}|_{\beta\to\infty} = \infty$;
(iv) the maximum span of f we determine as follows, $\Delta\alpha|_{|\beta|\to\infty} = \alpha^{\max} - \alpha^{\min} = \frac{1}{\ln N} \ln\left(\frac{\omega^{\max}}{\omega^{\min}}\right)$. We continue to use the simplified designation $\Delta\alpha = \Delta\alpha|_{|\beta|\to\infty}$;
(v) the following asymmetry factor can be used to determine the degree of asymmetry f, $\gamma|_{|\beta|\to\infty} = \frac{\alpha(\beta=0)-\alpha^{\min}}{\alpha^{\max}-\alpha(\beta=0)}$, where $\alpha(\beta = 0)$ is given by Eq. (11.16). We continue to use the simplified designation $\gamma = \gamma|_{|\beta|\to\infty}$.

It should be emphasized that, in general $f(\alpha = \alpha^{\min}, \alpha^{\max}) \neq 0$. This happens when at least one of the boundary values $\omega^{\min}, \omega^{\max}$ is degenerated. This is discussed in Sect. 11.3.2 for a four-group company market model. Additionally, we introduced time dependence ad hoc in Appendix A.21.

One can also analyze the asymmetry of f using the coefficient γ. If $\gamma > 1$, then we are talking about the advantage on the market of large companies, as opposed to the situation of $\gamma < 1$. The marginal case $\gamma = 1$ corresponds to the balanced situation.

11.3 Application to Reality

11.3.1 Multifractality of Real Company Market

As an example of the application of our method, we utilize the SPDR S&P500 ETF (SPY) index. This ETF index mimics the behavior of the S&P500 index and the weights of its constituents can be found online. For illustration, we use the results published in our earlier work [18]. The available empirical data cover approximately 70–80% of US stock market capitalization. These data directly provide daily market share values for individual companies, which are required for our analysis.

Let us characterize the market shares of companies using the Quetelet rank-size distribution (see Fig. 11.1), i.e., we construct a cumulative distribution

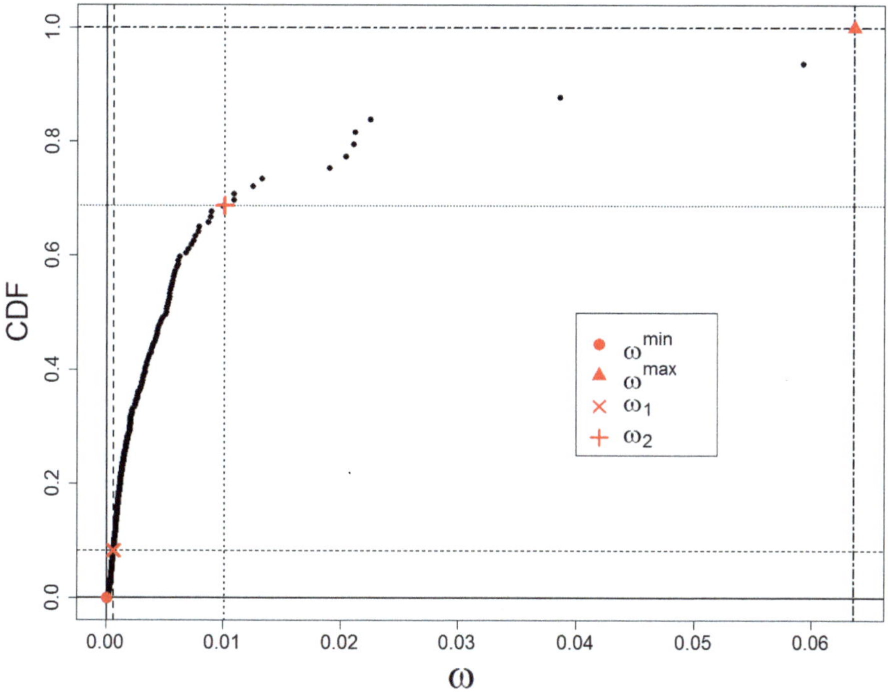

Fig. 11.1 Quetelet curve: the empirical relationship between the standardized rank of companies belonging to the S&P500 index, i.e., the cumulative distribution function (CDF) versus market company share, ω, plot. The border values define the CDF range, that is, $\omega^{\min} = 0.00007$ (red circle) and $\omega^{\max} = 0.06369$ (red triangle). In the main text, we describe (based on Eqs. (11.26) and (11.27)) how to obtain the remaining representative shares $[\omega_1 = 0.00065, \text{CDF}(\omega_1) = 0.08269]$ (red cross) and $[\omega_2 = 0.01001, \text{CDF}(\omega_2) = 0.68776]$ (red plus). To systematically analyze the empirical data, we employ the multifractal formalism. Reprinted under CC-BY-4.0 from [18]. © 2022, Chorowski and Kutner

function (CDF) versus a company market share plot. The span of the CDF is limited as follows: $\omega^{min} = 0.00007 \leq \omega \leq \omega^{max} = 0.06369$. For reasons we explain below, instead of using the full CDF, we will replace them with just four ω values, namely ω^{min}, ω_1, ω_2, and ω^{max}, where ω_1 and ω_2 are self-consistent solutions of multifractal market characteristics.

In such a scenario, the question of the actual dominance of companies in the market arises. Will small companies dominate large ones or vice versa? We utilize the multifractal analysis described in Sect. 11.2 to address this question.

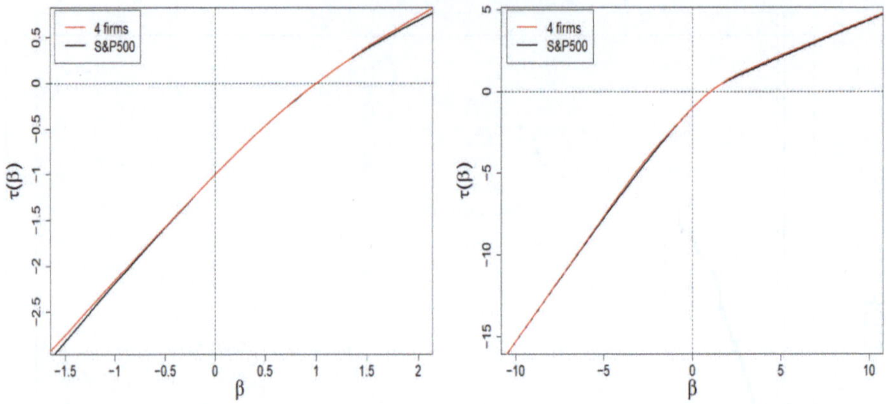

Fig. 11.2 The generalized (or global) scaling exponent, $\tau(\beta)$, dependence on β (basic) exponent for the company market from the S&P500 index. The plot on the left magnifies the range β within the interval $[-1.5, 2.0]$. The right plot displays the full β range from -10 to 10. At the assumed resolution for the entire (right) graph, it becomes impossible to distinguish between the results of the four-group company market model (red curve) and the empirical (black curve). In particular, both curves are asymptotically indistinguishable. In addition, at the points $\beta = 0$ and $\beta = 1$, both curves take on identical values, as shown below. Slight differences are visible only in the left graph. All these global features are reflected in the remaining more sensitive local characteristics of multifractality presented below. Reprinted under CC-BY-4.0 from [18]. © 2022, Chorowski and Kutner

In particular, multifractality would not be encountered if the CDF were constructed based on a power, exponential, or Gaussian distribution. In the first case, the generalized scaling exponent $\tau(\beta)$ would exhibit a linear relationship with β; in the second case, it would be logarithmic; and in the third case, it would be a linear combination of logarithmic and linear functions. For a discussion of this aspect, i.e., monofractality versus multifractality, see Sect. 3.7 .

We further explore the empirical relationship depicted in Fig. 11.1 using the multifractal approach outlined in Sect. 11.2. When employing Eqs. (11.1) and (11.5), we establish the dependence of $\tau(\beta)$ on β. However, we do not examine whether the market is stationary, i.e., we admit that the number of firms N may fluctuate around 500; we consider that shares may also fluctuate. We can apply it here because the method above is applicable to both stationary and non-stationary states

The dependence above of the generalized (or global) scaling exponent $\tau(\beta)$ on β exponent is illustrated in Fig. 11.2.

The depicted dependence is a non-linear function of β, enabling us to proceed with the subsequent steps of the method.

In Fig. 11.3, the dependence of the generalized exponent $\alpha(\beta)$ on the exponent β is presented. As one can see, this type of relationship allows us to reveal its subtle structure.

In Fig. 11.4, we illustrated the dependence of the generalized Hurst exponent $h(\beta)$ on the β exponent. Its range is adequate for one of the singularity spectra depicted

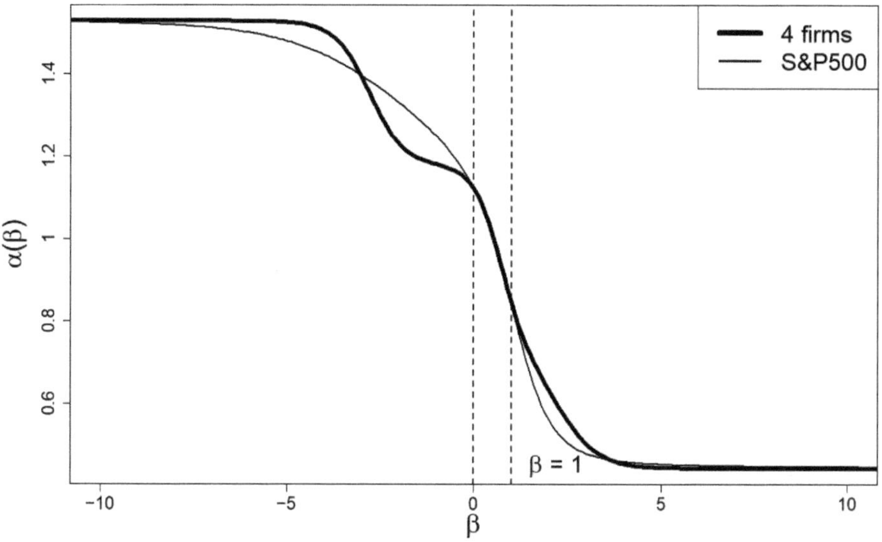

Fig. 11.3 The dependence of the generalized exponent $\alpha(\beta)$ on the β exponent is presented. The subtle/local structure of the theoretical (black bold) curve is visible. However, the fine structure is not discernible in Fig. 11.2 because the generalized scaling exponent is a global characteristic of multifractality. Only the local characteristic of the dependence $\alpha(\beta)\left(=\frac{d\tau(\beta)}{d\beta}\right)$ versus β, allows for its extraction. The black bold (theoretical) and gray thin (empirical) curves have identical values at the points $[\beta = 0,\ \alpha(\beta = 0)]$ and $[\beta = 1,\ \alpha(\beta = 1)]$. The curves are almost identical for asymptotically large values of $|\ \beta\ |$. The differences between both curves are more visible than in Fig. 11.2 for the curves $\tau(\beta)$ versus β, because here we are dealing with the corresponding local quantities. Reprinted under CC-BY-4.0 from [18]. © 2022, Chorowski and Kutner

in Fig. 11.5 (compared to the black bold curve) to establish a robust multifractality. Since the dependence of Rényi dimensions $D(\beta)$ versus β is very similar, we did not provide it here, leaving it to the interested reader.

11.3.2 Real Market Versus Four-Group Company Market

Now, we address the following question: How should companies' markets be grouped/organized to maintain their diversity, i.e., to replicate (at least approximately) empirical data?

First, we consider the spectrum of dimensions (illustrated in Fig. 11.5). This concerns its positioning and the fundamental shape determined by the following characteristic

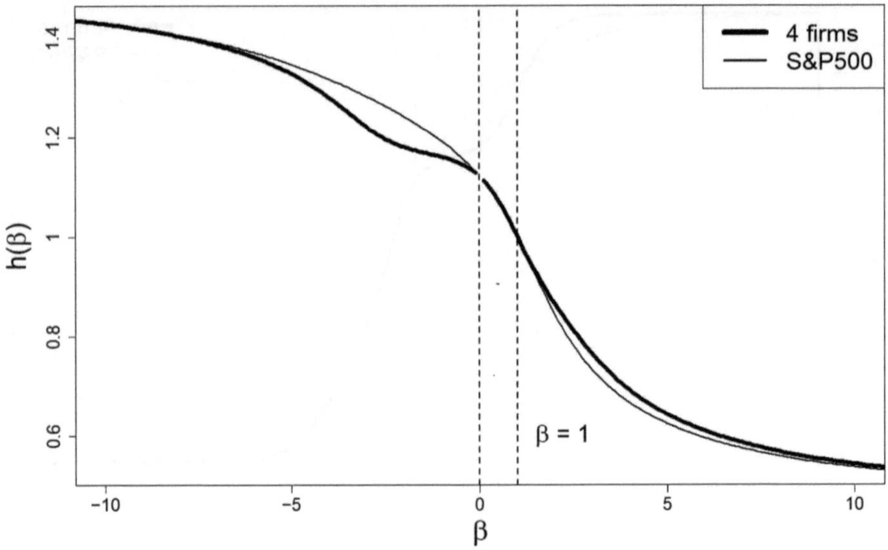

Fig. 11.4 The dependence of the generalized Hurst exponent, $h(\beta)$, on the β exponent is illustrated. Its range is adequate for one-dimensional spectra presented in Fig. 11.5 (having the same span) to establish a robust multifractality. There are slight local/intermediate differences between the curves in both figures (the gray thin empirical one and the black bold theoretical curve for the four-group company market). However, both curves have identical values at the points $[\beta = 0,\ h(\beta = 0)]$ and $[\beta = 1,\ h(\beta = 1)]$. The curves are almost identical for asymptotically large values of $|\ \beta\ |$. Similar to Fig. 11.3, the differences between the both curves are more visible here than in Fig. 11.2 since here we present the corresponding local quantities. Reprinted under CC-BY-4.0 from [18]. © 2022, Chorowski and Kutner

points: $(\alpha^{\min}, f^{\min})$, (α^1, f^1), $(\alpha^0, f^0 (= D^0))$, and $(\alpha^{\max}, f^{\max})$ (refer to Fig. 11.6 for further explanation).

For this purpose, we utilize the following expression for the generalized scaling exponent (derived from the multifractal formalism outlined in Sect. 11.2),

$$\tau(\beta) = -\frac{\ln Z^4_\beta(N)}{\ln N} = -\frac{1}{\ln N} \times \ln\left(M\left(\omega^{\min}\right)^\beta + K_1\omega_1^\beta + K_2\omega_2^\beta + L\left(\omega^{\max}\right)^\beta\right), \tag{11.19}$$

where $Z^4_\beta(N)$ denotes the partition function obtained from Eq. (11.1) for the four-group company market. This section demonstrates that such a division is sufficient to reproduce the localization and (approximately) the shape of the spectrum of dimensions and other multifractality characteristics, such as the generalized scaling exponent and generalized Hurst exponent. We can demonstrate that two- and three-group company markets are inadequate for describing the multifractality of real company markets. For instance, they cannot accurately replicate the location or range of the spectrum of dimensions.

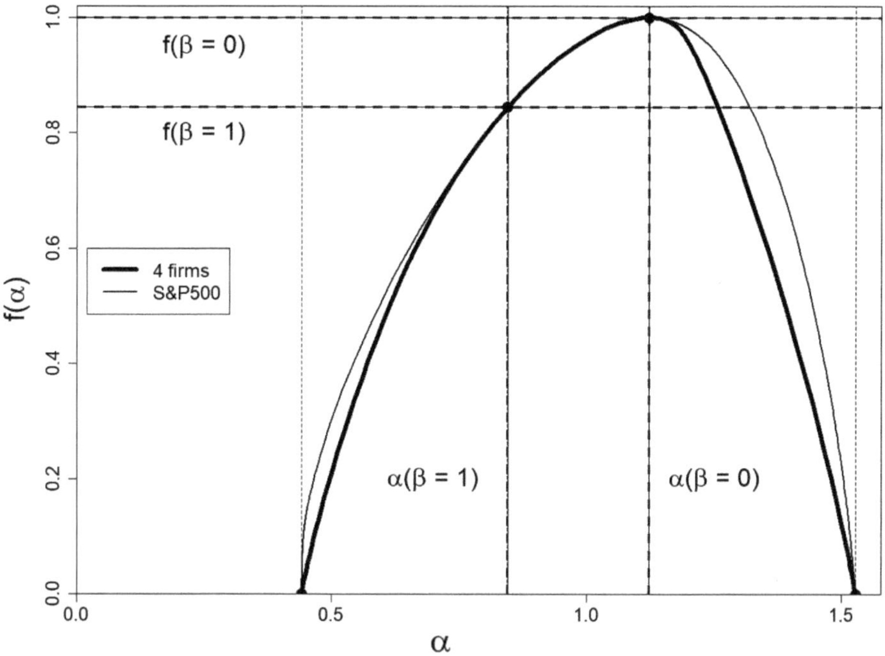

Fig. 11.5 Dependence of the spectrum of dimensions $f(\alpha)$ on α for the company market from the S&P500 index (gray thin curve). The asymmetry in f favoring large firms (smaller α values) is evident. For comparison, the spectrum of dimensions for the four-group company market is represented by the black bold curve. The key here is the fact that the theoretical curve coincides with the empirical one at the most important points: (denoted by the black half circles) $[\alpha(\beta = \infty) = \alpha^{\min},\ f^{min} = f(\alpha^{\min})]$, $[\alpha^{\max} = \alpha(\beta = -\infty),\ f^{\max} = f(\alpha^{max})]$, and (denoted black circles) $[\alpha(\beta = 0) = \alpha^0,\ f^0 = f(\alpha^0)]$, $[\alpha(\beta = 1) = \alpha^1,\ f^1 = f(\alpha^1)]$ (see Fig. 11.6 for further explanation of designations; here $f^{\min} = f^{\max} = 0$). This allows us to use our four-group approach for (at least rough) analysis of empirical data. The overlap of the empirical (gray) and theoretical (black) curves at these four points is a direct result of the corresponding overlap of the curves shown in Fig. 11.2. Thanks to this, we can say that the theoretical curve reproduces the empirical curve quite well. Side note: we leave the analysis of the important point of the spectrum $(\alpha(\beta') = (f^1)^{-1}, f^1)$ to the interested reader (we defined β' by $(f(\alpha(\beta')) = f^1)$, $\alpha(\beta') > \alpha^0 \Leftrightarrow \beta' < 0)$. Reprinted under CC-BY-4.0 from [18]. © 2022, Chorowski and Kutner

11.3.2.1 Specific Goal

Our specific goal is to determine eight unknowns: the size of each of the four groups of companies $M,\ K_1,\ K_2,\ L$ and their shares $\omega^{\min}, \omega_1, \omega_2, \omega^{\max}$. We can unequivocally determine the eight desired unknown variables for the four-group company market.

Figure 11.6 presents a schematic example illustration of the mature spectrum of dimensions. By a fully mature spectrum of dimensions, we define such a spectrum of dimensions that attains its limit values, $\alpha^{\min}$ and $\alpha^{\max}$. We remind that at these junctures $\frac{df}{d\alpha}|_{\alpha=\alpha^{\min,\max}} = \pm\infty$. It should be noted that parabolic multifractality, for instance, is not fully mature. The catalog of possible spectra is presented in Fig. 11.7.

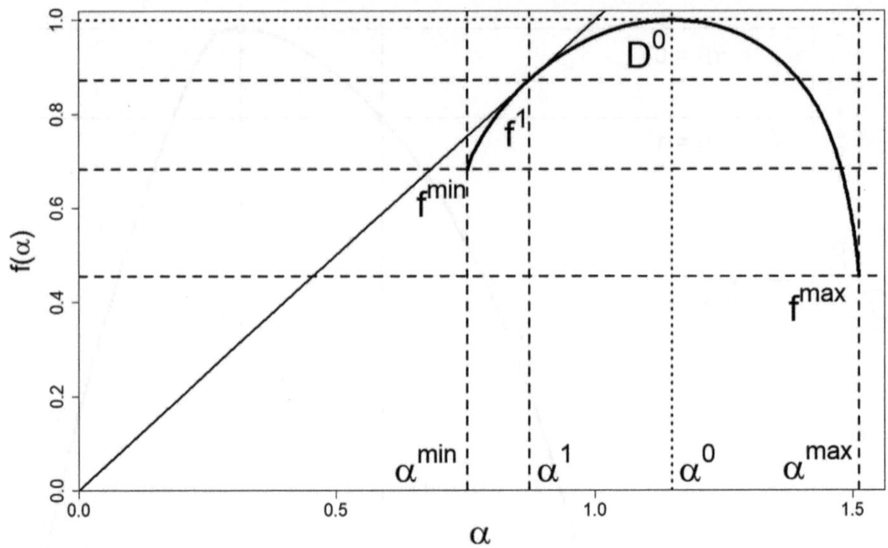

Fig. 11.6 The example of the schematic plot of the fully mature spectrum of dimensions f versus α is depicted. The catalog of schematic plots of possible spectra is presented in Fig. 11.7. The characteristic coordinates that we read from the plot define the conditions (discussed in the main text), which aid in determining the unknown variables M, K_1, K_2, L and $\omega^{\min}$, ω_1, ω_2, $\omega^{\max}$. Let's also recall that $f^0 = f(\alpha^0) = D^0$, where $\alpha^0 = \alpha(\beta = 0)$. We have limited the graph to the standardized situation, where $D^0 (= D(\beta = 0)) = 1$. We also introduced the notation: $f^{\min} = f(\alpha = \alpha^{\min})$ and $f^{\max} = f(\alpha = \alpha^{\max})$. This plot served to introduce the basic characteristics of the multifractal spectrum: ($\alpha^{\min}$, $f^{\min}$), (α^1, f^1), (α^0, f^0), ($\alpha^{\max}$, $f^{\max}$) used later. Reprinted under CC-BY-4.0 from [18]. © 2022, Chorowski and Kutner

The unknown variables we are after can be ascertained by interpreting the coordinates of particular points within this dimensional spectrum. In the subsequent phases, we illustrate this procedure.

The normalization condition takes the form,

$$Z^4_{\beta=1}(N) = M\omega^{\min} + K_1\omega_1 + K_2\omega_2 + L\omega^{\max} = 1, \qquad (11.20)$$

while the size of the market is fixed on N,

$$Z^4_{\beta=0}(N) = M + K_1 + K_2 + L = N. \qquad (11.21)$$

We derive N directly from empirical data.

We highlight Eqs. (11.20) and (11.21) represent the first two equations from the system of equations, enabling us to determine the unknowns mentioned above. Since the shares of $\omega^{\min}$ and $\omega^{\max}$ are directly obtained from empirical data (e.g., from the cumulative empirical data shown in Fig. 11.1), we need four additional equations to determine the remaining unknown variables, as discussed below.[2]

[2] A more optimal division into four groups would involve at least $\omega^{\min}$ not being directly obtained from empirical data. However, we must still utilize at least one more equation.

From Eq. (11.19), definitions (11.5) and (11.15) we get,

$$\alpha(\beta) = \frac{d\tau(\beta)}{d\beta} = -\frac{1}{\ln N}\frac{1}{Z_\beta^4(N)}$$
$$\times\ [M\left(\omega^{\min}\right)^\beta \ln \omega^{\min} + K_1\omega_1^\beta \ln \omega_1$$
$$+\ K_2\omega_2^\beta \ln \omega_2 + L\,(\omega^{\max})^\beta \ln \omega^{\max}]. \tag{11.22}$$

From the definition of the spectrum of dimensions (11.14), we obtain its boundary values for our case,

$$f^{\min} = f(\alpha^{\min} = \alpha(\beta \to \infty)) = \frac{\ln L}{\ln N},$$
$$f^{\max} = f(\alpha^{\max} = \alpha(\beta \to -\infty)) = \frac{\ln M}{\ln N}, \tag{11.23}$$

which can also be approximated from the empirical f shown in Fig. 11.5 (black curve). Thus, the number of unknowns is reduced to ω_1 and ω_2. It should be emphasized that only in the special case when M or L are equal to 1, i.e., when the marginal values of companies' market shares are nondegenerate, do the boundary values of the spectrum of dimensions (11.23) vanish. This occurs precisely in the case of the empirical data we use here.

Another needed quantity, which we read from the empirical f shown in Fig. 11.5 (black bold curve), is the location of the center of the peak f given by the formula,

$$\alpha(\beta = 0) = -\frac{1}{N \ln N}$$
$$\times\ \left(M \ln \omega^{\min} + K_1 \ln \omega_1 + K_2 \ln \omega_2 + L \ln \omega^{\max}\right). \tag{11.24}$$

The same applies to the point of contact $f(\alpha(\beta = 1)) = \alpha(\beta = 1)$. Therefore,

$$\alpha(\beta = 1) = -\frac{1}{\ln N}$$
$$\times\ [M\omega^{\min} \ln \omega^{\min} + K_1\omega_1 \ln \omega_1$$
$$+\ K_2\omega_2 \ln \omega_2 + L\omega^{\max} \ln \omega^{\max}]. \tag{11.25}$$

Both of the above equations have been obtained from Eq. (11.19) and definition (11.15), where $\alpha(\beta = 0)$ and $\alpha(\beta = 1)$ are determined indirectly from empirical data based on the definition (11.15), while N is obtained directly from the empirical data.

Table 11.1 Empirical data on the first and last components of the SPY index as of November 12, 2021

No.	Company	ω^{min}	ω^{max}	M	L
1	AAPL	–	0.06866056	–	1
505	NWS	0.00006948	–	1	–

Explanation of abbreviations: AAPL Apple Inc., NMS New Corporation Class B

Now we calculate the unknowns K_1 and K_2 from Eqs. (11.20) and (11.21) as functions of ω_1 and ω_2. We substitute the obtained quantities into Eqs. (11.24) and (11.25). Thus, we reduce our problem to two transcendental equations. For our case, $M = L = 1$, these equations can be reformulated as follows,

$$\alpha(\beta = 0)N \ln N + \ln\left(\omega^{\min}\omega^{\max}\right)$$
$$= (N-2)\frac{\omega_1 \ln \omega_2 - \omega_2 \ln \omega_1}{\omega_2 - \omega_1} + \Omega \frac{\ln\left(\frac{\omega_1}{\omega_2}\right)}{\omega_2 - \omega_1}, \tag{11.26}$$

and

$$\alpha(\beta = 1) \ln N + \omega^{\min} \ln \omega^{\min} + \omega^{\max} \ln \omega^{\max}$$
$$= (N-2)\frac{\omega_1\omega_2}{\omega_2 - \omega_1} \ln\left(\frac{\omega_2}{\omega_1}\right) + \Omega \frac{\omega_1 \ln \omega_1 - \omega_2 \ln \omega_2}{\omega_2 - \omega_1}, \tag{11.27}$$

(where $\Omega = 1 - \omega^{\min} - \omega^{\max}$), which are more suitable for a numerical solution. Thus, we have reduced our problem to the two transcendental equations above for two unknown variables ω_1 and ω_2.

Table 11.1 presents the empirical data needed regarding the first and last components of the SPY index of November 12, 2021, consisting (on this day) of $N = 505$ companies.

Based on these empirical data, we solve Eqs. (11.26) and (11.27) numerically and obtain $\omega_1 = 0.00065$ and $\omega_2 = 0.01001$. Hence, we have $K_1 = 439$ and $K_2 = 64$, while $M + K_1 + K_2 + L = N = 505$. So, in our case, we obtain nondegenerate share margins and strongly degenerate (though very different) intrinsic share values. The resulting spectrum of dimensions is presented in Fig. 11.5 using a red curve. Likewise, we have presented the remaining results in Figs. 11.2 and 11.4 by means of red curves.

We emphasize that the obtained result is universal in that, starting from the four-group market of companies, we obtain enough equations to define all characteristics of the multifractality, e.g., describing the location and shape of the multifractal characteristics.

11.3.3 Classification of Spectra of Local Dimensions

It's essential to understand how distributions give rise to multifractal structures. The focus is not on seeking out these structures but on contrasting them with one another and determining which structures exhibit more multifractality and which exhibit less. A preliminary step is to categorize them based on their symmetry and the degradation of their extreme/marginal proportions.

The asymmetry coefficient, $\gamma = \frac{\alpha^0 - \alpha^{min}}{\alpha^{max} - \alpha^0}$, dictates the extent of asymmetry in the multifractal structure. The symmetric multifractal structure is observed when $\gamma = 1$. Left asymmetry is present when $\gamma > 1$, while right asymmetry is observed for $\gamma < 1$.

The extent of degeneration of the extreme shares dictates the height of the spectral dimensions' edges: the left edge is influenced by the degeneration level of the maximum share, whereas the degeneration level of the minimum share influences the right edge.

In this way, we have categorized multifractal spectra into nine groups, where asymmetries and degenerations cause the corresponding spectra of dimensions to match each other, akin to the symmetry between left and right hands. See for illustration the corresponding pairs of spectra in Fig. 11.7: the first spectrum in the first column and the last spectrum in the third column (i.e., $f_{\gamma<1,\, \kappa<1},\ f_{\gamma>1,\, \kappa>1}$), the last spectrum in the first column and the first spectrum in the third column (i.e., $f_{\gamma<1,\, \kappa>1},\ f_{\gamma>1,\, \kappa<1}$), the second spectrum in the first column and the second spectrum in the third column (i.e., $f_{\gamma<1,\, \kappa=1},\ f_{\gamma>1,\, \kappa=1}$), and the first spectrum in the second column and the last spectrum in the same column (i.e., $f_{\gamma=1,\, \kappa<1},\ f_{\gamma=1,\, \kappa>1}$).

Classifying the multifractal spectra within each equivalence class would still be possible. However, this is not what we will deal with here, as the division into equivalence classes is already sufficient for our needs. Instead, we introduce a supporting phenomenological definition of a multifractal capacity, MC, exploring the definitions visualized in Fig. 11.6, we can introduce

$$MC \stackrel{\text{def.}}{=} \frac{\Delta\alpha}{\alpha^0\,\alpha^1}\,\frac{f^0\,f^1}{N^{f^{min}+f^{max}}}, \tag{11.28}$$

where for the relationships given in Eq. (11.23), denominator $N^{f^{min}+f^{max}} = L\,M$. We recall $\Delta\alpha = \alpha^{max} - \alpha^{min}$. The above form of MC is easier to interpret. In fact (by item (ii) in Sect. 11.2.5) f^1 and α^1 contract with each other.

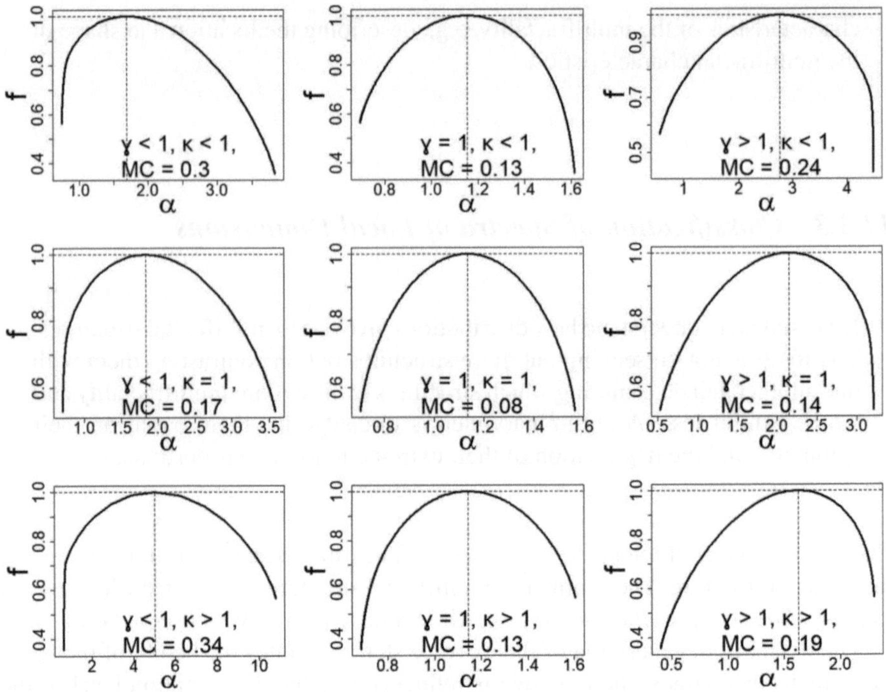

Fig. 11.7 The schematic two-parameter classification of fully mature standardized spectra of local dimensions is based on asymmetry $\gamma = \frac{\alpha^0 - \alpha^{min}}{\alpha^{max} - \alpha^0}$ and degeneration ratio $\kappa = M/L$. In other words, we present a matrix $\hat{F}$ whose elements are spectra of dimensions $f_{\gamma,\kappa}$ indexed by the appropriate ranges of parameter values γ and κ. In essence, each spectrum represents the equivalence class of spectra characterized by the pair of parameters mentioned. The highest coefficient value MC (for the graph in the lower left corner) is associated with the highest $\Delta\alpha$. On the other hand, the lowest MC coefficient value is found in symmetrical spectra with the smallest values of span $\Delta\alpha$. A more detailed classification is possible considering each class separately. Reprinted under CC-BY-4.0 from [18]. © 2022, Chorowski and Kutner

The definition of MC was intended to describe the intuition that spectra of dimensions have more multifractals when they:

(i) have a larger range $\Delta\alpha$
(ii) have higher points f^0 and f^1
(iii) contain more large fluctuations, i.e. they have smaller α^0 and α^1
(iv) have their ends of the spectrum of dimensions lower, i.e., smaller L and M.

The dependence on these effects separately is linear or the inverse of the linear dependence. Notably, our definition of multifractal capacity can be applied to any mature multifractality.

As one can see, the above definition of the multifractal capacity indicator is entirely based on the knowledge of the characteristic points of the spectrum of dimensions, answering the question of which spectrum contains more multifractality, i.e., is more multifractal. This indicator is more subtle than the width of the spectrum of dimensions $\Delta\alpha$. This indicator has a factorized form: the first factor, $\frac{\Delta\alpha}{\alpha^0\alpha^1}$ (which could be called the reduced width/range of the spectrum), directly related to the width of the spectrum its location and the contact point and the second factor, $\frac{f^0 f^1}{N^{f^{min}+f^{max}}}$, directly related to the height of the spectrum. Therefore, the multifractal capacity allows you to classify series/data according to the content of multifractality, i.e., the greater the value of $\mathcal{MC}$, the more multifractality the spectrum of a given time series contains. It should be added that it favors multifractality caused by large fluctuations because the first factor includes the spectral locations α_0 and α_1 in the denominator.

Notably, we can apply the approach to any array of securities, e.g., shares of turnover volumes on the stock exchange and shares of companies' quotations on the stock exchange. In short, the approach can be applied to any normalized series of positively defined elements, performing roles in coarse-grain probabilities (coarse-grain measure). Moreover, our approach makes it possible to examine the evolution of the multifractality of the company market, especially in the vicinity of crash regions. That is why studying the relationship between multifractality and criticality is essential.

Control & Review Questions and Exercises

11.1 Exercise
Prove for what β value the normalized generalized partition function leads to the disappearance of the generalized scaling exponent.

11.2 Exercise
Derive asymptotes of the following multifractality characteristics: τ, α, D, and h.

11.3 Exercise and Questions
Derive asymptotic forms of the generalized scaling exponent. Indicate the conditions under which multifractality disappears—how does it affect the spectrum of dimensions? What reduction does the spectrum of dimensions for mono-fractality undergo?

11.4 Exercise
Derive the formula for Rényi dimensions. Show under what conditions an appropriately chosen Rényi dimension is proportional to Shanon's information.

11.5 Exercise
Prove that $\alpha(\beta = 1) = D(\beta = 1)$.

11.6 Problem
Justify why at points $\beta = 0$, $\beta = 1$, $\beta = \mp\infty$ the characteristics of multifractality obtained within the four-group model and empirical multifractality may overlap.

11.7 Problem

Prove that the number of equations needed to determine all parameters characterizing multifractality within the four-group model is sufficient.

References

1. Zhi-Qiang Jiang, Wen-Jie Xie, Wei-Xing Zhou, Sornette, D.: Multifractal analysis of financial markets review. Reports on Progress in Physics 82(12), 125901 (2019). https://doi.org/10.1088/1361-6633/ab42fb
2. Detecting long-range correlations with detrended fluctuation analysis: Kantelhardt, J.W., Kościelny-Bunde, E., Rego, H.H.A., Havlin, S., A. Bunde, A. Physica A **295**, 441–454 (2001). https://doi.org/10.1016/S0378-4371(01)00144-3
3. Kwapień, J., Stanisław Drożdż, S.: Physical approach to complex systems. Phys. Rep. **515**, 115–226 (2012). https://doi.org/10.1016/j.physrep.2012.01.007
4. Perelló, J., Masoliver, J., Kasprzak, A., Kutner, R.: Model for interevent times with long tails and multifractality in human communications: An application to financial trading. Phys. Rev. E **78**, 036108 (2008). https://doi.org/10.1103/PhysRevE.78.036108
5. Kasprzak, A., Kutner, R., Perelló, J., Masoliver, J.: Higher-order phase transitions on financial markets. Eur. Phys. J. B **76**, 513–527 (2010). https://doi.org/10.1140/epjb/e2010-00064-y
6. Kantelhardt, J.W., Zschlegner, S.A., Koscielny-Bunde, E., Havlin, S., Bunde, A., H. Eugene Stanley, H.E.: Multifractal detrended fluctuation analysis of nonstationary time series. Physica A: Statistical Mechanics and its Applications 316(1-4), 87-114 (2002), https://doi.org/10.1016/S0378-4371(02)01383-3
7. Oświęcimka, P., Kwapień, J., Drożdż, D.: Wavelet versus detrended fluctuation analysis of multifractal structures. Phys. Rev. E **74**(6), 016103 (2006). https://doi.org/10.1103/PhysRevE.74.016103
8. Grech, D., Pamuła, G.: On the multifractal effects generated by monofractal signals. Physica A: Statistical Mechanics and its Applications **392**(23), 5845–5864 (2013). https://doi.org/10.1016/j.physa.2013.07.045
9. Grech, D., Mazur, Z.: On the scaling ranges of detrended fluctuation analysis for long-term memory correlated short data series. Physica A **392**, 2384–2397 (2013). https://doi.org/10.1016/j.physa.2013.01.049
10. Grech, D., Mazur, Z.: On the scaling range of power-laws originated from fluctuation analysis. Phys. Rev. E **87**, 052809 (2013). https://doi.org/10.1103/PhysRevE.87.052809
11. Oświęcimka, P., Drożdż, S., Forczak, M., Jadach, S., Kwapień, J.: Detrended cross-correlation analysis consistently extended to multifractality. Phys. Rev. E. **89**(2), 023305 (2014). https://doi.org/10.1103/PhysRevE.89.023305
12. Drożdż, S., Oświęcimka, P.: Detecting and interpreting distortions in the hierarchical organization of complex time series. Phys. Rev. E **91**(3), 030902(R) (2015). https://doi.org/10.1103/PhysRevE.91.030902
13. Drożdż, S., Kowalski, R., Oświęcimka, P., Rak, R., Gębarowski, R.: Dynamical Variety of Shapes in Financial Multifractality. Complexity Volume 218, ID 7015721 (2018). https://doi.org/10.1155/2018/7015721
14. Klamut, J., Kutner, R., Gubiec, T., Struzik, Z.R.: Multibranch multifractality and the phase transitions in time series of mean interevent times. Phys. Rev. E **101**(6), 063303 (2020). https://doi.org/10.1103/PhysRevE.101.063303
15. Stanley, H.E.: Fractals and Multifractals: The interplay of Physics and Geometry. Chapt. 1, pp. 1–56. In: Bunde, A., Havlin, S. (eds.) Fractals and Disordered Systems. Second Revised and Enlarged Edition. Springer, Berlin (1996)
16. Halsey, T.C., Jensen, M.H., Kadanoff, L.P., Procaccia, I., Shraiman, B.I.: Fractal measures and their singularities: The characterization of strange sets. Phys. Rev. **33A**, 1141 (1986)

17. Grassberger, P., Procaccia, I.: On the Characterization of Strange Attractors. Phys. Rev. Lett. **50**, 346–349 (1983). https://doi.org/10.1103/PhysRevLett.50.346
18. Chorowski, M., Kutner, R.: Multifractal Company Market: An Application to the Stock Market Indices. Entropy 2022 24, 130(1). In: Kutner, R., Schinckus, Ch., Stanley, H.E. (eds.) Three Risky Decades. A Time for Econophysics? pp. 553-568. MDPI, Basel (2022). https://doi.org/10.3390/e24010130

Chapter 12
Epilogue

Abstract In this chapter, we delineate the fundamental components of our textbook, which have been tested using examples from diverse fields. (This chapter is very important in its entirety, which is why we have given up on additional marking in grey). Specifically, we explore topics ranging from the transport and diffusion of charge carriers in amorphous materials to markets, such as financial and free markets with competing firms. Of course, all this is in the context of singular stochastics.

12.1 Fundamental Components

The principal focus of this textbook lies in the simultaneous asymptotic presence of both regular (i.e., Brownian and Gaussian, and singular domains/worlds—see Chap. 6 and 7 for detailed considerations[1]). The preponderance of either domain is contingent upon the specific environmental conditions or scenarios under which the phenomena manifest. Depending on these scenarios, either regularity or singularity may prevail. Both fundamentally different worlds are covered by a single model, namely the Weierstrass random wandering.

In other words, the textbook delves into the Weierstrass process, which integrates both domains (regular and singular). From a statistical perspective, this process is depicted as a self-similar hierarchical random walk, which is amenable to computer simulations, even in instances where the mean and variance of the process are divergent.

Singularities are particularly distinguished by their ability to disrupt pivotal limit theorems, notably the Central Limit Theorem (CLT) and Bernoulli's Law of Large Numbers (BLLN), thus deviating from traditional frameworks of probability theory, mathematical statistics, and stochastic process theory. Such deviations compel the introduction of the generalized (or Lévy–Khinchine) Central Limit Theorem (GCLT)

[1] In this context, it is worth recommending the basic bibliography given in Chap. 1, on which our textbook is based.

M. Chorowski et al., *Anomalous Stochastics*, Understanding Complex Systems,
https://doi.org/10.1007/978-3-031-80392-5_12

and the Maximum Limit Theorem (MLT), both of which are integral to this textbook and elucidated through examples, primarily from the realms of physics and financial markets.

We elucidated (in Chap. 2) the fundamental principles underlying stationary asymptotically power-law probability density functions (PDFs). This elucidation delineated the nexus between the Langevin equation of stochastic dynamics and the realm of stochastic processes, as formulated in the Itô representation. Consequently, this bridges the chasm between the foundational *ab initio* stratum and the sophisticated stochastic Fokker-Planck equation tier (refer to Appendix A.14).

Our textbook elucidates the intricate interconnections among various phenomena in statistical physics, with a particular emphasis on singular stochastic processes. It highlights the significant influence of superdiffusive or persistent characteristics of long-range autocorrelations in these processes (manifested by slowly decaying autocorrelations) on the variance of the process, as expounded in Chap. 2. This phenomenon exemplifies a fluctuation-correlation effect pivotal in statistical physics, particularly in statistical thermodynamics concerning critical phenomena.

It is crucial to recognize that the enhanced definition of anomalous diffusion given by Eq. (3.47) (elucidated in Chap. 3), through the integration of a revised scaling parameter, facilitates a more sophisticated classification and analysis of phenomena such as Brownian motion.

A salient feature of the textbook is its exhaustive delineation of random walks concerning Brownian and Gaussian processes. This taxonomy clarifies the extent to which the observed process diverges from the Wiener process (denoted in Table 12.1 as BG for the Brownian and Gaussian process, or succinctly as Bm for Brownian motion, due to the CLT). Although this classification is depicted schematically in Table 12.1, additional exegesis is required.

The discussion pertains to the 'Velocity AC' (Velocity Autocorrelation) and 'Multifractality' columns.

(i) The manifestation of the velocity autocorrelation function for the BG and BnG (Brownian and non-Gaussian) processes is contingent upon the inclusion of a modification within Eq. (3.47) to the scaling term $CS(t)$, which exhibits either a sublinear increment or a decrement over temporal progression.
(ii) The phenomenon of multifractality, frequently in an embryonic state (i.e., not fully developed), in BG is observed after statistical averaging of the process

Table 12.1 Classification of the random wandering *(The notation $\mp$ says that in some processes of a given type, the indicated feature occurs, and in others of this type, it does not.)*

Classification	Fat tail PDF	Process AC	Velocity AC	Multifractality
BG $\equiv$ Bm	$-$	$+$	$\mp$	$\mp$
BnG	$+$	$+$	$\mp$	$\mp$
nBG $\equiv$ nfBm	$+$	$+$	$+$	$\mp$
nBnG $\equiv$ fBm	$+$	$+$	$+$	$\mp$

distribution across the energy landscape of the substrate, particularly in scenarios involving dispersive transport and diffusion within amorphous or glossy materials (refer to Sect. 8.2 for further elucidation).

In this textbook, we address several pivotal inquiries, including exploring the origins of multifractality. In this context, it is worth noting Chap. 4, in which we show that stochasticity enters explicitly into the fractal dimension. This is important in building stochastic multifractality (what we show in the next Chap. 5).

We delineate several primary phenomena that are attributed to multifractality:

(i) stochastic processes engendering broadened or fat-tailed distributions pertinent to the process itself or its fluctuations,
(ii) stochastic processes that engender long-range or enduring autocorrelations, particularly non-linear manifestations,
(iii) the scaling form of the probability density function (PDF) characterized by a random Hurst exponent (refer to Chap. 3 for details),
(iv) stochastic processes typified by bi- or polynomial cascades (refer to Chap. 5 for details),
(v) a singular or invariant measure (considerations discussed in Chap. 5),
(vi) carrier dynamics within the Continuous-Time Random Walk (CTRW) valley framework, delineated by a stretched-exponential domain. Furthermore, this scenario includes an examination of a specific instance of parabolic multifractality (see Chap. 8). Parabolic multifractality has also been previously explored in Chap. 3 within the context of the stochastic parameter that characterizes the stationary solution of the Fokker-Planck equation.

Moreover, we analyzed the multifractality based on the fluctuation function (termed fluctuational multifractality, or MF-DFA) with that based on a nonlinear autocorrelation function (called nonlinear autocorrelation multifractality, or MF-DACA). This comparative study elucidates that both methodologies yield congruent outcomes concerning higher-order corrections (refer to the discussion in Chap. 5). Such equivalence may be conceptualized as a fluctuation-correlation theorem within the domain of multifractality, suggesting the fungibility of these analytical approaches. In essence, nonlinear autocorrelation multifractality invariably underpins fluctuational multifractality. Thus, our research underscores the pivotal role of nonlinear autocorrelations in the emergence of multifractality.

If a given random walk is subject to finite-size scaling (FSS), then the size of its trajectory does not affect its multifractality property. However, such an effect exists when the random walk is not subject to FSS—it is worth being aware of this, as it can lead to spurious multifractality.

It is well-recognized that the profound fluctuation and dissipation theorem (exemplified notably by Einstein's equation linking diffusion and friction coefficients) suggests that the dynamics of energy dissipation are intricately linked to fluctuations within a state of thermodynamic/statistical equilibrium.

An additional contemporaneous example of a similarly profound correlation is the nexus between fluctuations and disorder, a phenomenon explored and elucidated by Giorgio Parisi.[2] This textbook presents the interrelation between fluctuations and autocorrelations via multifractality as a substantive augmentation of the abovementioned theories.

At this pivotal moment, it is imperative to underscore that the generalized or comprehensive partition function constitutes the quintessential and inaugural component across a spectrum of multifractal formalisms. This methodology is universally applicable and harmonizes seamlessly with the principles of statistical thermodynamics.

Elucidating the relaxation mechanism of photocurrents, induced by pulse illumination in commercially utilized amorphous, glassy, and photosensitive materials, presented a formidable challenge in the early 1970s. This challenge was adequately addressed (in Chap. 6) by developing the Continuous-Time Random Walk (CTRW) formalism. This formalism is classified as a jump model (hopping), encompassing jump diffusion or hopping process. The theoretical underpinning of this formalism is the waiting-time distribution (WTD). This sharply defined conditional probability density delineates the survival time statistics of charge carriers within distinct nodes of the spatial network. The formalism employs an average version of this distribution, articulated through sophisticated statistical analyses. When integrated with empirically derived characterizations of amorphous materials or substrates, this leads to an asymptotic decay of the WTD that follows a power-law distribution. The decay power exponent, indicative of the substrate's properties and thermodynamic behavior, establishes a form of the fluctuation-dissipation theorem. In this formalism, carrier dynamics is modeled by a multistep propagator, which traces the trajectory of carriers from their origin to their present location. A notable achievement of this approach was the derivation of characteristic patterns in the Fourier-Laplace domain, which catalyzed the exploration of anomalous transport and diffusion phenomena.

The Continuous Time Random Walk formalism can generate a Weierstrass random wandering, exemplifying a fractal trajectory within fractional Brownian motion. Furthermore, this formalism has been integratively applied to fractal differential equations, which include various models of fractal relaxation and diffusion processes

As for the characteristic example of the Weierstrass random wandering in Chap. 7, we showed that it focuses "as in a lens" on the issues discussed in the book. The answer, based on computer simulations, especially tells the question of how theoretically unlimited variance manifests itself in the real world. Also, the consequences of this in the form of the CLT and BLLN violation are very characteristic, clearly revealing the role of the colloquially termed "black swans".

The inquiry into extending the Continuous-Time Random Walk (CTRW) framework to encompass multiscale/multifractal scenarios has been addressed. In Chap. 8,

---

[2] In 2021, the Nobel Prize "For breakthrough contributions in our understanding of complex systems," was awarded half to Professors Syukuru Manabe (Princeton University) and Klaus Hasselmann (University of Hamburg) "for physically modeling the Earth's climate, quantifying its variability and reliably predicting global warming" and a half to Professor Giorgio Parisi (Sapienza University of Rome) "for the discovery of the interdependence of disorder and fluctuations in physical systems from the atomic to the planetary scale."

we delineated two methodologies—both concerning waiting- or pausing times and not test-particle displacements. Initially, in Sect. 8.2, a Gaussian substrate was utilized instead of the conventional exponential substrate typical of the canonical CTRW; a concise summary of this adaptation is depicted in Fig. 8.2. Subsequently, in the approaches detailed in Sects. 8.3 and 8.3.3, a stretch-exponential substrate, motivated by empirical observations, was examined; the comparative analysis of this stretch-exponential CTRW against empirical data is illustrated in Fig. 8.2. This augmented variant of CTRW has been designated as the Multifractal Continuous-Time Random Walk (MF-CTRW). In this iteration, we have elucidated a non-matured spectrum of dimensions, specifically the left-sided spectrum as exhibited in Fig. 8.4. This extends beyond the conventional classifications of spectra of local dimensions, which are further discussed subsequently.

The paramount importance of extreme absolute values of increments within the framework of a random walk is elucidated by the Generalized Extreme Value Theory (refer to Chap. 9). This theory posits that the dynamics of any random walk are predominantly influenced by extreme events, termed "black swans."

Chapter 10 presents the first systematically studied example of a CLT breakdown in the context of the financial markets. It was one of the key examples that sparked a flood of interest from physicists in financial markets.

We have systematically delineated (in Chap. 11) fully matured, standardized spectral classifications of (local) dimensions, as encapsulated in Fig. 11.7. This classification consists of only three parameters. Such a parsimonious parameterization was feasible because of the employing a four-group model (specifically within the context of a corporate market) upon which the generalized partition function was constructed. This framework facilitated a comprehensive analytical elucidation of multifractality as a robust reference for more complex multifractal analyses. In this regard, the classification achieves universal applicability.

12.2 Instances of Singular Stochastics Used in the Textbook

It is imperative to underscore that our textbook encompasses instances of phenomena typically referred to as singular stochastics or at least constitute a starting point/base for such processes. Subsequently, we delineate the most salient among them.

- The pioneering Vilfredo Pareto power-law cumulative distribution function (CDF) characterization of phenomena (profits and wealth) within socio-economic frameworks (Chap. 2).
- The power-law probability distribution function (PDF) for securities' prices in the stock market represents a stationary solution to the corresponding Fokker-Planck equation (see Chap. 2). The exponent within this power law is not arbitrarily but systematically influenced by the securities' drift and variance parameters. This relationship exemplifies the fluctuation-dissipation theorem, facilitating the empirical derivation of drift and variance metrics.

- The log-periodic oscillations (LPO) modify the power law, discernible directly through the stock market quotes of securities/assets (Chap. 3). Upon meticulous observation, these oscillations enable the forecasting of pivotal occurrences, such as the precipitous declines characteristic of stock market crashes, particularly during the disintegration of market bubbles. In addition, this methodology facilitates the anticipation of recoveries in waning market trends after such crashes. Consequently, LPO augments econometric analysis through the integration of econophysical paradigms.
- The Petersburg paradox, as postulated by Bernoulli, and its subsequent elaboration into Weierstrass random wandering, encompassing flights and walks, are discussed in detail (see Chap. 3).
- Various intricate systems, including viscoelastic and glassy substances as well as the aging of materials, are governed by decelerated, non-Debye (nonexponential or fractional) relaxation dynamics (Chap. 3).
- Elucidation of multifractal characteristics in empirical time series and assorted entities (Chap. 5).
- Employing the Continuous-Time Random Walk paradigm to scrutinize empirical data sets exhibiting non-Debye transient photoconductivity in glass-like or amorphous materials (Chap. 6).
- Examination of empirical data sets using the Multifractal Continuous Time Random Walk (MF-CTRW) methodology, which constitutes an advanced elaboration of the conventional CTRW paradigm (Chap. 8).
- Analysis of empirical data from rival companies using a multifractal method and a generalized partition function based on the market shares of these firms (Chap. 11). This data can be directly obtained from appropriate publicly available sources.

The preceding discussion addresses singular stochastics, a nascent and rapidly evolving field within interdisciplinary physics and the intricate science of complexity. We hope this textbook will contribute to disseminating several significant advances in this field of scholarship.

Appendix
Complementary Mathematical Tools and Theoretical Frameworks

A.1 Fractal Derivative of Any Degree—Riemann-Liouville Definition

This Appendix presents the Riemann-Liouville (RL) definition of the fractal (i.e., fractional) derivative. The approach consists of three stages.

Stage one: definition of a negative integer derivative

As we already mentioned in Sect. 3.5, the negative derivative of degree -1 is simply a single Riemann integral (see formula (3.32)). Naturally, the derivative of the integer degree $-n$ (where n is the natural number) is the n-fold integral; therefore,

$$\begin{aligned}\frac{d^{-n} f(t)}{dt^{-n}} &= \int_0^t dt_{n-1} \int_0^{t_{n-1}} dt_{n-2} \ldots \int_0^{t_1} dt_0 f(t_0) \\ &= \frac{1}{(n-1)!} \int_0^t dt' \frac{f(t')}{(t-t')^{1-n}},\end{aligned} \tag{A.1}$$

which is the starting point for a more general definition.

Stage two: definition of a negative fractal derivative

A direct generalization of the derivative of degree $-n$ to the derivative of any negative degree $-\alpha$ (where α is a positive real number) is to replace in the last integral on the right-hand side of the expression (A.1) found there factorial $(n-1)!$ by the gamma Euler function $\boldsymbol{\Gamma}(\alpha)$ and the exponent n by α. Then, we can introduce the definition,

$$\frac{d^{-\alpha} f(t)}{dt^{-\alpha}} \stackrel{def.}{=} \frac{1}{\boldsymbol{\Gamma}(\alpha)} \int_0^t dt' \frac{f(t')}{(t-t')^{1-\alpha}}; \tag{A.2}$$

in the more general case, when the lower integration limit may be different from zero, the notation is used,

M. Chorowski et al., *Anomalous Stochastics*, Understanding Complex Systems,
https://doi.org/10.1007/978-3-031-80392-5

$$ {}_aD_t^{-\alpha} f(t) \stackrel{def.}{=} \frac{1}{\Gamma(\alpha)} \int_a^t dt' \frac{f(t')}{(t-t')^{1-\alpha}}, \tag{A.3} $$

where for $a = 0$, we have,

$$ {}_0D_t^{-\alpha} f(t) = \frac{d^{-\alpha} f(t)}{dt^{-\alpha}}. \tag{A.4} $$

Another representation[1] of the RL fractional derivative (A.3) is given through the limit of the following summation [1–4],

$$ {}_aD_t^{-\alpha} f(t) = \lim_{N\to\infty} \left(\left(\frac{t-a}{N} \right) \sum_{j=0}^{\alpha N-1} \frac{\Gamma(j+\alpha)}{\Gamma(\alpha) j!} f\left(\frac{Nt-(t-a)j}{N} \right) \right) \tag{A.5} $$

valid for arbitrary α.

Now, we can answer the question: How to define a fractal derivative of degree α?

Stage three: definition of the fractal derivative of a positive degree.

The answer to the question posed above is simple. It is enough to differentiate n times the fractal derivative of degree $\alpha - n (< 0)$,

$$ {}_aD_t^{\alpha} f(t) \stackrel{def.}{=} \frac{d^n}{dt^n} ({}_aD_t^{\alpha-n} f(t)), \tag{A.6} $$

which is the Riemann-Liouville definition of a fractal derivative (positive real degree). We will continue to use this derivative only for $a = 0$. Another special case is when $a = -\infty$ is called the Weyl fractal derivative. Of course, from this point of view, both the fractal relaxation equation (3.35) and the fractal diffusion equation (3.169) are ambiguous due to the freedom to choose the lower integration limit. The variable t is associated with time, but the definition (A.6) can also be expressed in the variable X associated with space,

$$ {}_aD_X^{\beta} f(X) \stackrel{def.}{=} \frac{d^n}{dX^n} \left({}_aD_X^{\beta-n} f(X) \right), \tag{A.7} $$

it can be bound by both fractalized time and fractalized space. The difference between the above operator and the given equation (A.6) is that the domain of the t variable is the non-negative semi-real axis. In contrast, the domain of the X variable is the entire axis.

[1] *De facto* is an extension of this definition to negative values of α. We only consider the actual values of α here for simplicity.

Useful properties of the Riemann-Liouvile differ-integration or fractal derivative

We now provide several properties of the fractal derivative that will enable us to operate this tool, which is very useful in analyzing non-Gaussian processes.

First, note that any order of the RL fractal derivative of the power function is,

$$ {}_0D_t^{\alpha} t^{\mu} = \frac{\Gamma(\mu+1)}{\Gamma(\mu+1-\alpha)} t^{\mu-\alpha} \tag{A.8} $$

hence,

$$ \begin{aligned} {}_0D_t^{\alpha} \exp t = & \; {}_0D_t^{\alpha} \sum_{m=0}^{\infty} \frac{t^m}{\Gamma(m+1)} = \sum_{m=0}^{\infty} \frac{t^{m-\alpha}}{\Gamma(m+1-\alpha)} \\ = & \; \frac{t^{-\alpha}}{\Gamma(1-\alpha)} \, {}_1F_1(1, 1-\alpha, t). \end{aligned} \tag{A.9} $$

By the way, it is worth noting that the Weyl fractal derivative gives (just like the regular derivative), for any real α,

$$ {}_{-\infty}D_t^{\alpha} \exp t = \exp t. \tag{A.10} $$

Secondly, we show what the Laplace transform of the fractal derivative (for $\alpha > 0$ and $a = 0$) equals to a simple form,

$$ \mathcal{L}\left[{}_aD_t^{-\alpha} f(t)\right](u) = u^{-\alpha} \tilde{f}(u), \tag{A.11} $$

where, as one can see, the entire dependency on α is in the power-law dependence of the Laplace variable u, and $(\tilde{\ldots})$ simply means (we remind) the Laplace transform $\mathcal{L}$. This is a beneficial formula that essentially reduces the calculation of the fractal derivative of a given function to the calculation of the Laplace transform of this function; the power-law factor here is universal, i.e., independent of this function.

Let us also note that moving from the t variable to the X variable means moving from the formula (A.11) to the following:

$$ \mathcal{F}\left[{}_aD_X^{\beta} f(X)\right](k) = \mid k \mid^{\beta} \tilde{f}(k), \tag{A.12} $$

where $\beta > 0, a = -\infty$, and $\mathcal{F}$ is the Fourier transform and k is the Fourier variable.

The family of Fox H-functions

The family of Fox H-functions fundamentally analyses anomalous diffusion using fractional partial differential equations (e.g., fractional-time or space diffusion equation). It is defined as follows using the inverse Mellin transform:

$$H_{pq}^{mn}(z) \overset{\text{def.}}{=} H_{pq}^{mn}\left[z \left| \begin{matrix} (a_1, A_1), (a_2, A_2), \ldots, (a_p, A_p) \\ (b_1, B_1), (b_2, B_2), \ldots, (b_q, B_q) \end{matrix} \right. \right]$$
$$\overset{\text{def.}}{=} \frac{1}{2\pi i} \int_{c-i\infty}^{c+i\infty} \mathcal{H}_q^m(s)\mathcal{H}_p^n(s) z^s ds, \qquad (A.13)$$

where

$$\mathcal{H}_q^m(s) \overset{\text{def.}}{=} \begin{cases} \frac{\Pi_{j=1}^m \Gamma(b_j - B_j s)}{\Pi_{j=m+1}^q \Gamma(1-(b_j - B_j s))}, & m > 0,\ q \geq m+1 \\ \frac{1}{\Pi_{j=m+1}^q \Gamma(1-(b_j - B_j s))}, & m = 0,\ q \leq m \end{cases} \qquad (A.14)$$

and

$$\mathcal{H}_p^n(s) \overset{\text{def.}}{=} \begin{cases} \frac{\Pi_{j=1}^n \Gamma(1-(a_j - A_j s))}{\Pi_{j=n+1}^p \Gamma(a_j - A_j s)}, & n > 0,\ p \geq n+1 \\ \frac{1}{\Pi_{j=n+1}^p \Gamma(a_j - A_j s)}, & n = 0,\ p \leq n \end{cases} \qquad (A.15)$$

whereby the actual number c in the second equality in (A.13) was chosen to enable the construction of a (closed) contour on the complex plane $\{z\}$ containing the poles of the integrand function, that is adapted to the calculation the above integral using the residual method (i.e., the Mellin-Barnes type path integral calculations).

In the solution (3.170) of the mono-fractal subdiffusion equation (3.169) (i.e., for $\alpha < 1$), we are dealing with a simplified Fox H-function H_{11}^{10}. Using the explicit form of this function, we obtain the desired solution in the following way:

$$P(X,t) = \frac{1}{\sqrt{4\pi D_\alpha t^\alpha}} H_{11}^{10}\left[\frac{|X|}{\sqrt{4D_\alpha t^\alpha}} \left| \begin{matrix} (1-\frac{\alpha}{2}, \frac{\alpha}{2}) \\ (0,1) \end{matrix} \right. \right].$$
$$= \frac{1}{\sqrt{4\pi D_\alpha t^\alpha}} \frac{1}{2\pi i} \int_{c-i\infty}^{c+i\infty} ds \left(\frac{|X|}{\sqrt{4D_\alpha t^\alpha}}\right)^s \frac{\Gamma(-s)}{\Gamma(1-\alpha/2-\alpha/2\, s)}$$
$$= \frac{1}{\sqrt{4\pi D_\alpha t^\alpha}} \sum_{n=0}^{\infty} \frac{(-1)^n}{n!\,\Gamma(1-(n+1)\alpha/2)} \left(\frac{|X|}{\sqrt{4D_\alpha t^\alpha}}\right)^n. \qquad (A.16)$$

Using the residual method, we represented the integral contained in the second equality as a sum of an alternating infinite sequence. It came down to using the (single) poles of the $\Gamma(-s)$ function, which are $s = s_l = l = 0, 1, 2, \ldots$, i.e., they lie on the nonnegative semireal axis. The infinitely swelling rectangular contour of the integration has been selected to cover these poles. The disappearance of the integral on the right side/edge of the contour occurs only when $\frac{|X|}{\sqrt{4D_\alpha t^\alpha}} < 1$. The integrals on the upper and lower sides/edges of the contour disappear, which results from the estimation (A.50) in Appendix A.7. The poles correspond to the residuals $(-1)^l/l!$, $l = 0, 1, 2, \ldots$ The sum of these residuals multiplied by the remaining part of the integrand function and the factor $2\pi i$ gave the infinite sum mentioned above.

Subsequent fractional Riemann-Liouville differ-integration feature of the Fox H-function

The following property of a Fox H function (which we give without proof) is crucial for solving fractional diffusion equations, e.g., of type (3.169). Namely,

$$\begin{aligned} &{}_0D_t^{\nu}\left(z^{\alpha}H_{pq}^{mn}\left[(az)^{\beta}\left|\begin{matrix}(a_1,A_1),(a_2,A_2),\ldots,(a_p,A_p)\\(b_1,B_1),(b_2,B_2),\ldots,(b_q,B_q)\end{matrix}\right.\right]\right)\\ &= z^{\alpha-\nu}H_{p+1q+1}^{mn+1}\left[(az)^{\beta}\left|\begin{matrix}(-\alpha,\beta),(a_1,A_1),(a_2,A_2),\ldots,(a_p,A_p)\\(b_1,B_1),(b_2,B_2),\ldots,(b_q,B_q),(\nu-\alpha,\beta)\end{matrix}\right.\right] \end{aligned} \tag{A.17}$$

where we used the mutual assignment of fractal differentiation operators $\frac{\partial^{1-\alpha}}{\partial t^{1-\alpha}} \equiv {}_0D_t^{\nu}$. Due to this, it is relatively easy to check that the distribution (3.170) is a solution of the fractal diffusion equation (3.169).

A.2 Fractal Structure Empty Almost Everywhere

In this Appendix, we show the paradoxical properties of fractal structures.

The formula (4.28) shows that

$$d_s = \frac{\ln(N(k))}{\ln(n(k))} = \frac{\ln(N(k))}{\ln(L(k)/l)} = \frac{\ln(n^d - m)}{\ln n} = d_s - \Delta_s,\ k = 1, 2, \ldots, \tag{A.18}$$

where, as for bounded fractals, the deficit of the self-similar dimension, Δ_s, is given by the formula

$$\Delta_s = \frac{\ln(1 - \frac{m}{n^d})}{\ln \epsilon}, \tag{A.19}$$

and as needed, it does not depend on the generation number k.

The difference between both types of structures (limited and unlimited) is visible only in the volume formulas, which in the first case decreases according to the formula (4.9) as one goes to older generations and growing in the second. Namely, it is

$$V(k) = l^d N(k) = l^d n(k)^{d_s} = l^d \left(\frac{L(k)}{l}\right)^{d_s},\ k = 1, 2, \ldots, \tag{A.20}$$

which is often written in the form,

$$V(k) \sim (L(k))^{d_s},\ k = 1, 2, \ldots \tag{A.21}$$

The volume $V(k)$ of a growing self-similar structure scales in power with its linear size $L(k)$. In addition,

$$d_s = \frac{\ln(V(k)/l^d)}{\ln(L(k)/l)}, \ k = 1, 2, \dots \tag{A.22}$$

and it doesn't depend on k as it should be.

Based on (A.20) we see that the number density, of the fractal mass

$$\rho = \frac{N(k)}{V(k)} = \frac{1}{l^d}, \ k = 1, 2, \dots, \tag{A.23}$$

i.e., it does not change from a regular structure to a fractal one, as it is for bounded structures.

It can be seen from (4.27), (4.29) and (A.20) that

$$\lim_{k\to\infty} \frac{V(k)}{(L(k))^d} = \lim_{k\to\infty} \frac{l^d N(k)}{l^d n(k)^d} = \lim_{k\to\infty} \left(1 - \frac{m}{n^d}\right)^k = 0, \tag{A.24}$$

or using an equivalent Eq. (4.28), and also Eqs. (4.29) and (A.20) we get,

$$\lim_{k\to\infty} \frac{V(k)}{(L(k))^d} = \lim_{k\to\infty} \frac{l^d N(k)}{l^d n(k)^d} = \lim_{k\to\infty} \frac{n(k)^{d_s}}{n(k)^d} = \lim_{k\to\infty} \frac{1}{n^{k(d-d_s)}} = 0. \tag{A.25}$$

The above result is valid only in the limit $k \to \infty$, showing that the limit self-similar structure is maximally defective because the area of holes equals in size the entire area of the structure, even though only a minimal amount has been removed at each granularity level ($m = 1$) squares needed to provide (non-trivial) self-similarity of the unlimited statistical Sierpiński carpet.

To better understand the result (A.24), we calculate it in the limit $k \to \infty$, the number of elementary squares $N(k)$ and elementary holes $M(k)$. According to (4.27),

$$\lim_{k\to\infty} \frac{N(k)}{n(k)^d} = \lim_{k\to\infty} \left(1 - \frac{m}{n^d}\right)^k = 0 \tag{A.26}$$

and

$$\lim_{k\to\infty} \frac{M(k)}{n(k)^d} = \lim_{k\to\infty} \left(1 - \left(1 - \frac{m}{n^d}\right)^k\right) = 1. \tag{A.27}$$

The result above is paradoxical because it becomes evident that in the limit $k \to \infty$, the number of holes equals (with an accuracy of zero measure) the number of all the squares that make up a homogeneous (non-defected) object.

A.3 Bi-fractal: Limitations

In this Appendix, we show, using one characteristic example, the relationship between an extreme fractal structure, here bi-fractal, and a multiscale fractal, i.e., multifractal.

Both plots shown in Fig. A.1 require discussion. These plots illustrate important properties of multifractality. Graph (a) shows the relative positioning, for a fixed value $0 < \gamma < 1$, of the line $(1-\gamma)^q$ versus q and $1-\gamma q$ versus q. They have the first common point at $q_0 = 0$, regardless of the value of the γ parameter.

The location of the second common point at $q_1 = 1$ is also independent of the value of the γ parameter, as illustrated in graph (b). In this plot, we present the course of the derivatives of the lines given in graph (a) at point q_0 depending on the parameter γ. As one can see, in the range, we are interested in $0 < \gamma < 1$, the derivative of the numerator of the weight $a(q)$ is greater than the derivative of the denominator, which gives (as needed) $0 \leq a(q) \leq 1$.

Therefore, we have explained that the domain q of the multifractality is limited to the range $0 \leq q \leq 1$ (and to $q = \infty$ as we talk about in Sect. 5.1.1). How does this translate into the relationship $f(\alpha)$ versus α we showed in Sect. 5.1.3.

A.4 Saddle-Point Approximation

The Saddle-Point Approximation/Method (SPA) [5] is also called the Steepest Descent Method.[2] This method allows for the approximate calculation of a broad class of integrals, which we present in the following.

Therefore, let the integral requires determination in a closed form,

$$I = \int \exp(-NF(x))dx, \tag{A.28}$$

for which the function $F(x)$ has a minimum at the saddle point[3] $x = x^*$, i.e., this point is defined using the equality for the first derivative $F^{(1)}(x^*) = 0$, with the second derivative at this point $F^{(2)}(x^*) > 0$, while $N(> 0)$ is for now an arbitrary multiplicative parameter.

We now expand the function $F(x)$ into a Taylor series at the point $x = x^*$, obtaining

---

[2] This method is also known as the Parabolic Approximation, sometimes called the Method of Stationary Phase.

[3] The point is that the minimum occurs only in the variable x; the other variables appear implicitly (if we are dealing with a multidimensional problem) and may have in this place, for example, a maximum, which creates a characteristic saddle; other shapes are also allowed, e.g., a gutter or an inflection point.

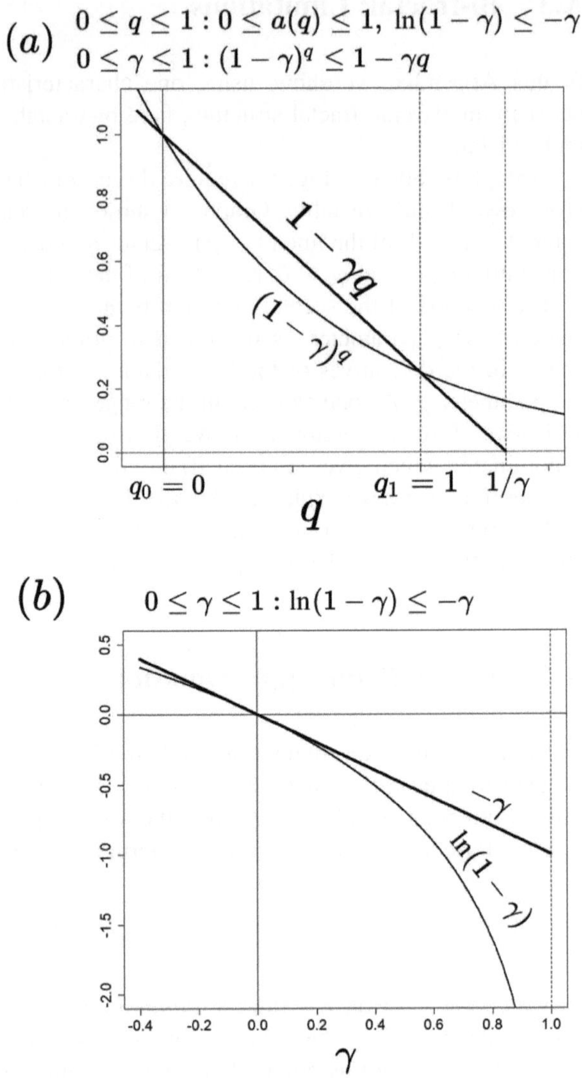

Fig. A.1 Schematic plots illustrating the origins of the constraints imposed on the variable q and the exponent γ. Plot **a** shows that in the interval $0 \leq q \leq 1$ the curve $(1-\gamma)^q$ versus q runs below the straight line $1-\gamma q$. That is, only in this range does $a(q)$ play the role of a (normalized) weight. Plot **b** presents the situation (including the range of the γ parameter) showing the course of the $-\gamma$ and $\ln(1-\gamma)$ lines versus γ, which are the derivatives at the point $q=0$ of both lines shown in plot (**a**). It is visible that for the entire previously defined range $0<\gamma<1$, the inequality $\ln(1-\gamma) \leq -\gamma$ is satisfied, which is a necessary condition for the straight line and curved line we have presented on plot (**a**)

$$I \approx \exp(-NF(x^*)) \int \exp[-NF^{(2)}(x^*)(x-x^*)^2/2]dx, \tag{A.29}$$

where we assume that $\mid F^{(j)}(x^*) \mid \ll \mid F^{(2)}(x^*) \mid$, and $j=3,4,\ldots$, numbers the degrees of derivatives. As one can see, we have reduced the calculation of the initial integral (A.28) to the calculation of the Gaussian integral.[4] Finally, for $N \gg 1$, we can write with a good (controlled) approximation that

$$I \approx \exp(\mp N F(x^*)) \sqrt{\frac{2\pi}{N \mid F^{(2)}(x^*) \mid}}, \tag{A.30}$$

where the '−' sign refers to a function that has a minimum at x^* and a '+' sign has a maximum.[5]

It can be seen that *the integral I calculated in this way is burdened with a minor error the more negligible the dispersion* $1/\sqrt{N \mid F^{(2)}(x^*) \mid}$ *of the function* $NF(x)$ *is.* That's why assuming the enormous value of N here was necessary.

The SPA formulation presented above is pragmatic and often used in physics due to its efficiency (i.e., a compromise between accuracy and simplicity). The systematic development of the narrowed SPA method (referring to PDF or CDF) is based on the cumulative expansion of the characteristic function, i.e., the moment-generating function derived from its moment expansion.

A.5 Remarks on the Legendre Transformation. A Canonical Approach

The Legendre transformation, sometimes called the Legendre-Fenchel (LF) transformation, plays a crucial role in multifractal analysis. Let us discuss this transformation. In its canonical form, the transformation applies exclusively to convex or concave functions.

Figure A.2 depicts a scenario in which the variable x is tuned such that for a constant value of $\alpha (= \tan\gamma)$, the difference $\alpha \cdot x - \tau(x)$ achieves a supremum, that is, x is a variable and α is a constant.

The presence of the extreme, which always exists under these conditions (since $\tau(x)$ is a convex function of the variable x), means that the derivative of the difference $\alpha \cdot x - \tau(x)$ vanishes at point $q \stackrel{\text{def.}}{=} x_{sup}$ which leads to the expression,

$$\alpha(q) = \frac{d\tau(q)}{dq} \stackrel{\text{def.}}{=} \frac{d\tau(x)}{dx} \mid_{x=q}, \tag{A.31}$$

that is, to a one-unique relationship between α and q. Using this, we can introduce,

$$f(\alpha(q)) \stackrel{\text{def.}}{=} q \cdot \alpha(q) - \tau(q). \tag{A.32}$$

---

[4] The Gaussian variant of the integral is a simplification. For example, when all odd PDF moments disappear, the approximation used here is more accurate the smaller the excess kurtosis.

[5] The reader can check the formula (A.30) by replacing in (A.28) the sign '−' with a '+' and then repeat the above procedure.

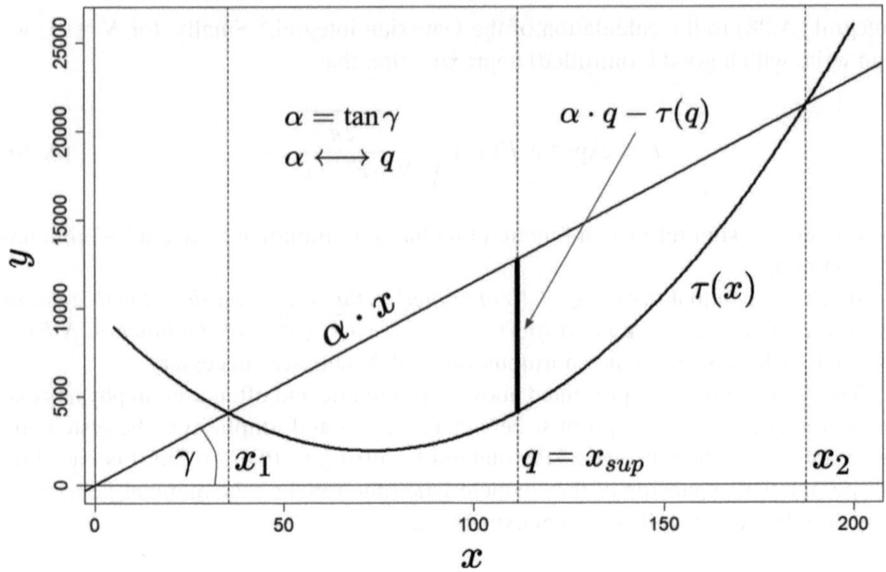

Fig. A.2 Presentation of the components of the Legendre transformation and how to build it. The difference $\alpha \cdot q - \tau(q)$ (represented by the vertical black bold segment) is the value of the expression $\alpha \cdot x - \tau(x)$ at $q = x_{sup}$, where the difference reaches its supremum. The point q uniquely corresponds to the slope $\alpha (= \tan \gamma)$ of the black tangent straight line $\alpha \cdot x$. The vertical axis corresponds to this tangent line, the convex function $\tau(x)$ (e.g., $\tau(x) = 1.5\,x^2 - 200\,x + 10^4$), and the above vertical bold segment. The analysis concerns a fragment of the domain (x_1, x_2), in which (as can be seen) the difference $\alpha \cdot x - \tau(x)$ certainly has a supremum. A similar analysis can be performed for the concave function $\tau(x)$, but then we are dealing with the infimum of the difference $\alpha \cdot x - \tau(x)$, i.e., instead of x_{sup}, we have $q = x_{inf}$. This leads to the definition of the so-called negative Legendre transformation

Differentiating the Eq. (A.32) by the variable q and using the Eq. (A.31) we get,

$$\frac{df(\alpha(q))}{d\alpha(q)} = q. \tag{A.33}$$

In the following, for simplicity of notation, we will drop q from the left side of this equation. The Eq. (A.32) together with the auxiliary equations (A.31) and (A.33) define the canonical form of the Legendre transformation.

Axiomatic extended approach

Departing from the canonical approach, the axiomatic extended approach is inherently formal. It relies on the following definitions,

$$
\begin{aligned}
f(\alpha(q)) &\stackrel{\text{def.}}{=} q\alpha(q) - \tau(q), \text{ if } q \text{ is treated as an independent variable, where} \\
\alpha(q) &\stackrel{\text{def.}}{=} \frac{d\tau(q)}{dq}, \text{ and this relationship is reversible, or} \\
f(\alpha) &\stackrel{\text{def.}}{=} q(\alpha)\alpha - \tau(q(\alpha)), \text{ if } \alpha \text{ is treated as an independent variable,}
\end{aligned}
\tag{A.34}
$$

i.e., on the assumption that the variable q depends on the variable α and vice versa. We emphasize that the term $q\alpha(q)$ is generally a nonlinear function of q, and similarly, the term $q(\alpha)\alpha$ is generally a nonlinear function of α. As for the function $\tau(q)$ and the remaining functions appearing in the Eq. (A.34), we do not assume anything more than they are the differentiable functions.[6] It is important to realize that the origin of the q variable is different from that of the x (see Fig. A.2 for explanation). Here, we explore the q variable based only on the axioms defined in Eq. (A.34). We emphasize that we do not answer the question here about the course of the function $\tau(x)$ versus x and only to the question about the dependence of $\tau(q)$ on q.

From the second and third equations in (A.34), it follows directly that

$$\frac{df(\alpha)}{d\alpha} = q(\alpha), \tag{A.35}$$

where (for simplicity of symbols) we left the dependence on q on the left side of the above equation. We also obtain,

$$\frac{df}{dq} = q\frac{d^2 f}{dq^2}. \tag{A.36}$$

The axioms (A.34) define the Legendre transformation used in this textbook in a broader sense because no knowledge of the $\tau(x)$ function is required—we only know $\tau(q)$.

The Hamiltonian function is an illustrative example of the Legendre transformation in classical mechanics. This function converts the generalized velocities of particles (i.e., the time derivatives of their generalized coordinates) into their corresponding generalized momenta through the Lagrangian function.

Maxwell relations provide another exemplary case in thermodynamics, which facilitates the transformation of one thermodynamic variable into another by utilizing thermodynamic potentials. We presented the relationship between thermodynamic and multifractal quantities in Sect. 5.1.6.

In summarizing both sections, the Legendre transformation consistently substitutes one set of variables with another. Treat both transformed sets impartially without favoring one over the other; essentially, it is a form of variable replacement.

---

[6] More specifically, we assume that the function $\tau(q)$ is at least three times differentiable concerning q, the function $f(\alpha)$ is a twice differentiable function of α, the quantity α is at least once differentiable function q and q is at least once differentiable function α.

A.6 The Role of the First Waiting and Flight

Let us dive into a more generalized scenario, where initial waiting and passing/jumping/hopping/flight events are characterized by a distinct distribution function (denoted as $h(x,t)$) compared to other events. The previous case, where $h = \psi$, refers to a situation in which the beginning of the process aligns with the beginning of its observation. Essentially, when a particle enters the system, both its wandering process and the observation of this process commence simultaneously. However, this may not be the norm in general; in most cases, the observation begins later, and the system already possesses a particular history. It is imperative to account for this history when describing the initial waiting event, which leads to modifications in the following manner.

(1) in the formula (6.27), defining the partial propagator $P_{\mathcal{E}}^{(0)}(X,t)$, the distribution $\Psi_{\mathcal{E}}(t)$ must be replaced by the more general $\Xi_{\mathcal{E}}(t)$, defined below;
(2) in the remaining formulas (6.28), (6.29) and (6.30) defining higher-order propagators $P^{(n)}_{\mathcal{E}_0,\mathcal{E}_1,\ldots,\mathcal{E}_n,\mathcal{E}}(X,t)$ (where $n \geq 1$) the distribution function $\psi_{\mathcal{E}_0}(x_1 - X_0, t_1 - t_0)$ should be replaced by more general $h_{\mathcal{E}_0}(x_1 - X_0, t_1 - t_0)$;

of course, between the distributions $h(x,t)$ and $\Xi(t)$ there is a relationship analogous to (6.11), i.e., the relationship between $\Psi_{\mathcal{E}}(t)$ and $\psi_{\mathcal{E}}(x,t)$

$$\Xi_{\mathcal{E}}(t) = \int_t^{\infty} dt' \int_{-\infty}^{\infty} dx h_{\mathcal{E}}(x,t') = 1 - \int_0^t dt' \int_{-\infty}^{\infty} dx h_{\mathcal{E}}(x,t'). \tag{A.37}$$

We can now rewrite the Eq. (6.39) in a more general form,

$$\begin{aligned}\tilde{\mathcal{P}}(k,s) &= \tilde{\Xi}(s) + \tilde{h}(k,s)\frac{\tilde{\Psi}(s)}{1-\tilde{\psi}(k,s)} \\ &= \frac{1}{s}\left[1 - \tilde{h}(k=0,s) + \tilde{h}(k,s)\frac{1-\tilde{\psi}(k=0,s)}{1-\tilde{\psi}(k,s)}\right].\end{aligned} \tag{A.38}$$

Similarly to Sect. 6.3.4, separability $h(x,t)(= q(x)\chi(t))$ and $\psi(x,t)(= p(x)\phi(t))$ simplifies the formula (A.38) to the form,

$$\tilde{\mathcal{P}}(k,s) = \frac{1}{s}\left[1 - \tilde{\chi}(s) + \tilde{q}(k)\tilde{\chi}(s)\frac{1-\tilde{\phi}(s)}{1-\tilde{p}(k)\tilde{\phi}(s)}\right]. \tag{A.39}$$

Of course, in the case when $h(x,t) = \psi(x,t)$ both above formulas change into the previously derived formulas (6.39) and (6.40), respectively. Generality of patterns (A.38) and (A.39) allows the study of systems both in a state of equilibrium and away from it.

Inhomogeneous generalized master equation

We will further show that, similar to Sect. 6.3.5, Eq. (A.38) can be written in differential-integral form. For this purpose, we will rewrite the first equality in (A.38) in the following intermediate form,

$$\frac{1-\tilde{\psi}(k,s)}{\tilde{\Psi}(s)}\tilde{\mathcal{P}}(k,s) = \{\tilde{\Xi}(s)\frac{1-\tilde{\psi}(k,s)}{\tilde{\Psi}(s)} + \tilde{h}(k,s)\}\tilde{P}(k,t=0) \tag{A.40}$$

which can be briefly written in a more convenient form for further transformations,

$$[s - \tilde{K}(k,s)]\tilde{\mathcal{P}}(k,s) = \tilde{\mathcal{P}}(k,t=0) + \tilde{\mathcal{I}}(k,s), \tag{A.41}$$

where the integral kernel of memory

$$\tilde{\mathcal{K}}(k,s) = s\frac{\tilde{\psi}(k,s) - \tilde{\psi}(k=0,s)}{1-\tilde{\psi}(k=0,s)}. \tag{A.42}$$

The Eq. (A.41) is a generalization of the analogous one introduced earlier for the separable case, adding inhomogeneity (see (6.41) and (6.42)),

$$\begin{aligned}\tilde{\mathcal{I}}(k,s) = &\ \frac{\tilde{h}(k,s) - \tilde{\psi}(k,s) - \tilde{h}(k=0,s) + \tilde{\psi}(k=0,s)}{1-\tilde{\psi}(k=0,s)} \\ + &\ \frac{\tilde{h}(k=0,s)\tilde{\psi}(k,s) - \tilde{h}(k,s)\tilde{\psi}(k=0,s)}{1-\tilde{\psi}(k=0,s)};\end{aligned} \tag{A.43}$$

as one can see, the heterogeneity disappears as it should when $h(x,t) = \psi(x,t)$. Finally, similarly to before, this equation can be written in the desired form,

$$\frac{\partial}{\partial t}\tilde{\mathcal{P}}(k,t) = \int_0^t dt'\tilde{\mathcal{K}}(k,t-t')\tilde{\mathcal{P}}(k,t') + \tilde{\mathcal{I}}(k,t). \tag{A.44}$$

This is the heterogeneous, generalized master's equation. In the separable case, when $h(x,t) = \psi(x,t)$, this equation becomes the previously derived equation (6.45).

A.7 Rigorous Waiting Time Distribution and its Asymptotic Form

In this Appendix, we present a very useful and rigorous calculation of the waiting-time distribution, especially its asymptotic form by using the Mellin transform [6].

The starting point is to represent the exponential function below using the Mellin transform,

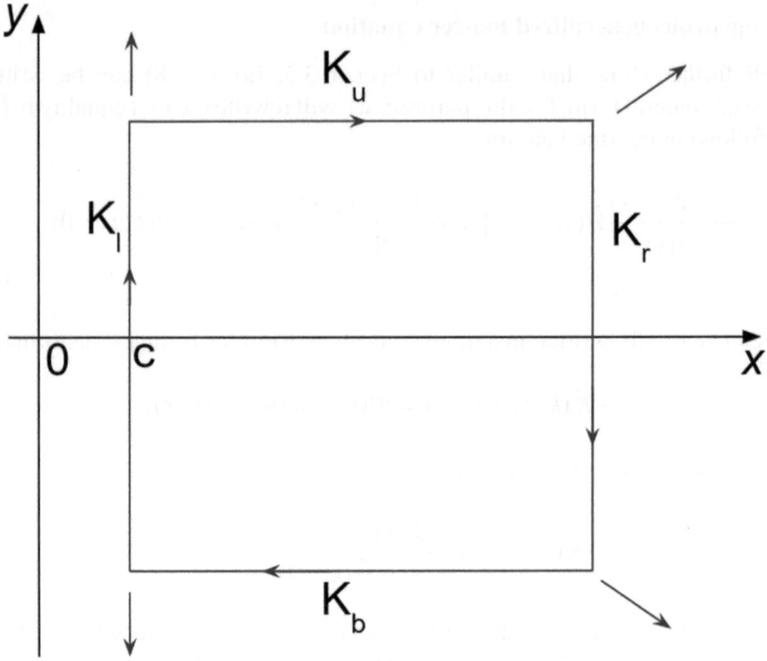

Fig. A.3 Rectangular, closed, diverging integration contour is defined in the complex plane $s = x + \imath, y$. As can be seen, the orientation of the contour is negative. The edges K_r, K_u, and K_b of this contour move infinitely away, while the edge K_l of this contour remains fixed at the point $(x = c, y = 0)$, increasing only its length upwards and downwards

$$\exp\left(-\gamma_0 \exp\left(-\frac{\mathcal{E}}{k_B \mathcal{T}}\right) t\right) = \frac{1}{2\pi i} \int_{c-i\infty}^{c+i\infty} ds (\gamma_0 t)^{-s} \mathbf{\Gamma}(s) \exp\left(\frac{\mathcal{E}}{k_B \mathcal{T}} s\right),$$
$$0 < c = \Re s < 1. \quad \text{(A.45)}$$

Then, substituting (A.45) into (6.67) we get,

$$\begin{aligned}\phi(t) &= \frac{1}{2\pi i} \gamma_0 \alpha \int_{c-i\infty}^{c+i\infty} ds (\gamma_0 t)^{-s} \mathbf{\Gamma}(s) \int_0^\infty d\left(\frac{\mathcal{E}}{k_B \mathcal{T}}\right) \exp\left(-\frac{\mathcal{E}}{k_B \mathcal{T}}(1 + \alpha - s)\right) \\ &= \frac{1}{2\pi i} \gamma_0 \alpha \int_{c-i\infty}^{c+i\infty} ds (\gamma_0 t)^{-s} \frac{\mathbf{\Gamma}(s)}{1 + \alpha - s}; \end{aligned} \quad \text{(A.46)}$$

we remind that $\alpha = \langle \mathcal{E} \rangle / k_B \mathcal{T}$. The mathematically rigorous expression (A.46) is the Mellin representation of the waiting time distribution $\phi(t)$.

We now introduce a rectangular contour $K = K_r \cup K_u \cup K_b \cup K_l$ (see Fig. A.3) whose right (K_r), top (K_u), and bottom (K_b) edges move infinitely away, while the left edge (K_l) is fixed but its length increases indefinitely.

Hence, the integral in (A.46) can be converted into a contour integral,

$$\int_{K_l^\infty} ds(\gamma_0 t)^{-s} \frac{\Gamma(s)}{1+\alpha-s} = \int_{c-i\infty}^{c+i\infty} ds(\gamma_0 t)^{-s} \frac{\Gamma(s)}{1+\alpha-s}$$
$$= \frac{1}{2\pi i}\gamma_0\alpha \oint_{K^\infty} ds(\gamma_0 t)^{-s} \frac{\Gamma(s)}{1+\alpha-s}, \quad \text{(A.47)}$$

where $K^\infty(= K_r^\infty \cup K_u^\infty \cup K_b^\infty \cup K_l^\infty)$ is the contour of K that has gone infinite magnified (let's also note that the orientation of the contour K and K^∞ is negative). Here, we took advantage of the fact that when $x \to \infty$ (where $s = x + iy$) then

$$\int_{K_r} ds(\gamma_0 t)^{-s} \frac{\Gamma(s)}{1+\alpha-s} \to 0, \quad \text{(A.48)}$$

because $(\gamma_0 t)^{-x-iy} \to 0$ (when $x \to \infty$) for $\gamma_0 t > 1$; i.e., the above integral on the right edge (side) of the contour (rectangle) K_r disappears when this edge moves away to infinity. Moreover, we took advantage of the property stating that when $| y |\to \infty$ then,

$$\int_{K_{u,d}} ds(\gamma_0 t)^{-s} \frac{\Gamma(s)}{1+\alpha-s} \to 0. \quad \text{(A.49)}$$

The proof of the above property follows directly from the asymptotic representation of the function $\Gamma(s(= x + iy))$ for $| y |\to \infty$ [6, 7]

$$| \Gamma(x+iy) |\to (2\pi)^{1/2} | y |^{x-1/2} \exp(-\frac{\pi}{2} | y |), \quad \text{(A.50)}$$

in which the exponential factor determines decay with growth $| y |$ with x fixed.

So, our task now is to evaluate the contour integral on the right side of the equality (A.47); this can be done using the residual method [8]. For this purpose, let us note that the only pole of the integrand function (let us denote it by F) that is located on the positive real axis is the zero of the denominator $1 + \alpha - s$, i.e., the pole of the function F is at the point $s_0 = 1 + \alpha$. Moreover, as one can see, this pole is of the first order. The method of residuals says that the contour integral mentioned above is

$$\oint_{K^\infty} ds(\gamma_0 t)^{-s} \frac{\Gamma(s)}{1+\alpha-s} = -2\pi i \; Res \; F(s_0), \quad \text{(A.51)}$$

where $Res\; F(s_0)$ is the residual of the integrand function F at the point s_0; however, the "−" sign appears on the right side of the above equation because the integration contour K and hence the limit integration contour K^∞ have a negative orientation. The residual method provides a recipe for determining the residual at a point, such as the pole of the first order, in the following limit.

$$Res\; F(s_0) = \lim_{s\to s_0} ((s - s_0)F(s)). \quad \text{(A.52)}$$

The computation of the above limit is immediate and gives

$$\lim_{s\to s_0}(s-s_0)F(s) = -(\gamma_0 t)^{-1-\alpha}\mathbf{\Gamma}(1+\alpha). \tag{A.53}$$

Finally, substituting (A.53) into (A.52), then (A.52) to (A.51) and we can finally write this to (A.47) using such a transformed expression (A.46) in the form (A.63)

$$\phi(t) \approx \gamma_0 \frac{\alpha\mathbf{\Gamma}(1+\alpha)}{(\gamma_0 t)^{1+\alpha}}, \tag{A.54}$$

taken out on an alternative route in the Appendix A.8.

A.8 Useful Properties of the Waiting Time Distribution Function

Initially, we analyze the scenario for brief temporal intervals, specifically when $\gamma_0 t \ll 1$. Then the exponential component within the integrand can be succinctly expanded into a truncated power series, encompassing merely the initial three terms,

$$\begin{aligned}
\phi(t) &\approx \frac{\gamma_0\alpha}{(\gamma_0 t)^{1+\alpha}}\int_0^{\gamma_0 t} y^\alpha\left(1-y+\frac{1}{2}y^2\right)dy \\
&= \gamma_0\alpha\left[\frac{1}{1+\alpha}-\frac{1}{2+\alpha}\gamma_0 t+\frac{1}{2}\frac{1}{3+\alpha}(\gamma_0 t)^2\right] \\
&\approx \gamma_0\frac{\alpha}{1+\alpha}\exp\left(-\frac{1+\alpha}{2+\alpha}\gamma_0 t\right).
\end{aligned} \tag{A.55}$$

One can see that $\phi(t=0)=\gamma_0\frac{\alpha}{1+\alpha}$. From (A.55) it follows that the third expansion term is much smaller than the second, which means that the function $\phi(t)$ decreases (to a good approximation) linearly with time for short times.

Secondly for $0\le\alpha\le 1$ and $y\ge 1$, the integrand function is easily evaluated,

$$\exp(-y)\le y^\alpha\exp(-y)\le y\exp(-y), \tag{A.56}$$

where for $\alpha\ne 0, 1$, both equalities hold at the same time only for extreme values $y=1$ or $y=\infty$. In general, for $m\le\alpha\le m+1,\ m=0,1,2,\ldots$, and $y\ge 1$ the estimation takes place,

$$y^m\exp(-y)\le y^\alpha\exp(-y)\le y^{m+1}\exp(-y). \tag{A.57}$$

Further, we perform calculations $m=0$, leaving the situation of any natural m to the interested reader. It should be noted that marginal cases $\alpha=m,\ m=0,1,2,\ldots$ can be calculated directly (see below, in the end, an example calculation for $\alpha=1$).

From the relation (A.56) we obtain an integral estimate for $\gamma_0 t \geq 1$,

$$\int_{\gamma_0 t}^{\infty} \exp(-y)dy \leq \int_{\gamma_0 t}^{\infty} y^{\alpha} \exp(-y)dy \leq \int_{\gamma_0 t}^{\infty} y \exp(-y)dy, \tag{A.58}$$

where for $\alpha \neq 0, 1$, both equalities hold at the same time only in the limit $\gamma_0 t \to \infty$; Eq. (A.58) gives the desired estimate,

$$\exp(-\gamma_0 t) \leq \int_{\gamma_0 t}^{\infty} y^{\alpha} \exp(-y)dy \leq 2\exp(-\gamma_0 t). \tag{A.59}$$

Moreover, according to the definition of the $\mathbf{\Gamma}$ function we have,

$$\int_{0}^{\gamma_0 t} y^{\alpha} \exp(-y)dy = \mathbf{\Gamma}(1+\alpha) - \int_{\gamma_0 t}^{\infty} y^{\alpha} \exp(-y)dy. \tag{A.60}$$

Continue using estimate (A.59) and equality (A.60), we get,

$$\begin{aligned}\mathbf{\Gamma}(1+\alpha) - 2\exp(-\gamma_0 t) \leq &\int_{0}^{\gamma_0 t} y^{\alpha} \exp(-y)dy \\ \leq &\ \mathbf{\Gamma}(1+\alpha) - \exp(-\gamma_0 t).\end{aligned} \tag{A.61}$$

Hence, for an asymptotically long time (i.e., $\gamma_0 t \gg 1$) we have a good approximation,

$$\int_{0}^{\gamma_0 t \gg 1} y^{\alpha} \exp(-y)dy \approx \mathbf{\Gamma}(1+\alpha). \tag{A.62}$$

Therefore, under these conditions, we obtain a good approximation that,

$$\phi(t) \approx \gamma_0 \alpha \mathbf{\Gamma}(1+\alpha)\frac{1}{(\gamma_0 t)^{1+\alpha}}. \tag{A.63}$$

Note that the relationship (A.63) holds for any, non-negative exponent α, and our main interest is still in the case of $\alpha < 1$.

Marginal case $\alpha = 1$. Integrating by parts of the standing integral (in the third equality) in expression (6.69) we get that

$$\int_{0}^{\gamma_0 t} y^{\alpha=1} \exp(-y)dy = 1 - \exp(-\gamma_0 t). \tag{A.64}$$

Therefore, for $\gamma_0 t \gg 1$, $\phi(t)$ takes on immediately asymptotic form

$$\phi(t) \approx \frac{\gamma_0}{(\gamma_0 t)^2}, \tag{A.65}$$

which also results from the expression (A.63) (obtained in a much longer way).

Third, we discuss the intermediate case when $\alpha \lesssim \gamma_0 t \lesssim 1$ (we continue to consider situations for which $\alpha < 1$). The formula (6.71) shows that the ϕ function is the product of a decreasing function (power-wise in time) and an increasing function (in time, given in integral form). Therefore, we can look for a time for which the ϕ distribution function has a local maximum.

A.9 Weierstrass Waiting Time Distribution Function

I. Direct method. We shall elucidate the rigorous formulation of the Weierstrass-Mandelbrot waiting-time distribution function employing the Mellin transform coupled with integration via the method of residues. This methodology facilitates the direct derivation of the power law decay characteristic of the distribution function $\phi(t)$ as a function of time t, obviating the necessity for dual applications of the Laplace transform—initially from the real space to the inverse space, and subsequently reverting to the real space after the analysis via the inverse Laplace transform. This novel strategy presents an alternative to the conventional approaches documented in the scholarly literature.

We will start our argument with the probability distribution function (PDF) given by the expression (6.120)

$$\phi(t) = \gamma_0 \left(1 - \frac{1}{M}\right) \sum_{j=0}^{\infty} \frac{1}{M^j} \gamma^j \exp(-\gamma_0 \gamma^j t). \tag{A.66}$$

Using the Mellin transform of the exponential function [18],

$$\exp(-\gamma_0 \gamma^j t) = \frac{1}{2\pi i} \int_{c-i\infty}^{c+\infty} ds (\gamma_0 t)^{-s} \mathbf{\Gamma}(s) \gamma^{-js}, \ 0 < \Re s = c < 1, \tag{A.67}$$

the distribution function given by the expression (A.66) can be transformed into the form,

$$\phi(t) = \gamma_0 \left(1 - \frac{1}{M}\right) \frac{1}{2\pi i} \int_{c-i\infty}^{c+\infty} ds (\gamma_0 t)^{-s} \mathbf{\Gamma}(s) \sum_{j=0}^{\infty} \left(\frac{\gamma^{1-s}}{M}\right)^j, \tag{A.68}$$

which we further transform using the fact that for $\gamma < 1$ and $M > 1$. One can always choose the number $0 < c < 1$ so that the value of the quotient of the subsequent components of the geometric series (occurring in (A.68)), i.e., $| \gamma^{1-s}/M | = \gamma^{1-c}/M$, was less than one. Therefore, the expression (A.68) takes the form,

$$\phi(t) = \gamma_0 \left(1 - \frac{1}{M}\right) \frac{1}{2\pi i} \int_{c-i\infty}^{c+\infty} ds (\gamma_0 t)^{-s} \frac{\mathbf{\Gamma}(s)}{1 - \frac{\gamma^{1-s}}{M}}, \tag{A.69}$$

which we continue to transform using the method residues [8].

To compute the integral delineated in Eq. (A.69), we use the contour previously delineated in Fig. A.3, characterized as a rectangular contour $K (= K_l \cup K_u \cup K_r \cup K_b)$.

$$\int_{c-i\infty}^{c+\infty} ds (\gamma_0 t)^{-s} \frac{\mathbf{\Gamma}(s)}{1 - \frac{\gamma^{1-s}}{M}} = - \oint_{K^\infty} ds (\gamma_0 t)^{-s} \frac{\mathbf{\Gamma}(s)}{1 - \frac{\gamma^{1-s}}{M}}, \tag{A.70}$$

where the integration on the right side of the above equality was performed along the contour K, the right side of which and the base and upper edge move away to infinity (hence the notation K^∞). Moreover, the orientation of the contour integral is negative (hence, the sign "$-$" in front of it).

First note that when $x \to \infty$ then

$$\int_{K_r} ds (\gamma_0 t)^{-s} \frac{\mathbf{\Gamma}(s)}{1 - \frac{\gamma^{1-s}}{M}} \to 0, \tag{A.71}$$

where $s = x + iy$ because $\gamma_0 t > 1$ and $(\gamma_0 t)^{-x-iy} \to 0$ when $x \to \infty$; i.e., the above integral on the right edge (side) of the contour (rectangle) K_r disappears when this edge moves away to infinity.

Secondly we will show that when $\mid y \mid \to \infty$ then

$$\int_{K_{u,d}} ds (\gamma_0 t)^{-s} \frac{\mathbf{\Gamma}(s)}{1 - \frac{\gamma^{1-s}}{M}} \to 0. \tag{A.72}$$

The proof of the above theorem follows directly from the asymptotic representation of the function $\mathbf{\Gamma}$ for $\mid y \mid \to \infty$ [9],

$$\mid \mathbf{\Gamma}(x + iy) \mid \to (2\pi)^{1/2} \mid y \mid^{x-1/2} \exp\left(-\frac{\pi}{2} \mid y \mid\right), \tag{A.73}$$

in which the exponential factor determines here decay with growth $\mid y \mid$ with x fixed.

So, our task now is to evaluate the contour integral on the right side of the equality (A.70); this can be done using the residual method. For this purpose, let us note that the only poles of the integrand function located on the positive side of the real axis are the zeros of the denominator $1 - \frac{\gamma^{1-s}}{M}$, i.e.,

$$1 - \frac{\gamma^{1-s}}{M} = 0 \Rightarrow s_n = 1 + \alpha - \frac{2\pi i}{\ln \gamma} n, \; n = 0, \pm 1, \pm 2, \ldots, \tag{A.74}$$

where by s_n, $n = 0, \pm 1, \pm 2, \ldots$, the searched poles are marked. Moreover, as one can see, these poles are of first order. The residual method says that the contour integral mentioned above

$$\oint_{K^\infty} ds(\gamma_0 t)^{-s} \frac{\boldsymbol{\Gamma}(s)}{1 - \frac{\gamma^{1-s}}{M}} = -2\pi i \sum_{n=-\infty}^{\infty} Res\ F(s_n), \tag{A.75}$$

where $Res\ F(s_n)$ is the residual of the integrand function, denoted here by F, at the point s_n. The residue method provides, among others, a recipe for determining the residual at a point that is the pole of the first order, that is, the following limit.

$$Res\ F(s_n) = \lim_{s \to s_n} [(s - s_n) F(s)]. \tag{A.76}$$

In our case, calculating this limit comes down to calculating the limit of the expression below,

$$\begin{aligned}
\lim_{s \to s_n} \frac{s - s_n}{1 - \frac{\gamma^{1-s}}{M}} &= \lim_{s \to s_n} \frac{s - s_n}{1 - \frac{1}{M} \exp((1 - s_n) \ln \gamma) \exp((s_n - s) \ln \gamma)} \\
&= \lim_{s \to s_n} \frac{s - s_n}{1 - \exp((s_n - s) \ln \gamma)} = \frac{1}{\ln \gamma}.
\end{aligned} \tag{A.77}$$

Hence and from (A.76), (A.75), (A.73) and (A.69) we finally get,

$$\begin{aligned}
\phi(t) &= \gamma_0 \left(1 - \frac{1}{M}\right) \oint_{K^\infty} ds(\gamma_0 t)^{-s} \frac{\boldsymbol{\Gamma}(s)}{1 - \frac{\gamma^{1-s}}{M}} \\
&= -\frac{\gamma_0}{\ln \gamma} \left(1 - \frac{1}{M}\right) (\gamma_0 t)^{-1-\alpha} \times \sum_{n=-\infty}^{\infty} \boldsymbol{\Gamma}(s_n) (\gamma_0 t)^{\frac{2\pi i}{\ln \gamma} n} \\
&= -\frac{\gamma_0}{\ln \gamma} \left(1 - \frac{1}{M}\right) (\gamma_0 t)^{-1-\alpha} \\
&\times \{\boldsymbol{\Gamma}(s_0) + \sum_{n=1}^{\infty} [\boldsymbol{\Gamma}(s_n) (\gamma_0 t)^{\frac{2\pi i}{\ln \gamma} n} + \boldsymbol{\Gamma}(s_{-n}) (\gamma_0 t)^{\frac{-2\pi i}{\ln \gamma} n}]\}.
\end{aligned} \tag{A.78}$$

It can be shown numerically that (for $\gamma_0 t > 1$),

$$\boldsymbol{\Gamma}(s_0) \gg \sum_{n=1}^{\infty} [\boldsymbol{\Gamma}(s_n) (\gamma_0 t)^{\frac{2\pi i}{\ln \gamma} n} + \boldsymbol{\Gamma}(s_{-n}) (\gamma_0 t)^{\frac{-2\pi i}{\ln \gamma} n}]. \tag{A.79}$$

Although its analytical proof, as yet, is unknown, we can write to a good approximation,

$$\phi(t) \approx -\frac{\gamma_0}{\ln(\gamma)}\left(1-\frac{1}{M}\right)\Gamma(1+\alpha)(\gamma_0 t)^{-1-\alpha}, \tag{A.80}$$

where we are reminding that $\alpha = -\frac{\ln M}{\ln \gamma} > 0$. Equation (A.80) is a prominent example of Eq. (A.54).

A.10 Abelian and Tauberin Theorems

The Abelian theorems we are considering belong to the category of Tauberin theorems. Its importance to the subject matter of this textbook must be regarded, so we present them with proofs, which are an instructive exercise.

Theorem J. 1 (Abelian) **Case** $\alpha > -1$

Suppose that for $t \to \infty$, the function is of the power-law form,

$$f(t) \approx t^{\alpha} F(t), \tag{A.81}$$

where $F(t)$ is an asymptotically homogeneous function or a slowly-varying function,

$$\lim_{t\to\infty} \frac{F(ct)}{F(t)} = 1, \tag{A.82}$$

for every $c > 0$ (e.g., $F(t) \propto \ln(t)$) and not decreasing faster than $t^{-(1+\alpha)}$. Then, the Laplace transform $\tilde{f}(s)$ of $f(t)$ for $s \to 0$ takes the form,

$$\tilde{f}(s) \approx \frac{\Gamma(1+\alpha)}{s^{1+\alpha}} F\left(\frac{1}{s}\right). \tag{A.83}$$

Proof of Theorem J. 1

It consists of two parts.

First, we write the Laplace transform of $\tilde{f}(s)$ as,

$$\begin{aligned}\tilde{f}(s) &= \int_0^{\infty} f(t)\exp(-st)dt \\ &\approx \int_0^{t_{\max}} [f(t) - t^{\alpha}F(t)]\exp(-st)dt + \int_0^{\infty} t^{\alpha}F(t)\exp(-st)dt,\end{aligned} \tag{A.84}$$

where $t_{\max}$ is the time above which the function $f(t)$ assumes, with a good approximation, its asymptotic form.

Secondly, we use the s variable to match condition $\Re\, s t_{\max} \ll 1$ (i.e., having a vanishingly small real part), then the above expression can be converted to the form,

$$\begin{aligned}\tilde{f}(s) &\approx \int_0^{t_{\max}} [f(t) - t^{\alpha} F(t)]dt - s \int_0^{t_{\max}} [f(t) - t^{\alpha} F(t)]tdt \\ &+ \frac{1}{s^{1+\alpha}} \int_0^{\infty} y^{\alpha} F\left(\frac{y}{s}\right) \exp(-y)dy \\ &\approx const_0 - s \times const_1 + \frac{1}{s^{1+\alpha}} F\left(\frac{1}{s}\right) \int_0^{\infty} y^{\alpha} \exp(-y)dy, \quad \text{(A.85)}\end{aligned}$$

where the constants $const_0$ (the first term in the third line) and $const_1$ (the integral expression standing in the second term also in the third line) are finite (which is an additional condition imposed on the F function) and depend on the parameter $t_{\max}$; moreover, we substituted $y = st$. We used the asymptotic form (A.81) of the function $f(t)$ and the property (A.82) (and also from the definition of the function Γ). Of course, for $s \to 0$ the expression (A.85) is dominated by the third term, so

$$\tilde{f}(s) \approx \frac{\Gamma(1+\alpha)}{s^{1+\alpha}} F\left(\frac{1}{s}\right). \quad \text{(A.86)}$$

Theorem J. 2 (Abelian) **Case** $\alpha = -1 - \beta,\ 0 < \beta < 1$

Suppose that for $t \to \infty$ the function f decays exponentially,

$$f(t) \approx t^{-(1+\beta)} F(t), \quad \text{(A.87)}$$

where $F(t)$ is an asymptotically homogeneous function,

$$\lim_{t\to\infty} \frac{F(ct)}{F(t)} = 1, \quad \text{(A.88)}$$

for each $c > 0$ (e.g., $F(t) \propto \ln(t)$) and a slower decreasing from $t^{-(1+\beta)}$; furthermore, for $t \to 0$, the $F(t)$ function decreases not slower than t. Then, the Laplace transform $\tilde{f}(s)$ of $f(t)$ for $s \to 0$ takes the form,

$$\tilde{f}(s) \approx const + s^{\beta}\Gamma(-\beta) F\left(\frac{1}{s}\right), \quad \text{(A.89)}$$

where $\Gamma(-\beta) = -\Gamma(1-\beta)/\beta$ for the β exponent in the above range.

Proof of Theorem J. 2

The first step of the proof is analogous to the previous case, so

$$\begin{aligned}\tilde{f}(s) &= \int_0^{\infty} f(t)\exp(-st)dt \\ &\approx \int_0^{t_{max}} [f(t) - t^{-(1+\beta)} F(t)] \exp(-st)dt + \int_0^{\infty} t^{-(1+\beta)} F(t) \exp(-st)dt,\end{aligned} \quad \text{(A.90)}$$

where t_{max} is the time above which the function $f(t)$ assumes, with a good approximation, its asymptotic form.

Secondly, we use the s variable to match condition $\Re s t_{\max} \ll 1$ (i.e., having vanishingly small real part). Then, the above expression can be converted (by using property (A.88)) to the form,

$$\begin{aligned}\tilde{f}(s) &\approx \int_0^{t_{max}} [f(t) - t^{-(1+\beta)} F(t)]dt - s \int_0^{t_{\max}} [f(t) - t^{-(1+\beta)} F(t)]t dt \\ &+ s^\beta \int_0^\infty y^{-(1+\beta)} F\left(\frac{y}{s}\right) \exp(-y) dy \\ &\approx const - s \times const_1 + s^\beta \boldsymbol{\Gamma}(-\beta) F\left(\frac{1}{s}\right) \\ &\approx const + s^\beta \boldsymbol{\Gamma}(-\beta) F\left(\frac{1}{s}\right),\end{aligned} \tag{A.91}$$

where the constants $const$ (the first term in the first row) and $const_1$ (the integral expression standing in the second term in the same row) are finite (which, as in the previous case, is an additional condition imposed on the F function) and depend on the parameter $t_{\max}$; moreover, we substituted $y = st$ and used the asymptotic form (A.81) of the function $f(t)$ and the property (A.82) (and also from the definition of $\boldsymbol{\Gamma}$ function). Of course, for $s \to 0$, the expression (A.91) is dominated by a zero-order term in s (or $const$) and a sublinear in s, damping the second linear term.

A.11 Useful Laplace Transforms

We will derive the formula for the Laplace transform $\tilde{\phi}(s)(\stackrel{\text{mark}}{=} \mathcal{L}_s(\phi(t)))$ of the waiting time distribution function. For this purpose, we will use the auxiliary relation,

$$\mathcal{L}_s\left(-\frac{d}{dt}\Phi(t)\right) = \Phi(0) - s\tilde{\Phi}(s), \tag{A.92}$$

satisfied for any differentiable function Φ having the Laplace transform $\tilde{\Phi}(s)$ [6].

From (6.71) we can see that for $t \to \infty$,

$$\begin{aligned}\phi(t) &\approx \frac{k_B \mathcal{T} \boldsymbol{\Gamma}(1+\alpha)}{\gamma_0^\alpha \bar{\mathcal{E}}} \frac{1}{t^{1+\alpha}} = -\frac{k_B \mathcal{T} \boldsymbol{\Gamma}(1+\alpha)}{\alpha \gamma_0^\alpha \bar{\mathcal{E}}} \frac{d}{dt}(t^{-\alpha} + const) \\ &= -\frac{d}{dt}\Phi(t),\end{aligned} \tag{A.93}$$

where the asymptotic form of the function

$$\Phi(t) \approx \frac{k_B \mathcal{T} \mathbf{\Gamma}(1+\alpha)}{\alpha \gamma_0^\alpha \bar{\mathcal{E}}} (t^{-\alpha} + const), \tag{A.94}$$

has for $\alpha < 1$ the Laplace transform given by [6],

$$\tilde{\Phi}(s) \approx \frac{k_B \mathcal{T} \mathbf{\Gamma}(1+\alpha)}{\alpha \gamma_0^\alpha \bar{\mathcal{E}}} \left(\frac{\mathbf{\Gamma}(1-\alpha)}{s^{1-\alpha}} + \frac{const}{s} \right), \tag{A.95}$$

From the relations (A.92), (A.93) and (A.95) it follows that

$$\tilde{\phi}(s) = \mathcal{L}_s \left(-\frac{d}{dt} \Phi(t) \right) = \Phi(0) - \frac{k_B \mathcal{T} \mathbf{\Gamma}(1+\alpha)}{\alpha \gamma_0^\alpha \bar{\mathcal{E}}} (\mathbf{\Gamma}(1-\alpha) s^\alpha + const). \tag{A.96}$$

The constant $const$ should be chosen so that the distribution function ϕ met the normalization condition (6.66), which can be expressed in the form

$$\tilde{\phi}(s=0) = 1. \tag{A.97}$$

Hence and from (A.96) it follows that

$$const = (\Phi(0) - 1) \left(\frac{k_B \mathcal{T} \mathbf{\Gamma}(1+\alpha)}{\alpha \gamma_0^\alpha \bar{\mathcal{E}}} \right)^{-1} \tag{A.98}$$

From the above and from (A.96) we finally get that

$$\tilde{\phi}(s) = 1 - \frac{\pi \alpha}{\sin(\pi \alpha)} \left(\frac{s}{\gamma_0} \right)^\alpha, \tag{A.99}$$

where we tacitly used auxiliary patterns

$$\alpha = \frac{k_B \mathcal{T}}{\bar{\mathcal{E}}}, \ \mathbf{\Gamma}(1+\alpha) = \alpha \mathbf{\Gamma}(\alpha), \ \mathbf{\Gamma}(\alpha) \mathbf{\Gamma}(1-\alpha) = \frac{\pi}{\sin(\pi \alpha)}. \tag{A.100}$$

By introducing a dimensionless coefficient

$$\gamma_f' = \frac{\sin(\pi \alpha)}{\pi \alpha}, \tag{A.101}$$

we can write (A.99) as follows,

$$\tilde{\phi}(s) = 1 - \frac{1}{\gamma_f'} \left(\frac{s}{\gamma_0} \right)^\alpha. \tag{A.102}$$

Using normalization condition (6.66), we get

$$\Phi(t) = \int_t^{\infty} dt' \phi\left(t'\right), \tag{A.103}$$

thus becoming the survival probability density of the molecule in some potential valley, at least for time t. Note that $\Phi(0) = 1$, which by Eq. (A.98) gives self-consistently $const = 0$. Distribution $\Phi(t)$ is a function that plays a supportive, important role in the model of the molecule's wandering in continuous time. From (A.102) and (A.103) it follows directly that the Laplace transform [6],

$$\tilde{\Phi}(s) = \frac{1 - \tilde{\phi}(s)}{s} = \frac{1}{\gamma_0 \gamma_f'} \left(\frac{\gamma_0}{s}\right)^{1-\alpha}. \tag{A.104}$$

From the above and from (6.110), we can see that the average time expectations

$$\langle t \rangle = \tilde{\Phi}(s = 0) \tag{A.105}$$

is, for $\alpha < 1$, infinite.

A.12 Spherical Weierstrass Flights

For spherical Weierstrass flights in discrete time, the spatial part $p(\mathbf{x})$ of the probability density of the particle moving by the vector $\mathbf{x}$ as a result of a single flight has the property of spherical symmetry,

$$p(\mathbf{x}) = \frac{1}{S_d \mid \mathbf{x} \mid^{d-1}} p_0(\mid \mathbf{x} \mid), \tag{A.106}$$

satisfying the normalization condition

$$\int d\mathbf{x} p(\mathbf{x}) = \int_0^{\infty} d|\mathbf{x}| p_0(\mid \mathbf{x} \mid) = 1 \tag{A.107}$$

where $S_d = 2\pi^{d/2}/\Gamma(d/2)$ is the area of the $d-1$ dimensional hypersphere of unit radius (where d is the dimension of Euclidean space). Let's note that $S_{d=3} = 4\pi$, $S_{d=2} = 2\pi$, $S_{d=1} = 2$. We continue to assume, similarly to one-dimensional Weierstrass flights, that

$$p_0(\mid \mathbf{x} \mid) = \frac{N-1}{N} \sum_{j=0}^{\infty} \frac{1}{N^j} \delta\left(\mid \mathbf{x} \mid - b^j\right); \tag{A.108}$$

it can be seen that in the special case $d = 1$, $p(x) = p_0(x)/2$ (where for simplicity, we have omitted the " " notation).

As an instructive example, we determine the structural factor of spherical Weierstrass flights for $d = 3$. By definition,

$$\begin{aligned}\tilde{p}\,(\mathbf{k}) &= \int d\mathbf{x} \exp\left(-i\mathbf{k}\cdot\mathbf{x}\right) p\,(\mathbf{x}) \\ &= \frac{2\pi}{S_d}\int_0^{\infty} d\mid\mathbf{x}\mid \int_0^{\pi} d\vartheta \sin(\vartheta) \exp\left(-i \mid\mathbf{x}\mid\mid\mathbf{k}\mid\cos(\vartheta)\right) p_0\left(\mid\mathbf{x}\mid\right) \\ &= \int_0^{\infty} d\mid\mathbf{x}\mid \left(\frac{\pi}{2\mid\mathbf{x}\mid\mid\mathbf{k}\mid}\right)^{1/2} J_{1/2}\left(\mid\mathbf{x}\mid\mid\mathbf{k}\mid\right) p_0(\mid\mathbf{x}\mid),\end{aligned} \tag{A.109}$$

where $J_{1/2}(z)$ denotes, as usual, the Bessel function (cylindrical of the first kind).

However, for the general situation (d-dimensional), we get,

$$\begin{aligned}\tilde{p}\,(\mathbf{k}) &= \mathbf{\Gamma}\left(\frac{d}{2}\right)\int_0^{\infty} d\mid\mathbf{x}\mid \left(\frac{1}{2}\mid\mathbf{x}\mid\mid\mathbf{k}\mid\right)^{1-d/2} J_{d/2-1}\left(\mid\mathbf{x}\mid\mid\mathbf{k}\mid\right) p_0\left(\mid\mathbf{x}\mid\right) \\ &= \frac{N-1}{N}\mathbf{\Gamma}\left(\frac{d}{2}\right)\sum_{j=0}^{\infty}\frac{1}{N^j}\left(\frac{\mid\mathbf{k}\mid b^j}{2}\right)^{1-d/2} J_{d/2-1}\left(\mid\mathbf{k}\mid b^j\right),\end{aligned} \tag{A.110}$$

where $J_{d/2-1}(z)$ is the Bessel function (cylindrical of the first kind).

At this point, as in the case of the one-dimensional case, a question arises **about the conditions under which the form of the structural factor obtained above can be presented in a closed form**. To answer this question, let us note that $\tilde{p}\,(\mathbf{k})$ satisfies a non-uniform scaling equation of the form,

$$\tilde{p}(b\mathbf{k}) = N\tilde{p}(\mathbf{k}) - (N-1)\mathbf{\Gamma}\left(\frac{d}{2}\right)\left(\frac{1}{2}\mid\mathbf{k}\mid\right)^{1-d/2} J_{d/2-1}\left(\mid\mathbf{k}\mid\right). \tag{A.111}$$

Solving this equation (just like we did for Eqs. (7.20) in Chap. 7) we are looking for in the form of a sum

$$\tilde{p}\,(\mathbf{k}) = \tilde{p}_{reg}\,(\mathbf{k}) + \tilde{p}_{sing}\,(\mathbf{k}), \tag{A.112}$$

where the regular (normal, general) solution $\tilde{p}_{reg}\,(\mathbf{k})$ satisfies the non-homogeneous equation (A.111) and the singular solution $\tilde{p}_{sing}\,(\mathbf{k})$ the homogeneous part of this equation. As usual, the form of the general solution is dictated by heterogeneity.

Regular solution

Therefore, using the expansion of the Bessel function of the first kind

$$J_{d/2-1}(\mid\mathbf{k}\mid) = \frac{\mid\mathbf{k}\mid^{d/2-1}}{2^{d/2-1}}\sum_{l=0}^{\infty}(-1)^l\frac{\mid\mathbf{k}\mid^{2l}}{2^{2l}l!\mathbf{\Gamma}\left(\frac{d}{2}+l\right)} \tag{A.113}$$

we can assume the solution $\tilde{p}_{reg}(\mathbf{k})$ in the form of a power series also containing only even powers of the variable $|\mathbf{k}|$, namely,

$$\tilde{p}_{reg}(\mathbf{k}) = 1 + \sum_{j=1}^{\infty}(-1)^j \frac{1}{j!} a_j \mid \mathbf{k} \mid^{2j}, \tag{A.114}$$

where $a_j,\ j = 1, 2, \ldots,$ are the coefficients we are looking for (slightly anticipating our considerations, we assumed $a_0 = 1$). Substituting (A.113) and (A.114) into (A.111) and comparing the coefficients with the same powers of $|\mathbf{k}|$, we obtain (after simple algebraic transformations) the expression sought for the coefficient.

$$a_j = \frac{1}{2^{2j}} \frac{\Gamma\left(\frac{d}{2}\right)}{\Gamma\left(\frac{d}{2}+j\right)} \frac{1-\frac{1}{N}}{1-\frac{b^{2j}}{N}},\ j = 0, 1, 2, \ldots. \tag{A.115}$$

For example, by limiting ourselves to at most square terms in $|\mathbf{k}|$, we get,

$$\tilde{p}_{reg}(\mid \mathbf{k} \mid) \approx 1 - D' \mid \mathbf{k} \mid^2,\ D' = a_1 = \frac{1}{2d} \frac{1-\frac{1}{N}}{1-\frac{b^2}{N}}, \tag{A.116}$$

which is a generalization of the expressions (7.22) and (7.23) valid in one dimension to any number dimensions $d \geq 1$.

Singular solution

The singular solution, $\tilde{p}_{sing}(\mathbf{k})$, can be immediately written in a form analogous to (7.24) with an auxiliary expression (7.26) (where $|k|$ should be replaced by $|\mathbf{k}|/2$) and the exponent given by the formula (7.27), because it formally satisfies the equation identical to (7.21), The coefficients A_j (for the flight process of Weierstrass taking place in a space with any number of dimensions $d \geq 1$) are determined below, using a general method using the Mellin transform and integration by residuals.

In the special case of limiting ourselves to only the constant coefficient A_0, we obtain

$$\tilde{p}_{sing}(\mathbf{k}) \approx -D'_f \mid \mathbf{k} \mid^{\beta},\ D'_f = A_0, \tag{A.117}$$

which is formally identical to the second term in the Eq. (7.29) (for $\beta < 2$).

The complete form of a singular solution

The strategy is first to perform the Mellin transform of the function (A.110) and then (after performing the appropriate calculations) to calculate the inverse Mellin transform. In this way, we will extract the complete singular solution from the function (A.110).

Let us recall at the beginning that the Mellin transform (TM) of the function $f(x)$ is defined as follows

$$\tilde{f}(s)\,(\equiv \mathcal{M}[f;s]) \stackrel{\text{def.}}{=} \int_0^{\infty} f(x)x^{s-1}dx, \tag{A.118}$$

where s is a complex variable. However, the inverse Mellin transform (TM^{-1}) becomes more intense a complex but beneficial character

$$f(x)\left(\equiv \mathcal{M}^{-1}[\tilde{f};x]\right) \stackrel{\text{def.}}{=} \frac{1}{2\pi\imath}\int_{c-\imath\infty}^{c+\imath\infty} \tilde{f}(s)x^{-s}ds,\ 0<c<1,\ 0<\Re s\le c \tag{A.119}$$

Note that we will apply TM to the function $f(x) = x^{-\nu/2}J_\nu(2x^{1/2})$, where $J_\nu(\cdots)$ is a Bessel function (cylindrical of the first kind). The Mellin transform of this function is $\tilde{f} = \mathbf{\Gamma}(s)/\mathbf{\Gamma}(\nu - s + 1),\ 0 < \Re s < 2^{-1}\Re\nu + 3/4$. In our case, the following substitution applies: $\nu = d/2 - 1,\ 2x^{1/2} = \mid \mathbf{k} \mid b^j \equiv x = \left(\frac{1}{2}\mid \mathbf{k}\mid\right)^2 b^{2j} \equiv \left(x^{1/2}\right)^{1-d/2} = \left(\frac{1}{2}\mid \mathbf{k}\mid b^j\right)^{1-d/2}$. Using the above, we can express our function (A.110) as follows:

$$\tilde{p}(\mathbf{k}) = \frac{1}{2\pi\imath}\int_{c-\imath\infty}^{c+\imath\infty} \frac{N-1}{N}\frac{1}{\left(\left(\frac{1}{2}\mid\mathbf{k}\mid\right)^2\right)^s}\frac{\mathbf{\Gamma}\left(\frac{d}{2}\right)\mathbf{\Gamma}(s)}{\mathbf{\Gamma}\left(\frac{d}{2}-s\right)}\frac{1}{1-\left(Nb^{2s}\right)^{-1}}ds, \tag{A.120}$$

where we used the commutativity of summation (occurring in the formula (A.110)) and integration in the formula (A.119)) because both the integral and the sum converge. Moreover, instead of summing the geometric series term by term, we inserted the appropriate sum of the geometric series. Let us emphasize that the integration in formula (A.120) will be performed using residuals, thanks to locating the poles of the integrand function in the left half-plane (see Fig. A.4) and assuming (still) that $\frac{1}{2}\mid \mathbf{k}\mid < 1$.

We will now show that the integral in (A.120) vanishes on the left edge of the contour and its lower and upper edges. Its disappearance on the left bank results directly from the above assumption. However, the pattern

$$\mid \mathbf{\Gamma}(s = x + \imath\, y)\mid \to \sqrt{2\pi}\mid y\mid^{x-1/2}\exp\left(-\frac{\pi\mid y\mid}{2}\right), \mid y\mid \to \infty, \tag{A.121}$$

shows directly how both the Eulerian gamma ratio vanishes $\frac{\mathbf{\Gamma}(s)}{\mathbf{\Gamma}(\frac{d}{2}-s)} \to \mid y\mid^{2x-d/2}$ appearing in the integrand function in the expression (A.120), followed by the full integral function and hence the integral when the upper and lower edges of the contour move away infinitely.

Therefore, we can write that

$$\tilde{p}(\mathbf{k}) = \sum_{s_0} \mathcal{R}es\ F(s_0), \tag{A.122}$$

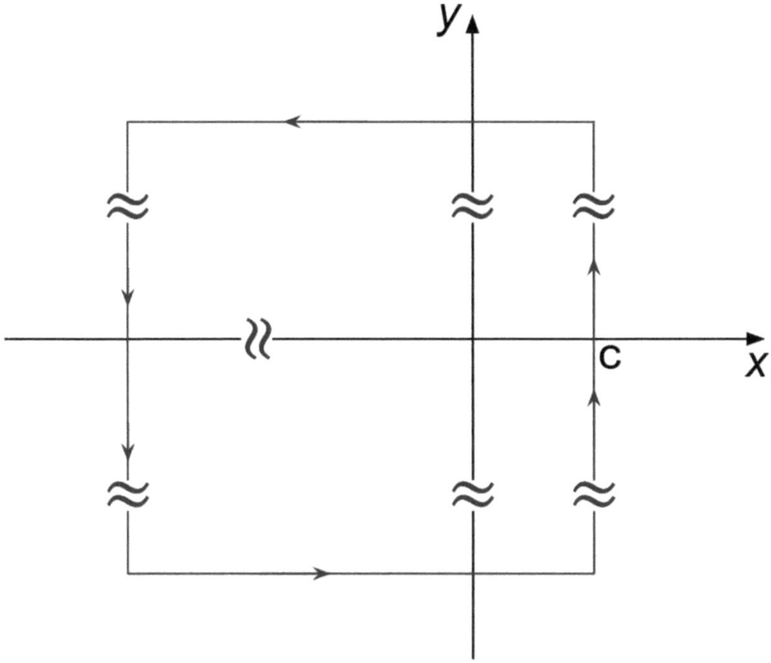

Fig. A.4 A rectangular, closed, diverging integration contour represented in the complex plane $s = x + \imath\, y$. This contour covers all poles of the structural factor $\tilde{p}(\mathbf{k})$ of the spherical Weierstrass random flight

the right side of the above equality means the sum of residuals associated with all poles of the integrand function $F(s)$.

We will now deal with determining these poles and their associated residuals.

Poles and residuals

The poles of the integrand function are of two types:

(a) negative integers along with zero, which are the poles of the function $\mathbf{\Gamma}(s)$,
(b) complex (except for one real one), constituting the poles of the function $1/\left(1 - \left(Nb^{2s}\right)^{-1}\right)$.

In item (a), the poles can be easily found by recalling that $\mathbf{\Gamma}(s) = \pi/(\mathbf{\Gamma}(1 - s)\sin(\pi s))$. Hence, $s_0 = -l,\ l = 0, 1, 2, \ldots$.

In item (b), we determine the poles from the equality $N = \left(b^2\right)^{-s_0} = \exp(-s_0 \ln b^2 + 2\pi\imath\, n),\ n = 0, 1, 2, \ldots$. That is, $s_0 = -\frac{1}{2}\beta + \frac{\pi\imath\, n}{\ln b},\ n = 0, \pm 1, \pm 2, \ldots$.

Now let us recall that the residual (of the first order—but only such residuals are what we are dealing with here) of any complex function $F(s)$ of the complex variable s is determined from a simple formula

$$\mathcal{R}es\, F(s_0) = \lim_{s \to s_0} F(s)(s - s_0). \tag{A.123}$$

Therefore,

$$\mathcal{R}es\, \mathbf{\Gamma}(s_0) = \frac{(-1)^l}{l!},\ l = 0, 1, 2, \dots . \tag{A.124}$$

These residuals lead directly to the expression (A.114) with the auxiliary (A.115), i.e., to the regular solution—so we cannot take it into account here.

Now consider case (b). Using the formula (A.123) we get that

$$\begin{aligned}
\mathcal{R}es \frac{1}{1 - \left(Nb^{2s_0}\right)^{-1}} &= \lim_{s \to s_0} \frac{s - s_0}{1 - \frac{b^{-2s_0}}{N} b^{-2(s-s_0)}} = \lim_{s \to s_0} \frac{s - s_0}{1 - b^{-2(s-s_0)}} \\
&= \lim_{s \to s_0} \frac{s - s_0}{1 - \exp\left(-2(s - s_0)\ln b\right)} = \frac{1}{2\ln b},
\end{aligned} \tag{A.125}$$

where we used the equality $\frac{b^{-2s_0}}{N} = 1$, present above in the explanation of case (b).

In the light of the above, the full singular solution takes the form,

$$\begin{aligned}
\tilde{p}_{sing}(\mathbf{k}) = &\left(1 - \frac{1}{N}\right) \frac{1}{2\ln b} \mathbf{\Gamma}\left(\frac{d}{2}\right) \left(\frac{1}{2} \mid \mathbf{k} \mid\right)^{\beta} \\
&\times \sum_{n=-\infty}^{\infty} \frac{\mathbf{\Gamma}\left(-\frac{1}{2}\beta + \frac{\pi \iota n}{\ln b}\right)}{\mathbf{\Gamma}\left(\frac{d}{2} + \frac{1}{2}\beta - \frac{\pi \iota n}{\ln b}\right)} \exp\left(-2\pi \iota\, n \frac{\ln\left(\frac{1}{2} \mid \mathbf{k} \mid\right)}{\ln b}\right),
\end{aligned} \tag{A.126}$$

which, combined with the full regular solution (A.114) and the expression (A.115), gives the full solution $\tilde{p}(\mathbf{k})$.

A.13 Exact Structural Factor for One-dimensional Weierstrass Flights

We derive a strict form of the structural factor for the one-dimensional Weierstrass walk using the Mellin transform and integration by residuals, i.e., proceeding analogously to spherical flights Weierstrass (see Appendix A.12) but without the initial distribution of the solution sought into regular and singular parts.

Our starting formula is (7.19). Since the Mellin transform of $\cos(\mid k \mid b^j)$ is $\left(b^j\right)^{-s} \mathbf{\Gamma}(s) \cos\left(\frac{\pi}{2} s\right)$, so the sought structural factor can be presented in the form

$$
\begin{aligned}
\tilde{p}(k) &= \frac{N-1}{N}\frac{1}{2\pi\imath}\sum_{j=0}^{\infty}\int_{c-\imath\infty}^{c+\imath\infty} |\, k \,|^{-s}\, \boldsymbol{\Gamma}(s)\cos\left(\frac{\pi}{2}s\right)\left(Nb^{s}\right)^{-j} ds \\
&= \frac{N-1}{N}\frac{1}{2\pi\imath}\int_{c-\imath\infty}^{c+\imath\infty} |\, k \,|^{-s}\, \boldsymbol{\Gamma}(s)\cos\left(\frac{\pi}{2}s\right)\frac{1}{1-(Nb^{s})^{-1}} ds. \qquad \text{(A.127)}
\end{aligned}
$$

This form allows us to calculate the integral using the $|\, k \,|< 1$ residual method. Let us note that the second equality was obtained by replacing summation with integration, which was possible because both converge.

For this purpose, as in Appendix A.12, we surround the poles of the integrand function (let us denote it, as in the previous Appendix, by $F(s)$) located in the left half-plane, with a rectangular diverging contour (see Fig. A.4). The disappearance of the integral on the lower and upper edges of the contour results directly from the formula (A.70), while on the left edge, it results from the fact that we carry out calculations assuming $|\, k \,|< 1$.

We have (as before) two types of pole: the first comes from the product of the function $\boldsymbol{\Gamma}(s)\cos\left(\frac{\pi}{2}s\right)$ and the second one comes from the function $\frac{1}{1-(Nb^{s})^{-1}}$. The first set of poles is given by the expression (A.124), where l (thanks to $\cos\left(\frac{\pi}{2}s\right)$) covers only even numbers, while the second expression (analogous to the one for case (b) in Appendix A.12) $s_0 = -\beta + \frac{2\pi\imath\, n}{\ln b}$, $n = 0, \pm 1, \pm 1, \ldots$. The only problem left is finding residuals for both types of poles. Proceeding similarly to Appendix A.12, we obtain $\frac{(-1)^l}{(2l)!}$, $l = 0, 1, 2, \ldots$, and $\frac{1}{\ln b}$. Ultimately, the full form of the structural factor consists of two fundamentally different expressions

$$
\begin{aligned}
\tilde{p}(k) &= \sum_{j=0}^{\infty}\frac{(-1)^j}{(2j)!}\frac{1-\frac{1}{N}}{1-\frac{b^{2j}}{N}} |\, k \,|^{2j} + |\, k \,|^{\beta}\, \frac{1-\frac{1}{N}}{\ln b} \\
&\times \sum_{n=-\infty}^{\infty}\boldsymbol{\Gamma}\left(-\beta+\frac{2\pi\imath\, n}{\ln b}\right)\cos\left(\frac{\pi}{2}\left(-\beta+\frac{2\pi\imath\, n}{\ln b}\right)\right) \\
&\times \exp\left(-\frac{2\pi\imath\, n}{\ln b}\ln |\, k \,|\right), \qquad \text{(A.128)}
\end{aligned}
$$

explicitly containing the regular part (based on the first infinite series) and the singular part (based on the second).

The above formula and the formula for $\tilde{p}(\frac{1}{2} \mid \mathbf{k} \mid)$ from the Appendix A.12 are crucial for Weierstrass flights (both one- and multidimensional).

A.14 From the Non-linear Langevin Equation to the Fokker-Planck One

We present a vital link/bridge between stochastic processes and stochastic dynamics/statistical mechanics. This bridge is so important because it makes it possible to build the domain of stochastic processes on the foundation *ab initio* of the Langevin equation.

Derivation

We derive the Fokker-Planck equation from the non-linear Langevin equation. This derivation consists of several steps [10].

In the first step, we integrate the Langevin equation of stochastic dynamics,[7]

$$\frac{dX(t)}{dt} = A(X(t), t) + C(X(t), t)\eta(t) \tag{A.129}$$

ruled by white noise η, to Wiener's stochastic dynamics equation represented by Itô,

$$dX(t) = A(X(t), t)dt + C(X(t), t)dW, \tag{A.130}$$

where $dW = \eta(t)dt$ is the Wiener process, and dW is the so-called stochastic differential (see e.g., [11]). As one can see, the above equation is a direct generalization of Eq. (2.89) with constant coefficients.

From the definition of a Wiener process, a typical value of $(dW)^2$ is of the order of dt. In other words, a random variable dW is drawn from a symmetric Gaussian PDF with variance proportional to dt.

In the second step, we use the Itô's lemma, which boils down to the following expansion of function $h(X)$ that is, at least twice differentiable,

$$\begin{aligned} dh(X) &= \frac{dh(X)}{dX}dX + \frac{1}{2}\frac{d^2h(X)}{dX^2}(dX)^2 \\ &= A(X, t)\frac{dh(X)}{dX}dt + B(X, t)\frac{d^2h(X)}{dX^2}dt + \frac{dh(X)}{dX}C(X, t)dW, \end{aligned} \tag{A.131}$$

where X is a time-dependent variable, $B(X, t) = C^2(X, t)/2$ and along the way we used the Eq. (A.130) to the first equality in (A.131) and simply assumed that $(dW)^2 = dt$.

In the third step, we average the above equation, obtaining,

[7] We introduce the Langevin equation in a more general form than that given in Van Kampen's cult book [12], because we allow here an explicit dependence of the equation's coefficients on time. This approach is possible because we use the Itô representation here, not the Stratonovich one, for simplicity. Nevertheless, both representations give the stationary PDF solution as a power law. Furthermore, we consider the overdamped Langevin equation containing the friction/drift component without memory or retardation.

$$\frac{\langle dh(X)\rangle}{dt} = \frac{d\langle h(X)\rangle}{dt} = \left\langle A(X,t)\frac{dh(X)}{dX}\right\rangle + \left\langle B(X,t)\frac{d^2h(X)}{dX^2}\right\rangle =$$
$$\int_{-\infty}^{\infty}\left[A(X,t)\frac{dh(X)}{dX} + B(X,t)\frac{d^2h(X)}{dX^2}\right]P(X,t)dX, \quad \text{(A.132)}$$

where the term proportional to dW is set to zero (as $\langle dW\rangle = 0$) and $P(X,t)$ is a probability distribution function.

Integrating by parts the right side of the above equation[8] and using the vanishing of the probability distribution function, $P(X,t)$, on the boundary of the domain $\mid X \mid \rightarrow \infty$ for any time t, we get,

$$\frac{d\langle h(X)\rangle}{dt} = \int_{-\infty}^{\infty} h(X)\left\{\frac{\partial}{\partial X}\left[-A(X,t)P(X,t)\right] + \frac{\partial^2}{\partial X^2}\left[B(X,t)P(X,t)\right]\right\}dX, \quad \text{(A.133)}$$

where we assumed vanishing of $h(X,t)$ for $\mid X \mid \rightarrow \infty$, i.e., on the boundary of the domain X.

On the other side, the time derivative of the mean value is,

$$\frac{d\langle h(X)\rangle}{dt} = \frac{d}{dt}\int_{-\infty}^{\infty} h(X)P(X,t)dX = \int_{-\infty}^{\infty} h(X)\frac{\partial}{\partial t}P(X,t)dX. \quad \text{(A.134)}$$

In the last/fourth step, comparing the right sides of the Eqs. (A.133) and (A.134) because they are equal for a comprehensive class of functions $h(X)$, we finally get,

$$\frac{\partial}{\partial t}P(X,t) = \frac{\partial}{\partial X}\left[-A(X,t)P(X,t)\right] + \frac{\partial^2}{\partial X^2}\left[B(X,t)P(X,t)\right], \quad \text{(A.135)}$$

which is the sought-after Fokker-Planck equation. This equation is used many times in this textbook.

To get *the stationary solution* $P(X,t) = P_{stat}(X)$ it is enough to note that the Eq. (A.135) is the continuity equation for the distribution $P(X,t)$, where the density of the probability's flux,

$$j(X,t) = j_{drift}(X,t) + j_{diff}(X,t), \quad \text{(A.136)}$$

where the drift/convection/transport flux density, $j_{drift}(X,t) \stackrel{\text{def.}}{=} A(X,t)P(X,t)$ and the diffusion/fluctuation flux density, $j_{diff}(X,t) \stackrel{\text{def.}}{=} -\frac{\partial}{\partial X}\left[B(X,t)P(X,t)\right]$. The stationary solution is obtained when $j(X,t) = 0$ that is, $j_{drift}(X,t) = -j_{diff}(X,t)$. Hence,

[8] Whereas the second term in square brackets in Eq. (A.132) is integrated by parts twice.

$$P_{stat}(X) = \frac{const}{B(X)} \exp\left[\int_{X_0}^{X} \frac{A(X')}{B(X')} dX'\right], \tag{A.137}$$

where the following conditions must be met: $A(X,t) = A(X,t) < 0$, $B(X,t) = B(X,t) > 0$, and the ratio $A(X,t)/B(X,t)$ is time-independent quantity; otherwise stationary solution does not exist.

Note also that the Fokker-Planck equation (A.135) is a one-dimensional continuity equation of the form,

$$\frac{\partial}{\partial t} P(X,t) + \frac{\partial}{\partial X} j(X,t) = 0, \tag{A.138}$$

which can be treated as the law of conservation of the "mass" of the process.

Additional remarks

The Langevin equation (both in the form (A.129) and its equivalent form (A.130)) has a physical representation/interpretation depending on the definition of the random variable $X(t)$. We report two separate cases. The first one is when $X(t)$ is identified with the current speed, $V(t)$, of the test (Brownian) particle, and then we are dealing with the original/canonical Langevin equation. Second, when $X(t)$ is defined as the current position of the Brownian particle, we deal with the overdamped Langevin equation called the Smoluchowski equation. A hybrid case outside the Eqs. (A.129) and (A.130)) is when both the position, $X(t)$, and the velocity, $V(t)$, co-occur in the force acting on the test particle.

The most general form of the Langevin equation (the Generalized Langevin Equation, GLE), which is a generalization of the hybrid case, introduces a retarded drag force (retarded friction) and keeps (in general) the kinematic term containing the acceleration of the Brownian particle. Retarding means that the friction force includes a friction kernel (memory kernel). Defining the memory kernel as a scale-invariant opened Pandora's box of problems studied to this day, and GLE itself defined the foundation of statistical mechanics and the foundation of the field of stochastic processes [12–15].

The GLE describes various physical processes: (i) the tracer particle random walk in a viscoelastic complex fluid [16–18], (ii) the erratic motion of a tagged particle attached to a polymer chain [19–21], and (iii) the stochastic dynamics of the distance between two protein residues as experimentally found [22].

A.15 Asymptotics of Lévy Distribution

Let there be a symmetric (i.e., without drift) Lévy PDF for $0 < \alpha < 2,\ \gamma, a > 0$,

$$
\begin{aligned}
P_L(x;\alpha,\gamma) &= \lim_{a\to 0} P(x;\alpha,\gamma;a) \\
&= \lim_{a\to 0}\frac{1}{2\pi}\int_{-\infty}^{\infty}\exp(-a\mid q\mid)\exp\left(-\gamma\mid q\mid^{\alpha}\right)\cos(\mid x\mid\mid q\mid)dq \\
&= \lim_{a\to 0}\frac{1}{\pi}\int_{0}^{\infty}\exp(-a\,q)\exp\left(-\gamma\,q^{\alpha}\right)\cos(\mid x\mid q)dq \\
&= \frac{1}{\pi}\sum_{k=0}^{\infty}\frac{(-1)^k}{k!}\lim_{a\to 0}\int_{0}^{\infty}\exp(-a\,q)\,\gamma^k q^{k\alpha}\cos(\mid x\mid q)dq,
\end{aligned}
\tag{A.139}
$$

where (for simplicity) we put $\Delta t = 1$ and have expanded the function $\exp(-\gamma\, q^{\alpha})$ into a series, with $\exp(-a \mid q \mid)$ being the convergence factor of the integrals. Then, we calculate the limit of the integrals,

$$
\begin{aligned}
\lim_{a\to 0} I_k(x;\alpha,\gamma;a) &= \lim_{a\to 0}\int_{0}^{\infty}\exp(-a\,q)\,\gamma^k q^{k\alpha}\cos(\mid x\mid q)dq \\
&= \gamma^k\boldsymbol{\Gamma}(1+k\alpha)\lim_{a\to 0}\left(a^2+x^2\right)^{-(1+k\alpha)/2}\cos\left[(1+k\alpha)\arctan\left(\frac{\mid x\mid}{a}\right)\right] \\
&= \gamma^k\boldsymbol{\Gamma}(1+k\alpha)\cos\left[\frac{\pi}{2}(1+k\alpha)\right]\frac{1}{\mid x\mid^{1+k\alpha}},\quad k=1,2,\ldots,
\end{aligned}
\tag{A.140}
$$

and insert into the formula (A.139). This leads in the limit $\mid x \mid\to\infty$ to the expression,

$$
P_L(\mid x\mid\to\infty,\alpha,\gamma) = \frac{\gamma}{\pi}\boldsymbol{\Gamma}(1+\alpha)\sin\left(\frac{\pi}{2}\alpha\right)\frac{1}{\mid x\mid^{1+\alpha}}.
\tag{A.141}
$$

constituting the algebraic tail of the Lévy PDF. We obtained this limit transition in this form because the series '$P_L(x;\alpha,\gamma)$' is alternating in sign, i.e., the sum of all neglected terms is of the order of the first neglected term. On the other hand, the first neglected term is '$\mid x\mid^{\alpha}$' times smaller than the first non-neglected term of the series.

A.16 Thermodynamic Representation of the Parabolic Multifractality

The relationship between multifractality and thermodynamic formalism is based on selecting variables such that the spectrum of dimensions or singularities is shaped into a convex function with downward-facing arms, similar to how entropy depends on energy in thermodynamic formalism. To solidify this link, we apply the subsequent linear transformation to the independent variable q:

$$\tilde{q} = 2 - q. \tag{A.142}$$

This leads to the following partition function,

$$\tilde{Z}_{\tilde{q}} = \frac{\left\langle \left(\frac{1}{\gamma_0 t} \right)^{\tilde{q}-1} \right\rangle}{\mathbf{\Gamma}(2 - \tilde{q})} = \gamma^{\tilde{\tau}(\tilde{q})}, \tag{A.143}$$

where

$$\tilde{\tau}(\tilde{q}) = -\tau(q) = -\tau(2 - \tilde{q}) = (\tilde{q} - 1)\left(\frac{\bar{\mathcal{E}}}{\sigma} - \frac{1}{2}\hat{\beta}\sigma(\tilde{q} - 1) \right), \tag{A.144}$$

and $\mathbf{\Gamma}$ is the gamma Euler function; here we used the formula (8.15) with the auxiliary Eq. (8.16).

In Eq. (A.143), we transitioned from time scaling to frequency scaling (the inverse of time). This shift ensures that the transformed partition function maintains its normalization, as indicated by the equality $\tilde{Z}_{\tilde{q}=1} = 1$. This holds true because $\tilde{\tau}(\tilde{q} = 1) = 0$.

Thanks to the transformation (A.142) and (A.144), the Rényi dimensions take the form,

$$\tilde{D}(\tilde{q}) = \frac{\tilde{\tau}(\tilde{q})}{\tilde{q} - 1} = \frac{\bar{\mathcal{E}}}{\sigma} - \frac{1}{2}\hat{\beta}\sigma(\tilde{q} - 1), \tag{A.145}$$

where the scope of the $\tilde{q}$ variable is limited

$$1 - \frac{2}{\hat{\beta}\sigma}\frac{\bar{\mathcal{E}}}{\sigma} = \tilde{q}_- \leq \tilde{q} \ll \tilde{q}_+ = 1 + \frac{1}{\hat{\beta}\sigma}\left(\frac{\bar{\mathcal{E}}}{\sigma} - 1 \right). \tag{A.146}$$

Recall that the $\sigma \ll \bar{\mathcal{E}}$ holds. It's worth noting that $\tilde{D}(\tilde{q})$ exhibits a linearly decreasing pattern concerning $\tilde{q}$. Consequently, it effectively characterizes the central, straight-line segment of the Rényi exponent plot—albeit with a reasonable approximation.

With the transformed global multiscale exponent $\tilde{\tau}(\tilde{q})$ at our disposal, we can deduce both the transformed local scaling exponent $\tilde{\eta}(\tilde{q})$ and the transformed local dimension spectrum $\tilde{f}(\tilde{\eta})$—specifically,

$$\tilde{\tau}(\tilde{q}) = (\tilde{q} - 1)\left(\frac{\bar{\mathcal{E}}}{\sigma} - \frac{1}{2}\hat{\beta}\sigma\,(\tilde{q} - 1) \right) = \tilde{q}\tilde{\eta}(\tilde{q}) - \tilde{f}(\tilde{\eta}(\tilde{q})), \tag{A.147}$$

where

$$\tilde{\eta}(\tilde{q}) = \frac{d\tilde{\tau}(\tilde{q})}{d\tilde{q}} = -\frac{d\tau(q)}{dq}\frac{dq}{d\tilde{q}} = \eta(q) = \eta(2-\tilde{q}) = \frac{\bar{\mathcal{E}}}{\sigma} - \hat{\beta}\sigma\,(\tilde{q}-1)\,,\ \frac{dq}{d\tilde{q}} = -1, \tag{A.148}$$

and

$$\begin{aligned}\tilde{f}(\tilde{\eta}(\tilde{q})) = & -f(\tilde{\eta}(\tilde{q})) + 2\tilde{\eta}(\tilde{q}) = -\frac{1}{2}\frac{1}{\hat{\beta}\sigma}(\tilde{\eta}(\tilde{q}) - \tilde{\eta}(\tilde{q}=2))^2 + 2\tilde{\eta}(\tilde{q}) - \tilde{D}(\tilde{q}=2)\\ = & -\frac{1}{2}\frac{1}{\hat{\beta}\sigma}\left(\tilde{\eta} - \frac{\bar{\mathcal{E}}}{\sigma}\right)^2 + \tilde{\eta},\end{aligned} \tag{A.149}$$

where Rényi's fractal dimension $\tilde{D}(\tilde{q}=2)$ is defined in (A.153) below. Moreover, important relations follow from (A.149), (A.148) and the third equality in (A.153),

$$\begin{aligned}\tilde{f}(\tilde{\eta}(\tilde{q}=0)) = & \ \tilde{D}(\tilde{q}=0),\\ \tilde{f}(\tilde{\eta}(\tilde{q}=1)) = & \ \tilde{\eta}(\tilde{q}=1) = \tilde{D}(\tilde{q}=1).\end{aligned} \tag{A.150}$$

It also takes place (as necessary)

$$\frac{d\tilde{f}(\tilde{\eta})}{d\tilde{\eta}} = \tilde{q} \Rightarrow \frac{d\tilde{f}(\tilde{\eta})}{d\tilde{\eta}}|_{\tilde{q}=1} = 1. \tag{A.151}$$

The following restrictions apply,

$$\frac{\bar{\mathcal{E}}}{\sigma} - \hat{\beta}\sigma \le \tilde{\eta}(\tilde{q}) \le 2\frac{\bar{\mathcal{E}}}{\sigma} \tag{A.152}$$

As observed, the transition from quantities without a wavelet to their wavelet-marked counterparts follows a linear pattern. Naturally, the constraints (A.152) limit the scope of multifractality, yet its presence remains evident.

It is also worth mentioning the following useful properties of Rényi's fractal dimensions, namely:

(a) $\tilde{D}(\tilde{q}) = \frac{1}{2}\tilde{\eta}(\tilde{q}) + \frac{1}{2}\frac{\bar{\mathcal{E}}}{\sigma}$
(b) $\tilde{D}(2\tilde{q}) = \tilde{\eta}(\tilde{q}) + \tilde{D}'(\tilde{q})$, where $\tilde{D}'(= -\frac{1}{2}\hat{\beta}\sigma)$ means derivative $\tilde{D}(\tilde{q})$ with respect to $\tilde{q}$
(c) if $\tilde{q}_2 \ge \tilde{q}_1$ then $\tilde{D}(\tilde{q}_1) \ge \tilde{D}(\tilde{q}_2)$
(d) within the ranges determined by the inequalities (A.146) Rényi's dimensions $\tilde{D}(\tilde{q}) > 0$.

Hence, in important special cases, we get:

$$\tilde{D}\left(\tilde{q}=1-\frac{1}{\hat{\beta}\sigma}\cdot\frac{\bar{\mathcal{E}}}{\sigma}\right)=\frac{3}{2}\frac{\bar{\mathcal{E}}}{\sigma}$$
$$\tilde{D}(\tilde{q}=0)=\frac{\bar{\mathcal{E}}}{\sigma}+\frac{1}{2}\hat{\beta}\sigma$$
$$\tilde{D}(\tilde{q}=1)=\frac{\bar{\mathcal{E}}}{\sigma}$$
$$\tilde{D}(\tilde{q}=2)=\frac{\bar{\mathcal{E}}}{\sigma}-\frac{1}{2}\hat{\beta}\sigma$$
$$\tilde{D}\left(\tilde{q}=1-\frac{1}{\hat{\beta}\sigma}\cdot\left(1-\frac{\bar{\mathcal{E}}}{\sigma}\right)\right)=\frac{1}{2}\left(1+\frac{\bar{\mathcal{E}}}{\sigma}\right). \tag{A.153}$$

Our objective remains the analysis of the singularity spectrum as given by (A.149). It is crucial to note that the parameters that define the spectrum, as well as the other two functions $\tilde{\tau}(\tilde{q})$ and $\tilde{\eta}(\tilde{q})$, are dimensionless quantities: $\frac{1}{\hat{\beta}\sigma}$ and $\frac{\bar{\mathcal{E}}}{\sigma}$. This choice helps avoid the effect of the ultimate size. Essentially, as the system size increases (that is, the average depth of potential minima $\bar{\mathcal{E}}$), both σ and consequently $\tilde{\beta}$ should be adjusted proportionally to maintain the values of these dimensionless parameters constant.

A.17 General Derivation of the Partition Function

It was determined in Sect. 8.3 that the (multi) q-moment of the inter-event times scales nonlinearly with the volume fluctuations. This raises the question of whether the generalized partition function exhibits analogous scaling. The objective of this section is to derive, at least approximately, this partition function [23, 24] in a closed form within the MF-CTRW model. Furthermore, significant multifractal properties of the inter-event times structure, complementary to those discussed in Sect. 8.3, are considered.

As an initial step, we utilize the following generalized partition function form, analogous to the grand partition function in thermodynamics.

$$Z(q')=\frac{1}{\theta^{\,q'-1}}Z_0(q')=\langle\left(\frac{p_{\underline{i}}}{\theta}\right)^{q'}\rangle,\; q'=q+1, \tag{A.154}$$

Here, we define the analog of the canonical partition function

$$Z_0(q')=\sum_{\underline{i}}p_{\underline{i}}^{q'}, \tag{A.155}$$

and a two-component index $\underline{i}=(i_1,i_2)$, which denotes a rectangle or plaquette (with side lengths Δ_t and Δ_ξ, respectively) to which a coarse-grained probability (measure)

$p_{\underline{i}}$ is attributed. By using such rectangles, our phase space is partitioned. In our approach, based on the MF-CTRW model, we consider the support of the measure, which is flat, involving a half-space continuum defined by time $t \geq 0$ and volume ξ of futures traded that are bought or sold, which implies $| \xi | < \infty$. Our measure vanishes only for $t \to \infty$ and/or $| \xi | \to \infty$. The positive quantity θ (defined in a self-consistent manner later on) relates to the chemical potential of an open system.

Both partition functions are given by Eq. (A.154) satisfy, as required, the normalization condition

$$Z(q' = 1) = Z_0(q' = 1) = 1. \tag{A.156}$$

This condition implies that orders q' and q (used in the definition of the q-moment, Eq. (8.28)) are shifted by 1, i.e., $q' = q + 1$, which establishes the mutual calibration of both orders.

Moreover, one also has

$$Z(q' \to 0) = Z_0(q' \to 0) \to \infty, \tag{A.157}$$

since the substrate (defined by the Cartesian product of inter-event time and futures volume) is unrestricted. This is the reason why, in our case, also $Z(q' < 0) = Z_0(q' < 0) = \infty$; therefore, we need to restrict our considerations only to $q' > 0$.

Note that for non-vanishing integers q', the partition function $Z_0(q')$ represents the probability of observing q' transactions in any placket $\underline{i}$. As q' remains fixed and possibilities $p_{\underline{i}}$ vary, the partition function $Z_0(q')$ increases for more heterogeneous values. Therefore, $q' \geq 2$ signifies the likelihood of how closely q' transactions are localized. It can gauge the degree of structural heterogeneity or correlations between transactions. For instance, according to the definition (A.155), in the case of full heterogeneity or localization at some plaquette $\underline{j}$ (where $p_{\underline{i}} = \delta_{\underline{i},\underline{j}}$), the partition function reaches its maximum value $Z_0(q') = 1$. Furthermore, with fixed $p_{\underline{i}} < 1$ and varying power q', the partition function decreases as the exponent increases, indicating a decrease in correlations between transactions or localization as the number of transactions increases, as expected.

In the following discussions, we establish connections between probabilities $p_{\underline{i}}$ and the joint and conditional waiting-time distributions defined in our extended MF-CTRW formalism. Subsequently, we introduce the generalized partition function, $Z(q')$, and perform the necessary calculations

$$Z(q') \stackrel{\text{def.}}{=} \frac{1}{\theta^{\,q'-1}} \sum_{i_1=0}^{\infty} \sum_{i_2=-\infty}^{\infty} \left[\int_{t_{i_1}}^{t_{i_1}+\Delta_t} dt \int_{\xi_{i_2}}^{\xi_{i_2}+\Delta_\xi} d\xi\, \psi(t \mid \xi)\, \rho(\xi) \right]^{q'}$$
$$\approx \frac{1}{(\theta / \Delta_t\, \Delta_\xi)^{q'-1}} \sum_{i_1=0}^{\infty} \left[\psi(t_{i_1} \mid \xi_{i_2})\right]^{q'} \Delta_t \times$$
$$\times \sum_{i_2=-\infty}^{\infty} \left[\rho(\xi_{i_2})\right]^{q'} \Delta_\xi. \tag{A.158}$$

Here, Δ_ξ and Δ_t represent the surface area of the elementary rectangle (plaquette) with side lengths $\Delta_t (\ll \tau_0)$ and $\Delta_\xi (\ll \sigma)$—both of which are arbitrarily small positive numbers. Meanwhile, $t_{i_1} = i_1\, \Delta_t$ and $\xi_{i_2} = i_2\, \Delta_\xi$.

In Eq. (A.158), the sum of the variable i_1 is constrained from the bottom, while the summation over i_2 is unrestricted. This limitation implies that examining the generalized partition function scaling with side lengths is not feasible. However, determining its scaling with volume fluctuation remains possible, which is this section's primary focus.

So far, by comparing expressions (A.158) and (A.154), we find a joint probability,

$$p_{\underline{i}} \approx \psi(i_1\Delta_t \mid i_2\Delta_\xi)\, \rho(i_2\Delta_\xi)\, \Delta_t\Delta_\xi, \tag{A.159}$$

which relates to a grid of rectangles of equal size. Note that the first sum over t_{i_1} in expression (A.158) can be performed explicitly

$$\sum_{i_1=0}^{\infty} \left[\psi(t_{i_1} \mid \xi_{i_2})\right]^{q'} = \left[\tau(\xi_{i_2})\right]^{-q'} \times$$
$$\times \sum_{i_1=0}^{\infty} \exp(-i_1\, q'\, \Delta_t\, / \tau(\xi_{i_2}))$$
$$= \frac{\left[\tau(\xi_{i_2})\right]^{-q'}}{1 - \exp(-q'\, \Delta_t\, / \tau(\xi_{i_2}))} \approx \frac{\left[\tau(\xi_{i_2})\right]^{1-q'}}{q'\, \Delta_t}. \tag{A.160}$$

By introducing expression (A.160) into (A.158), we obtain (also changing the sum over i_2 in the last approximate equality of (A.158) by integration) the partition function in the form

$$\begin{aligned}
Z(q') &\approx \left(\frac{\tau_0\,\sigma}{M}\right)^{q'-1} \frac{1}{q'} \int_{-\infty}^{\infty} d\xi\ [\rho(\xi)]^{q'}\ [\tau(\xi)]^{1-q'} \\
&= \left(\frac{\sigma}{M}\right)^{q'-1} \frac{1}{q'} \int_{-\infty}^{\infty} d\xi\ [\rho(\xi)]^{q'}\ \exp(\beta\,(1-q')\,\xi) \\
&= N_{q'}\ \exp(\beta\mu\,(1-q'))\ J(q'),
\end{aligned} \tag{A.161}$$

where integral

$$J(q') = \int_{-\infty}^{\infty} dy \exp\left(-\frac{q'}{2} \mid y \mid^{\alpha} + \beta\sigma(1-q')y\right) \tag{A.162}$$

and factor

$$N_{q'} = \frac{1}{q'} M^{1-2q'}, \tag{A.163}$$

were obtained after performing a simple change of variables $y \overset{\text{def.}}{=} (\xi - \mu)/\sigma$ in the integrals in the first and second lines in expression (A.161). The (dimensionless) denominator θ in expression (A.158) is $\theta = M\Delta_t\,\Delta_\xi/\tau_0\,\sigma$, where the coefficient $M \overset{\text{def.}}{=} 2^{1+1/\alpha}\Gamma(1+1/\alpha)$. In the following section, we discuss the scaling form of the generalized partition function; we also consider useful Rényi dimensions.

A.18 Approximate Evaluation of I(q)

This Appendix considers two complementary cases: q far from zero and q small. This approach is crucial because it delineates the boundaries of multifractality, highlighting that the extent of multifractality, i.e., the range of q defining its scope, is finite.

Case q far from zero

We want to evaluate the integral (8.35) which we write in the form

$$I(q) = \lambda^{1/\alpha} \int_{-\infty}^{\infty} e^{-\lambda h(x)} dx, \tag{A.164}$$

where $\lambda \overset{\text{def.}}{=} (\beta\sigma)^{\alpha/(\alpha-1)}$

$$h(x) = \frac{1}{2}|x|^{\alpha} - qx, \tag{A.165}$$

where $\alpha > 1$. For λ large, we can employ the saddle-point approximation or Laplace's method [6, 25]. Expanding $h(x) = h(x_0) + (1/2)h''(x_0)(x-x_0)^2 + \mathrm{O}[(x-x_0)^3]$ and performing the resulting Gaussian integral, we obtain

$$I(q) = \lambda^{1/\alpha} \left[\frac{2\pi}{\lambda h''(x_0)} \right]^{1/2} e^{-[\lambda h(x_0) + O(\lambda^{-1/2})]}, \tag{A.166}$$

where x_0 is the minimum of $h(x)$. That is, x_0 is the solution to

$$h'(x_0) = \frac{1}{2}\alpha |x_0|^{\alpha-1} \mathrm{sgn}(x_0) - q = 0,$$

i.e.,

$$\frac{1}{2}\alpha |x_0|^{\alpha-1} \mathrm{sgn}(x_0) = q. \tag{A.167}$$

But $q = |q|\mathrm{sgn}(q)$ and we rewrite Eq. (A.167) as

$$\frac{1}{2} \frac{\alpha |x_0|^{\alpha-1}}{|q|} = \frac{\mathrm{sgn}(q)}{\mathrm{sgn}(x_0)}.$$

for $q \neq 0$; obviously, for $q = 0$ also $x_0 = 0$. Since the right-hand side of this equation is positive (recall that $\alpha > 0$) then necessarily $\mathrm{sgn}(x_0) = \mathrm{sgn}(q)$. Therefore,

$$|x_0| = \left(\frac{2|q|}{\alpha} \right)^{1/(\alpha-1)}, \tag{A.168}$$

and the two extremes of $h(x)$ are

$$x_0 = (2q/\alpha)^{1/(\alpha-1)}$$

when $q > 0$, and

$$x_0 = -(-2q/\alpha)^{1/(\alpha-1)}$$

when $-1 < q < 0$.

Taking into account that the second derivative $h''(x_0) = \frac{1}{2}\alpha(\alpha - 1)|x_0|^{\alpha-2}$, i.e.,

$$h''(x_0) = \frac{1}{2}\alpha(\alpha - 1) \left(\frac{2|q|}{\alpha} \right)^{(\alpha-2)/(\alpha-1)} > 0 \tag{A.169}$$

is always positive we conclude that x_0 is indeed a minimum of $h(x)$. We must realize that the applicability of the parabolic approximation of the function $h(x)$ is limited, and the better the larger $h''(x_0)$ is. In other words, $|\, x_0 \,|$ must be far from zero, and the variance of $1/h''(x_0)$ must be small. The complementary case is considered in the Section below.

On the other hand, recalling that x_0 has the same sign as q we can write $h(x_0)$ in the form

$$h(x_0) = \frac{1}{2}|x_0|^{\alpha} - |q||x_0|,$$

and using Eq. (A.168) we have

$$h(x_0) = -2^{1/(\alpha-1)}(\alpha-1)\left(\frac{|q|}{\alpha}\right)^{\alpha/(\alpha-1)}. \tag{A.170}$$

Collecting terms, we finally obtain

$$I(q) \simeq A(q)\exp\left(b|q|^{\alpha/(\alpha-1)}\right), \tag{A.171}$$

where factor

$$A(q) = \lambda^{1/\alpha}\left[\frac{4\pi}{\lambda\alpha(\alpha-1)}\right]^{1/2}\left(\frac{2|q|}{\alpha}\right)^{(2-\alpha)/2(\alpha-1)}, \tag{A.172}$$

is a slowly-varying function of $| q |$ (in comparison with the remaining factor in Eq. (A.171)), and (see Eq. (8.36))

$$b = 2^{1/(\alpha-1)}\frac{\lambda(\alpha-1)}{\alpha^{\alpha/(\alpha-1)}} = 2^{1/(\alpha-1)}(\alpha-1)\left(\frac{\beta\sigma}{\alpha}\right)^{\alpha/(\alpha-1)}. \tag{A.173}$$

Case |q| small

The starting point for the following considerations is the series expansion of the expression (8.33),

$$\begin{aligned} I(q) &= 2\int_0^{\infty}\exp\left(-y^{\alpha}/2\right)\cosh\left(q\sigma\beta y\right)dy \\ &= 2^{1+1/\alpha}\frac{1}{\alpha}\sum_{n=0}^{\infty}\frac{\Gamma((2n+1)/\alpha)}{(2n)!}\left(2^{1/\alpha}q\sigma\beta\right)^{2n}, \end{aligned} \tag{A.174}$$

which for $q \ll \frac{1}{2^{1/\alpha}\sigma\beta}$ takes the approximate form,

$$I(q) \approx 2^{1+1/\alpha}\frac{1}{\alpha}\left[1+\frac{1}{2}\Gamma(3/\alpha)\left(2^{1/\alpha}q\sigma\beta\right)^2\right], \tag{A.175}$$

that leads to

$$\langle\Delta t^q\rangle \approx 2^{1/\alpha}\frac{\Gamma(1+q)}{\Gamma(1/\alpha)}\tau_0 l^q\left[1+\frac{1}{2}\Gamma(3/\alpha)\left(2^{1/\alpha}q\sigma\beta\right)^2\right]. \tag{A.176}$$

As one can see, $\ln\langle\Delta t^q\rangle$ is a linear function of q, where the quadratic term in square brackets is a vanishing correction to Eq. (A.176).

A.19 Calculation of the Integral J(q')

In this Appendix, we calculate (similarly to Appendix A.18) the basic integral using the saddle-point approximation.

$$J(q') = \int_{-\infty}^{\infty} dy \exp\left(-\frac{q'}{2} \mid y \mid^{\alpha} + \beta\sigma(1 - q')y\right) \tag{A.177}$$

and hence the generalized partition function

$$Z(q') \approx N_{q'} \exp(\beta\,(1 - q')\mu)\, J(q'), \tag{A.178}$$

where factor

$$N_{q'} = \frac{1}{q'} M^{1-2q'}. \tag{A.179}$$

and coefficient

$$M = 2^{1+1/\alpha}\boldsymbol{\Gamma}(1 + 1/\alpha). \tag{A.180}$$

To evaluate the integral (A.177) we first change the integration variable $y = \kappa\, x$, (where $\kappa > 0$), which results in

$$J(q') = \kappa \int_{-\infty}^{\infty} \exp\left[-f(x)\right] dx \tag{A.181}$$

where

$$f(x) = \frac{q'}{2} \mid \kappa x \mid^{\alpha} - sgn(1 - q')\kappa\beta\sigma \mid 1 - q' \mid x; \tag{A.182}$$

by assuming that coefficients at both powers of variable x are equal, we obtain

$$\kappa = \left[2\frac{\beta\sigma \mid 1 - q' \mid}{q'}\right]^{1/(\alpha-1)}. \tag{A.183}$$

Hence, integral $J(q')$ takes the form

$$J(q') = \kappa \int_{-\infty}^{\infty} \exp\left[-\frac{q'\kappa^{\alpha}}{2} h(x)\right] dx, \tag{A.184}$$

where

$$h(x) = \mid x \mid^{\alpha} - sgn(1 - q')x. \tag{A.185}$$

Now, we are ready to perform both steps of the SPA for the calculation of integral $J(q')$:

(i) approximation of the function $h(x)$ by the parabolic expansion around its minimum x_0,

$$h(x) \approx h(x_0) + \frac{1}{2}h''(x_0)(x - x_0)^2, \tag{A.186}$$

where minimum

$$x_0 = \frac{sgn(1 - q')}{\alpha^{1/(\alpha-1)}} \tag{A.187}$$

while

$$\begin{aligned} h(x_0) &= \frac{1}{\alpha^{\alpha/(\alpha-1)}} - \frac{1}{\alpha^{1/(\alpha-1)}} \\ &= -\frac{\alpha - 1}{\alpha^{\alpha/(\alpha-1)}} < 0, \\ h''(x_0) &= \alpha(\alpha - 1) \mid x_0 \mid^{\alpha-2} \\ &= (\alpha - 1)\, \alpha^{1/(\alpha-1)} > 0, \end{aligned} \tag{A.188}$$

and next

(ii) the explicit calculation of the Gaussian integral (based on the quadratic part of $h(x)$ function) as an approximation of $J(q')$, namely

$$\begin{aligned} J(q') \approx\ & \kappa \exp\left[-\frac{q'}{2}\kappa^{\alpha}\, h(x_0)\right] \times \\ & \times \int_{-\infty}^{\infty} \exp\left[-\frac{q'\kappa^{\alpha}}{4}\, h''(x)(x - x_0)^2\right] dx \\ =\ & \frac{2\kappa\sqrt{\pi}}{\sqrt{q'\kappa^{\alpha}h''(x_0)}} \exp\left[-\frac{q'\kappa^{\alpha}}{2} h(x_0)\right], \end{aligned} \tag{A.189}$$

Finally, by introducing expressions (A.188) into (A.189) and using definition of κ (A.183) we obtain

$$\begin{aligned} J(q') = & \left(\frac{2\pi}{\alpha - 1}\right)^{1/2} \left[\frac{2}{\alpha}(\beta\sigma)^{2-\alpha}\right]^{1/2(\alpha-1)} \times \\ & \times \left(\frac{\mid 1 - q' \mid^{2-\alpha}}{q'}\right)^{1/2(\alpha-1)} \times \\ & \times \exp\left[b\left(\frac{\mid 1 - q' \mid^{\alpha}}{q'}\right)^{1/(\alpha-1)}\right], \end{aligned} \tag{A.190}$$

By using the last row in expression (A.161), we obtain the formula for the generalized partition function

$$\begin{aligned} Z(q') &\approx N_{q'} \left(\frac{2\pi}{\alpha - 1}\right)^{1/2} \left[\frac{2}{\alpha}(\beta\sigma)^{2-\alpha}\right]^{1/2(\alpha-1)} \times \\ &\times \left(\frac{\mid 1-q' \mid^{2-\alpha}}{q'}\right)^{1/2(\alpha-1)} \times \\ &\times \exp\left[\beta\mu(1-q') + b\left(\frac{\mid 1-q' \mid^{\alpha}}{q'}\right)^{1/(\alpha-1)}\right]. \end{aligned} \tag{A.191}$$

To return to normalized $Z(q')$ (i.e., to the partition function obeying (A.156)), we have to neglect a part of the prefactor of Eq. (A.191); thus our final, crude approximation for the partition function is

$$\begin{aligned} Z(q') &\approx M^{1-q'} q'^{(1-2\alpha)/2(\alpha-1)} \times \\ &\times \exp\left[\beta\mu(1-q') + b\left(\frac{\mid 1-q' \mid^{\alpha}}{q' + q'_0}\right)^{1/(\alpha-1)}\right], \end{aligned} \tag{A.192}$$

where the correction constant q'_0 should be estimated from the requirement of self-consistency of the approach, while its exact value should be obtained by comparing the prediction of the approach with the corresponding empirical data; this is further considered in Sect. 8.3.3. Besides the normalization obeyed by expression (A.192), it takes (for $\alpha = 2$) the form defined by the Gaussian kernel $\rho(\varepsilon)$, which independent exact calculations can derive without applying the SPA.

A.20 Properties of Rényi Dimensions and Generalized Hurst Exponent

Here, with the time-dependent Rényi dimensions denoted as $D(\beta, t)$ and generalized Hölder exponent/function as $h(\beta, t)$, we can define their time-dependent differences $\Delta D(\beta, t) \stackrel{\text{def.}}{=} D(-\beta, t) - D(\beta, t)$ and $\Delta h(\beta, t) \stackrel{\text{def.}}{=} h(-\beta, t) - h(\beta, t)$, respectively. Identical horizontal time-dependent asymptotes limit the former two quantities:

$$\begin{aligned} &D(\beta \to \mp\infty, t) = h(\beta \to \mp\infty, t) = \\ &= \begin{cases} D^{\max}(t) = h^{\max}(t) = -\frac{\ln \omega^{\min}(t)}{\ln \Lambda(t)}, \beta < 0, \\ D^{\min}(t) = h^{\min}(t) = -\frac{\ln \omega^{\max}(t)}{\ln \Lambda(t)}, \beta > 0, \end{cases} \end{aligned} \tag{A.193}$$

while

$$\begin{aligned} \Delta D(t) &\stackrel{\text{def.}}{=} D^{\max}(t) - D^{\min}(t) = \\ &= \Delta h(t) \stackrel{\text{def.}}{=} h^{\max}(t) - h^{\min}(t) = \\ &= \ln\left(\frac{\omega^{\max}(t)}{\omega^{\min}(t)}\right). \end{aligned} \tag{A.194}$$

Equations (A.193) and (A.194) are directly derived from the asymptotic scaling exponent properties provided by Eq. (11.10) and by Eqs. (11.6) and (11.7), respectively.

A.21 Properties of Spectrum of Dimensions

We present in the following useful quantities and properties, which characterize the spectrum of singularities,

(i) $f^0(t) = f(\alpha^0(t) \stackrel{\text{def.}}{=} \alpha(\beta = 0, t), t) = D^0(t) = D(\beta = 0, t)$, which results from Eqs. (11.7), (11.14), and (11.18), and moreover we get $\frac{\partial f(\alpha,t)}{\partial \alpha}|_{\alpha(\beta=0,t)} = 0$;

(ii) for $\beta = 1$ we immediately get from Eq. (11.18) $\frac{\partial f(\alpha,t)}{\partial \alpha}|_{\alpha(\beta=1,t)} = 1$ for any time, hence $f^1(t) \stackrel{\text{def.}}{=} f(\alpha(\beta = 1, t)) = \alpha^1(t) \stackrel{\text{def.}}{=} \alpha(\beta = 1, t)$;

(iii) for $\beta \to -\infty$ we get from Eqs. (11.14) and (11.17), that $f^{\max}(t) = f(\alpha(t) = \alpha^{\max}(t) = -\frac{\ln \omega^{\min}(t)}{\ln \Lambda(t)}) = \frac{\ln K_{\min}(t)}{\ln N(t)}$ and $\frac{\partial f(\alpha,t)}{\partial \alpha}|_{\alpha(\beta\to-\infty,t)} = -\infty$; similarly for $\beta \to \infty$ we get $f^{\min}(t) = f\left(\alpha(t) = \alpha^{\min}(t) = -\frac{\ln \omega^{\max}(t)}{\ln \Lambda(t)}\right) = \frac{\ln K_{\max}(t)}{\ln \Lambda(t)}$ and $\frac{\partial f(\alpha,t)}{\partial \alpha}|_{\alpha(\beta\to\infty,t)} = \infty$;

(iv) the maximum span of f we determine as follows, $\Delta\alpha(t)|_{|\beta|\to\infty} = \alpha^{\max}(t) - \alpha^{\min}(t) = \frac{1}{\ln \Lambda(t)} \ln\left(\frac{\omega^{\max}(t)}{\omega^{\min}(t)}\right)$. We continue to use the simplified designation $\Delta\alpha(t) = \Delta\alpha(t)|_{|\beta|\to\infty}$;

(v) the following asymmetry factor can be used to determine the degree of asymmetry f, $\kappa(t)|_{|\beta|\to\infty} = \frac{\alpha(\beta=0,t)-\alpha^{\min}(t)}{\alpha^{\max}(t)-\alpha(\beta=0,t)}$, where $\alpha(\beta = 0, t)$ is given by Eq. (11.16). We continue to use the simplified designation $\kappa(t) = \kappa(t)|_{|\beta|\to\infty}$;

(vi) we proved, using Eqs. (11.14), (11.15) and (11.18), that $\frac{\partial f(\alpha(t),t)}{\partial t} = -\frac{\partial \tau(\beta(t),t)}{\partial t}$.

From item (iii) follows that $f(\alpha = \alpha^{\min}(t), t) \neq 0$ and $f(\alpha = \alpha^{\max}(t), t) \neq 0$. It happens when the boundary values $\omega^{\max}(t)$ and $\omega^{\min}(t)$ are degenerated, respectively.

A.22 Deterministic Binomial Cascade: A Generic Approach

Presumably, the simplest version of the cascade model is the multiplicative deterministic binomial de Wijs cascade [26]. The starting point ($n = 0$) is the interval/segment

[0, 1] with a uniformly distributed mass $\mu_0(0)(= 1)$. In the subsequent steps of the construction procedure, each segment is partitioned into two subsegments of equal size; however, the mass relocation between these segments occurs according to the following rule.

In the first step/generation ($n = 1$), we divide the segment [0, 1] into two equal parts (thus, the scale in the first generation is $s_{n=1} = 1/2^{n=1}$), assigning to the left subsegment a fraction of mass $\mu_{n=1}(0) = p\mu_0(0) = p$ and to the right subsegment $\mu_{n=1}(1) = (1-p)\mu_0(0) = 1-p$, ensuring that the total mass is conserved ($0 < p < 1$). The law of mass conservation is satisfied for each generation separately.

Consequently, in the second generation ($n = 2$, scale $s_2 = 1/2^2$), we have $\mu_2(0) = p\mu_1(0) = p^2$, $\mu_2(1) = (1-p)\mu_1(0) = (1-p)p$, $\mu_2(2) = p\mu_1(1) = p(1-p)$, and $\mu_2(3) = (1-p)\mu_1(1) = (1-p)^2$. This is generally expressed as the product $p^k(1-p)^{n-k}$, where the variable $0 \le k \le n$ tells us how many events led to the appearance of p in this product (more about these events in the discussion of equations in (A.196) below). We proceed analogously in subsequent generations. As we can see, the number of subsegments with identical masses in generation $n(= 0, 1, 2, 3, \dots,)$ is given by the binomial coefficient $\binom{n}{k}$; the scale factor is given by $s_n = 1/2^n$.

Thus, a binomial tree of subsegments with masses $\mu_n(j)$ is created, where $n = 0, 1, 2, 3, \dots$ denotes the generation number and the index $0 \le j \le 2^n - 1$ denotes the subsegment number in a given generation. The following constitutive partial recurrence equations relate to these masses:

$$\begin{aligned} \mu_{n+1}(2j) &= p\,\mu_n(j), \\ \mu_{n+1}(2j+1) &= (1-p)\,\mu_n(j). \end{aligned} \tag{A.195}$$

We emphasize that j indexes subsegments in generation n, while the even index $2j$ and the odd index $2j+1$ index subsegments in the subsequent generation $n+1$.

Equations in (A.195) introduce a binary intergenerational coupling/dependence conditioned by the law of mass conservation and binary symmetry $p \Leftrightarrow (1-p)$ at every generation level. There is a stochastic variant of this cascade, which we will discuss next.

Although the relative mass asymmetry between neighboring nodes $2j$ and $2j+1$, defined as $\Delta\mu_{n+1}(2j)/\mu_{n+1}(j) \overset{\text{def.}}{=} \left[\mu_{n+1}(2j+1) - \mu_{n+1}(2j)\right]/\mu_{n+1}(j) = 1 - 2p$, depends solely on p and disappears only in the case of symmetry ($p = 1/2$), the mass distribution in a given generation additionally relies on the position of the mass in the binomial tree (see the example in Fig. A.5 for $n = 10$)[9].

Our goal here is to analyze the equations in (A.195) and their consequences, which will allow us to relate the magnitude of the mass distribution in a given generation to the numbers of subsegments and the multiplicity/degeneracy $\binom{n}{k}$ of the occurrence of a given mass in the spectrum. This will enable us to construct a complete mass spectrum. For example, for $n = 3$, we obtain the following mass spectrum:

---

[9] Of course, a related example of solely p dependence here is the ratio $\mu_{n+1}(2j+1)/\mu_{n+1}(2j) = (1-p)/p$.

$$\mu_3(0) = \mu_0(0)ppp = \mu_0(0)M(-1,-1,-1)$$
$$\mu_3(1) = \mu_0(0)pp(1-p) = \mu_0(0)M(-1,-1,+1)$$
$$\mu_3(2) = \mu_0(0)p(1-p)p = \mu_0(0)M(-1,+1,-1)$$
$$\mu_3(3) = \mu_0(0)p(1-p)(1-p) = \mu_0(0)M(-1,+1,+1)$$
$$\mu_3(4) = \mu_0(0)(1-p)pp = \mu_0(0)M(+1,-1,-1)$$
$$\mu_3(5) = \mu_0(0)(1-p)p(1-p) = \mu_0(0)M(+1,-1,+1)$$
$$\mu_3(6) = \mu_0(0)(1-p)(1-p)p = \mu_0(0)M(+1,+1,-1)$$
$$\mu_3(7) = \mu_0(0)(1-p)(1-p)(1-p) = \mu_0(0)M(+1,+1,+1). \quad \text{(A.196)}$$

For example, the mass $\mu_0(0)p^2(1-p)$ is triply degenerate, with masses of this value distributed throughout the spectrum. The same holds for the mass value $\mu_0(0)p(1-p)^2$, which arises from binary symmetry. The structure of each mass is multiplicative. Specifically, the total multiplier is given by $M(\eta_1, \eta_2, \ldots, \eta_n) = m(\eta_1)m(\eta_2)\ldots m(\eta_n)$, where $\eta_j = \mp 1$. Each individual multiplier is defined as $m(-1) = p$ or $m(+1) = 1-p$. In this context, the vector $(\eta_1, \eta_2, \ldots, \eta_n)$ represents an intergenerational dichotomous directed random walk on a binomial tree, where -1 corresponds to a displacement to the left and $+1$ to the right. The variable k (the same one we mentioned above when discussing the mass distribution, for example, for generation $n = 2$) measures the number of displacements to the left in the currently considered generation (here $n = 3$). It is given by the binomial relation $k = \frac{n-\sum_{j=1}^{n}\eta_j}{2}$, where $\sum_{j=1}^{n}\eta_j$ represents the net displacement on the binomial tree after n steps. This random walk is unbiased only when $p = \frac{1}{2}$. For $p \neq \frac{1}{2}$, we encounter an anisotropic random walk characterized by broken symmetry. In Fig. A.5, we present the mass spectrum that illustrates the hierarchical and inhomogeneous nature of the cascade[10].

Recall that we consistently utilize (various representations of) a generalized multiscale partition function to extract the multifractality of a given process, which, in this case, is generated by a multiplicative cascade. This partition function is defined as follows:

---

[10] The equations in (A.195) are solved using the generating (or z) function defined as $G_n(z) \stackrel{\text{def.}}{=} \sum_{j=0}^{2^n-1} z^j \mu_n(j)$, along with the resulting equality $\mu_n(j) = \frac{1}{j!}\frac{d^j}{dz^j}G_n(z)\big|_{z=0}$, $0 \le j \le 2^n - 1$. Therefore, the equations in (A.195) can be transformed into a more convenient form,

$$G_{n+1}(z) = [p + (1-p)z]\, G_n(z^2), \; n = 0, 1, 2, 3, \ldots$$
$$G_0(z) = \mu_0(0), \quad \text{(A.197)}$$

by using the generating function mentioned above. The solution to this equation is of the form:

$$G_{n+1}(z) = \mu_0(0)\prod_{l=0}^{n}\left[p + (1-p)z^{2^l}\right], \; n = 0, 1, 2, 3, \ldots \quad \text{(A.198)}$$

Notice that the first equation in (A.197) is unusual because the variable in the z function depends on z^2 rather than on the z variable in the right-hand side of this equation.

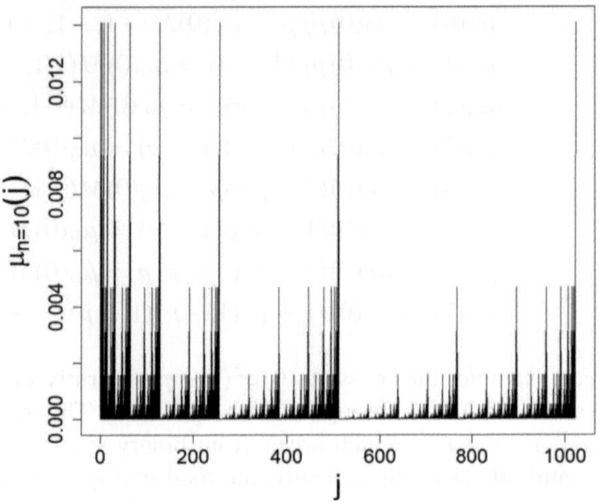

Fig. A.5 An illustration of the binomial cascade, as defined by the equations in (A.195), is provided for the case where $\mu_0(0) = 1$, $p = 1/4$, and $n = 10$. The mass distribution heterogeneity and its hierarchical structure are evident. The term 'heterogeneity' here indicates that translational symmetry is broken in each generation

$$
\begin{aligned}
Z_q(s_n) &= \sum_{j=0}^{s_n^{-1}-1} [\mu_n(j)]^q = \sum_{k=0}^{n-1} [p^k(1-p)^{n-k}]^q \binom{n}{k} \\
&= \left[p^q + (1-p)^q\right]^n = s_n^{\tau(q)}.
\end{aligned} \tag{A.199}
$$

The transition (in the above formula) from the first summation to the second one corresponds to the transition from summation over nodes $0 \leq j \leq s_n^{-1}(= s_1^{-n})1$ of the binomial tree in a given generation n to summation over various masses indexed by the variable k - hence a factor appears that determines the degeneracy (multiplicity of occurrence) of the masses. At the same time, this allows us to introduce the general scaling exponent $\tau(q)$, which serves as the starting point of the multifractal formalism.

From Eq. (A.199), it follows that the general scaling exponent of the mass distribution is given by:

$$
\tau(q) = -\frac{\ln\left(p^q + (1-p)^q\right)}{\ln 2}. \tag{A.200}
$$

Hence, we get the singularity index α,

$$
\begin{aligned}
\alpha(q) &= \frac{\partial \tau(q)}{\partial q} = -\frac{1}{\ln 2} \frac{p^q \ln p + (1-p)^q \ln(1-p)}{p^q + (1-p)^q} \\
&= -\frac{1}{\ln 2}[\xi(q) \ln p + (1-\xi(q)) \ln(1-p)],
\end{aligned} \tag{A.201}
$$

where we introduced the escort probability

$$\xi(q) = \frac{p^q}{p^q + (1-p)^q}, \tag{A.202}$$

and the spectrum of dimensions $f(\alpha)$,

$$\begin{aligned} f(\alpha) &= q\alpha(q) - \tau(q) = -\frac{\xi(q)\ln p^q + (1-\xi(q))\ln(1-p)^q}{\ln 2} + \frac{\ln\left(p^q + (1-p)^q\right)}{\ln 2} \\ &= -\frac{1}{\ln 2}[\xi(q)\ln\xi(q) + (1-\xi(q))\ln(1-\xi(q))] = \frac{\chi(\alpha(q))}{\ln 2} = \hat{\chi}(\alpha(q)). \end{aligned} \tag{A.203}$$

Here, $\chi(\alpha(q))$ represents the dimensionless Boltzmann-Gibbs-Shannon entropy density, expressed in a standardized form $\hat{\chi}(\alpha(q))$, which achieves a maximum value of 1 for $q = 0$ and $\alpha(q = 0) = -\ln(p(1-p))/2\ln 2$. The expression f as a function of α indicates that we first invert the last equality in (A.201), thereby determining $\xi(q)$ as a function of $\alpha(q)$, and then substitute this conditional $\xi(\alpha(q))$ into the last equality in (A.203). Therefore, the dependence of α on q is mediated by the escorting probability, which serves as an intermediate or auxiliary variable in this context.

As observed, a heterogeneous structure has been created, generated by intergenerational multiplicative (long-range) dependencies, often called the "domino effect." An essential feature of this structure is its multifractality.

The dependencies of τ and α on q (top plots) and $f(\alpha)$ on α (bottom plot) are illustrated in Fig. A.6 for three values of the parameter: $p = 0.25$, 0.5 and 0.75.

There is a clear dependence of the location of the spectrum of dimensions/singularities and its span on p, which is consistent with expectations[11].

Thanks to the entropic interpretation of f, the first equality in (A.203) can be formally treated as an analog of the thermodynamic identity, where q plays the role of the reciprocal of the temperature (assuming that the Boltzmann constant is set to 1), $\alpha(q)$ represents the specific energy and $\tau(q)$ denotes the specific free energy per unit temperature (see Table 5.2 in Sect. 5.1.6).

The transition from this deterministic cascade to the stochastic one involves reinterpreting the masses of individual subsegments. In the stochastic cascade, we refer to the measures of subsegments as their masses. Thus, the mass of a given subsegment is the average over the statistical ensemble of individual realizations (replicas) of the stochastic cascade. This approach enables the application of the dually interpreted binomial cascade formalism for both deterministic and statistical issues.

[11] The span of the spectrum of dimensions $\Delta\alpha = \alpha_{\max} - \alpha_{\min} = \ln\left(p/(1-p)\right)/\ln 2$, where $\alpha_{\max} = \alpha(q = -\infty) = -\ln p/\ln 2$ and $\alpha_{\min} = \alpha(q = \infty) = -\ln(1-p)/\ln 2$. Consequently, for $p = 1/2$, the span vanishes, and f reduces to a single point at $(\alpha = 1, f = 1)$. Moreover, $f(\alpha)$ for a given p is equal to $f(\alpha)$ for its counterpart $1 - p$, which follows from the easily provable equality that $\xi(q)$ for a given p equals $1 - \xi(q)$ for its counterpart $1 - p$.

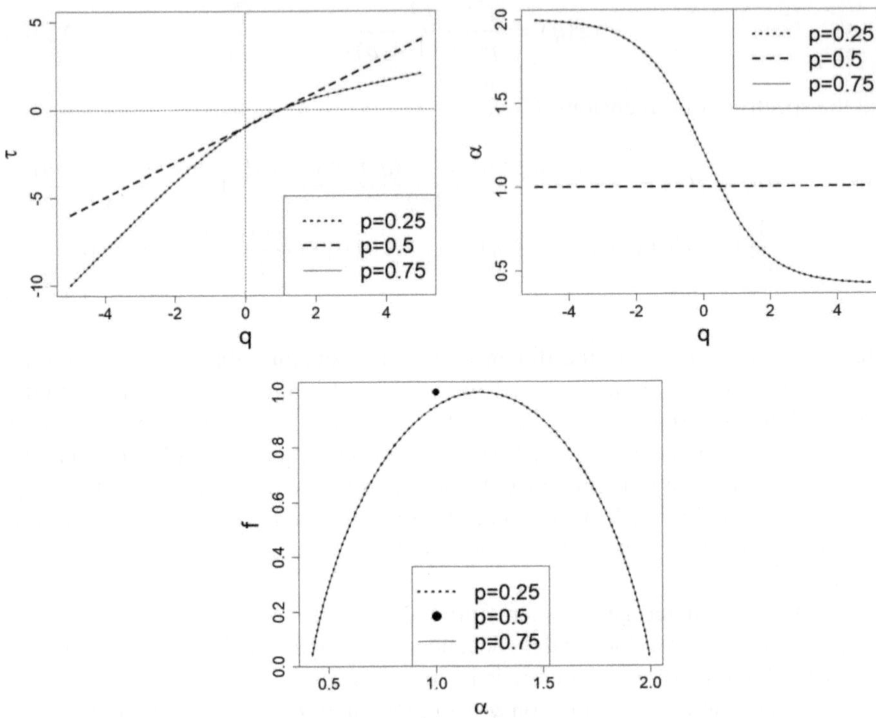

Fig. A.6 An illustration of Eqs. (A.200), (A.201), and (A.203) for three values of p highlights the essential characteristics of multifractality. The (vertical) span of α corresponds to the (horizontal) span of f in the bottom plot. Multifractality disappears for $p = 0.5$, while it collapses for the counterparts indexed by p and $1 - p$. Thus, the disappearance of the intergenerational mass drift (proportional to $(1 - 2p)$) in the system corresponds to the loss of multifractality (and vice versa). Additionally, the disappearance of the drift eliminates inhomogeneity in the system, which is already evident at the equations level in (A.195). Furthermore, these plots illustrate the multiparticle analogy of the binomial multiplicative cascade discussed in Sect. A.22.1 below.

A.22.1 *Multiparticle Analogy of the Binomial Multiplicative Cascade*

We now outline the functional analog of the previously presented binomial cascade derived from the statistical thermodynamics of gases.

We consider a volume container Ω containing a classical ideal gas under standard conditions of identical (but distinguishable) molecules n. We mentally designate a connected volume region V within this container. The *a priori* probability of finding a single molecule in this volume is given by $p = V/\Omega$. We demonstrate that Eq. (A.199) is formally unchanged. Therefore, we can write,

$$\begin{aligned} Z_q(s_n) &= \sum_{\{config\}} p^{qn_V}(1-p)^{q(n-n_V)} \\ &= \sum_{n_v=0}^{n} \sum_{\{config(n_V)\}} p^{qn_V}(1-p)^{q(n-n_V)} \\ &= \sum_{n_v=0}^{n} \binom{n}{n_V} p^{qn_V}(1-p)^{q(n-n_V)} \\ &= [p^q + (1-p)^q]^n = s_n^{\tau(q)}, \end{aligned} \tag{A.204}$$

where $\sum_{config} \cdots$ denotes the summation over all possible spatial configurations of gas molecules (integrated over all particle momenta). In contrast, $\sum_{config(n_V)} \cdots$ denotes the summation only in those configurations of all molecules that correspond to a fixed number of molecules n_V in volume V (the variable n_V here is equivalent to k used in Eq. (A.199)). Note that by subsequent configurations, we mean configurations corresponding to the exchange of molecule locations between regions V and $\Omega - V$ rather than within them. Moreover, we are not interested in the specific coordinates of the molecules but only in their presence in one of the two mentioned regions (which makes the possibility of using the scale $s_1 = 1/2$ and consequently the scale $s_n = 1/2^n$). This approach (e.g., to construct the Boltzmann configurational entropy) is called coarse-grained. Since our further considerations are analogous to those in Sect. A.22, there is no need to repeat them here.

A.22.2 Multifractality in the Thermodynamic Limit

In the thermodynamic limit, that is, in the asymptotic case, when the number density of the gas $n/\Omega \rightarrow_{n\to\infty} = \rho = \text{const}$, it is more convenient to use the transformed/asymptotic form of Z_q. Namely,

$$\begin{aligned} Z_q = \exp\left(-q\rho V + (\rho V)^q\right) = \exp(-\tau(q)) \Longleftrightarrow \ln Z_q &= -\tau(q), \\ \tau(q) &= q\rho V - (\rho V)^q, \end{aligned} \tag{A.205}$$

where, along the way, we used the fact that for asymptotically large n, the elementary probability is given by $p \approx p(n) = \rho V/n$, and then we performed the limit as $n \rightarrow \infty$. Let us add that ρV denotes the average number of particles in volume V.

From the above[12], we obtain that

---

[12] The first equality in (A.205) was obtained after taking the limit as $n \rightarrow \infty$ in the penultimate equality in (A.204). In this step, we used the relation that the probability p is inversely proportional to n. This allowed us to transform the binomial distribution present in the penultimate equality of (A.204) into the corresponding Poisson distribution.

$$\alpha(q) = \rho V - (\rho V)^q \ln(\rho V) \Longleftrightarrow q(\alpha) = \frac{1}{\rho V} \ln\left(\frac{\rho V - \alpha}{\ln(\rho V)}\right), \ \alpha(q) \le \rho V, \ \rho V \ge 1, \tag{A.206}$$

and

$$\begin{aligned} f(\alpha(q)) &= q\rho V - q(\rho V)^q \ln(\rho V) - q\rho V + (\rho V)^q \\ &= [1 - q \ln(\rho V)](\rho V)^q \\ &= \frac{\rho V - \alpha(q)}{\ln(\rho V)} - \ln\left(\frac{\rho V - \alpha(q)}{\ln(\rho V)}\right). \end{aligned} \tag{A.207}$$

We now present a list of properties satisfied by the fundamental characteristics of multifractality in the thermodynamic limit. Thus,

(i) $\tau(q = 0) = -1$ and $\tau(q = 1) = 0$.
(ii) $\alpha(q = 0) = \rho V - \ln(\rho V)$.
(iii) $\alpha(q \to -\infty) = \rho V = \alpha_{\max}$, $\alpha(q \to \infty) = -\infty$,
(iv) $f(\alpha(q = 0)) = 1$ and $f(\alpha(q = 1)) = \alpha(q = 1)$.
(v) $f(\alpha(q \to -\infty)) \to 0$, $f(\alpha(q \to \infty)) \to -\infty$, $f(\alpha(q = q_{root})) = 0$, where $q_{root} = \frac{1}{\ln(\rho V)}$ and $\alpha_{\min} = \alpha(q_{root}) = \rho V - (\rho V)^{1/\ln(\rho V)} \ln(\rho V)$.

A.22.3 *Multifractality Directly Via Grid Boxes Representation*

According to what we discussed above in Sect. A.22, we divide the initial interval [0, 1] then descend to obtain subintervals, always dividing into two equal parts. For $n = 1$, we have $s_1 = s_1(0) = s_1(1) = 1/2^1$. For any generation n, we divide the initial interval into equal parts of length $s_n = s_n(i) = (s_1)^n = 1/2^n$ each (where $0 \le i \le 2^n - 1$). We emphasize that the multiplicity of the cascade lies in the aforementioned power dependence $s_n = (s_1)^n$. For each such subinterval, we assign a mass, which we can express using the general formula: $\mu_n(\xi_n) = \left[p^{\xi_n}(1-p)^{1-\xi_n}\right]^n$, where $\xi_n \stackrel{\text{def.}}{=} k/n$, where k still has the same interpretation as k in the equation (A.199). We continue our search for such an exponent α (the so-called crowding index, 'local scaling exponent,' or 'local dimension') of a given subinterval, which leads to the following key equality:

$$\mu_n(\xi_n) = (s_n)^{\alpha(\xi_n)},$$

where the exponent may depend on ξ_n. Hence, we obtain directly,

$$\alpha(\xi_n) = -\frac{1}{\ln 2} [\xi_n \ln p + (1 - \xi_n) \ln(1 - p)], \tag{A.208}$$

which is formally identical to the last relation in (A.201).

We now perform a technical step by expressing Newton's factor as a fraction ξ_n. Thus, we obtain the following after simple algebraic manipulations using Stirling's approximation, $\ln n! \approx n \ln n - n$ for $n \gg 1$

$$\ln \binom{n}{k} = -n \left[\xi_n \ln \xi_n + (1 - \xi_n) \ln(1 - \xi_n)\right].$$

Next, we look for a distribution of the f of α exponents that satisfies the second key equation

$$\binom{n}{k} = (s_n)^{-f(\alpha(\xi_n))}.$$

Combining the two equalities above, we obtain

$$f(\alpha(\xi_n)) = -\frac{1}{\ln 2} \left[\xi_n \ln \xi_n + (1 - \xi_n) \ln(1 - \xi_n)\right]. \tag{A.209}$$

We can now determine the total mass of the $\binom{n}{k}$ subintervals for fixed n and k. Namely,

$$\mu_n(\xi_n) \binom{n}{k} = (s_n)^{\alpha(\xi_n)} (s_n)^{-f(\alpha(\xi_n))} = (s_n)^{\tau(\xi_n)},$$

where the summarized exponent is given by

$$\tau(\xi_n) = \alpha(\xi_n) - f(\alpha(\xi_n)),$$

whose definition is formally identical to the first equality in (A.203) for $q = 1$.

However, the approach presented in this subsection is simplified, yet it provides a straightforward and convenient means to directly address the question of the presence of multifractality in the system.

It should be noted that extending the binomial cascade to the multinomial cascade is straightforward, albeit tedious. The multifractality presented here is derived from the multinomial theorem.

$$Z_q(s_n) = \left(\sum_{j=1}^{r} p_j^q\right)^n = \sum_{k_1+k_2+\ldots+k_r=n} \binom{n}{k_1, k_2, \ldots, k_r} \prod_{j=1}^{r} \left(p_j^q\right)^{k_j} = s_n^{\tau(q)}, \tag{A.210}$$

where integers $k_j \geq 0$, $\sum_{j=1}^{r} p_j = 1$ and

$$\binom{n}{k_1, k_2, \ldots, k_r} \stackrel{\text{def.}}{=} \frac{n!}{\prod_{j=1}^{r} k_j!}, \tag{A.211}$$

where the first equality (in the above sequence of equalities) is an extension of the third equality in (A.199), the second equality is an extension of the second one, and the last equalities are formally the same. However, the scale s_n used here is $s_n = r^{-n}$ and the global scaling exponent

$$\tau(q) = -\frac{\ln \sum_{j=1}^{r} p_j^q}{\ln r}.$$

This is a straightforward extension of that given by Eq. (A.200).

Equation (A.210) describes the expansion of the power of the sum of individual components in terms of the powers of these components. This equation directly generalizes the binomial theorem from binomials to multinomials.

However, the following sequence of equations defining the multinomial cascade,

$$\mu_{n+1}(rj + l) = p_{l+1}\, \mu_n(j)$$
$$j = 0, 1, 2, \ldots, r^n - 1,\ l = 0, 1, 2, \ldots, r - 1,\ \text{integer}\ r \geq 2, \tag{A.212}$$

generalizes Eqs. (A.195); each index j corresponds to the complete set of all the values of the index l (above given).

Consequently, we encourage interested readers to derive the characteristics of the multifractality associated with the multinomial cascade.

References

1. Oldham, K., Spanier, J.: The Fractional Calculus Theory and Applications of Differentiation and Integration to Arbitrary Order. Dover, New York (2006)
2. Ross, B. (ed.): Fractional Calculus and its Applications. Lecture Notes in Mathematics, vol. 457. Springer, Berlin (1975)
3. Miller, K.S., Ross. B.: An Introduction to the Fractional Calculus and Fractional Differential Equations. Wiley, New York (1993)
4. Hilfer, R. (ed.): Applications of Fractional Calculus in Physics. World Scientific, Singapore (2000). https://doi.org/10.1142/3779
5. Butler, R.W.: Saddlepoint approximations with applications. Cambridge University Press, Cambridge (2007). https://doi.org/10.1017/CBO9780511619083
6. Bateman, H., Erdélyi, A.: Tables of Integral Transforms, vol. 1. McGraw-Hill, New York (1954)
7. Zwillinger, D. (ed.): Table of Integrals, Series, and Products, 8th edn. Academic Press, Amsterdam (2014)
8. Howie, J.M.: Complex Analysis. Springer, Berlin (2003)
9. Gradshteyn, I.S., Ryzhik, I.M.: Table of Integrals, Series, and Products. In: Jeffrey, A., Daniel Zwillinger, D. (eds.) 7th edn. Elsevier & Academic Press, Amsterdam (2007)
10. Jacobs, K.: Stochastic Processes for Physicists: Understanding Noisy Systems. Cambridge University Press, Cambridge (2010)
11. Bosq, D., Nguyen, H.T.: A Course in the Theory of Stochastic Processes. Stochastic Models and Statistical Inference, McGraw-Hill, New York (1981)

12. Van Kampen, N.G.: Stochastic Processes in Physics and Chemistry, 3rd edn. Elsevier, Amsterdam (2007)
13. Sornette, D.: Critical Phenomena in Natural Sciences: Chaos, Fractals, Self-Organization, and Disorder: Concepts and Tools. In: Springer Series in Synergetics: Springer Complexity, 2nd edn. Springer, Berlin (2006). https://doi.org/10.1007/3-540-33182-4
14. Kubo, R., Toda, M., Hashitsume, N.: Statistical Physics II: Nonequilibrium Statistical Mechanics. Springer, Berlin (1985)
15. Erdélyi, A.: Asymptotic Expansions. Dover, New York (1956)
16. Mason, T.G., Weitz, D.A.: Optical measurements of frequency-dependent linear viscoelastic moduli of complex fluids. Phys. Rev. Lett. **74**, 1250 (1995). https://doi.org/10.1103/PhysRevLett.74.1250
17. Gisler, T., Weitz, D.A.: Tracer microrheology in complex fluids. Curr. Opin. Colloid Interface Sci. **3**, 586–592 (1998). https://doi.org/10.1016/S1359-0294(98)80084-8
18. Mason, T., Van Zanten, J., Ganesan, K., Wirtz, D., Kuo, S.C.: Particle tracking microrheology of complex fluids. Phys. Rev. Lett. **79**, 3282 (1997). https://doi.org/10.1103/PhysRevLett.79.3282
19. Panja, D.: Anomalous polymer dynamics is nonmarkovian: memory effects and the Generalized Langevin Equation formulation. J. Stat. Mech. - Theor. Exp. **2010**, P06011 (2010). https://doi.org/10.1088/1742-5468/2010/06/P06011
20. Panja, D.: Generalized Langevin equation formulation for anomalous polymer dynamics. J. Stat. Mech. - Theor. Exp. **2010**, L02001 (2010). https://doi.org/10.1088/1742-5468/2010/02/L02001
21. Bullerjahn, J.T., Sturm, S., Wolff, L., Kroy, K.: Monomer dynamics of a wormlike chain. Europhys. Lett. **96**, 48005 (2011). https://doi.org/10.1209/0295-5075/96/48005
22. Min, W., Luo, G., Cherayil, B.J., Kuo, S.C., Xie, X.S.: Observation of a power-law memory kernel for fluctuations within a single protein molecule. Phys. Rev. Lett. **94**, 198302 (2005). https://doi.org/10.1103/PhysRevLett.94.198302
23. Halsey, Th.H., Jensen, M.H., Kadanoff, L.P., Procaccia, I., Shraiman, B.I.: Fractal measures and their singularities: the characterization of strange sets. Phys. Rev. A **33**(2), 1141–1151 (1986); Erratum: Phys. Rev. A **34**(2), 1601 (1986). https://doi.org/10.1103/PhysRevA.33.1141
24. Beck, C., Schlögl, F.: Thermodynamics of Chaotic Systems: An Introduction. Cambridge Nonlinear Science Series, Series Number 4. Cambridge University Press, Cambridge (1995)
25. Erdélyi, A.: Asymptotic Expansions. Dover, New York (1956)
26. Chang, Q.: Generalized binomial multiplicative cascade processes and asymmetrical multifractal distributions. Nonlin. Processes Geophys. **21**, 477 (2014). https://doi.org/10.5194/npg-21-477-2014

Index

M. Chorowski et al., *Anomalous Stochastics*, Understanding Complex Systems,
https://doi.org/10.1007/978-3-031-80392-5

MIX
Papier aus verantwortungsvollen Quellen
Paper from responsible sources
FSC® C105338

If you have any concerns about our products,
you can contact us on
ProductSafety@springernature.com

In case Publisher is established outside the EU,
the EU authorized representative is:
Springer Nature Customer Service Center GmbH
Europaplatz 3, 69115 Heidelberg, Germany

Printed by Libri Plureos GmbH
in Hamburg, Germany